AF361786

Advances in Heat and Mass Transfer in Micro/Nano Systems

Advances in Heat and Mass Transfer in Micro/Nano Systems

Editors

Junfeng Zhang
Ruijin Wang

MDPI • Basel • Beijing • Wuhan • Barcelona • Belgrade • Manchester • Tokyo • Cluj • Tianjin

Editors
Junfeng Zhang
Bharti School of Engineering
and Computer Science
Laurentian University
Sudbury
Canada

Ruijin Wang
School of Mechanical
Engineering
Hangzhou Dianzi University
Hangzhou
China

Editorial Office
MDPI
St. Alban-Anlage 66
4052 Basel, Switzerland

This is a reprint of articles from the Special Issue published online in the open access journal *Micromachines* (ISSN 2072-666X) (available at: www.mdpi.com/journal/micromachines/special_issues/Heat_Mass_Transfe).

For citation purposes, cite each article independently as indicated on the article page online and as indicated below:

LastName, A.A.; LastName, B.B.; LastName, C.C. Article Title. *Journal Name* **Year**, *Volume Number*, Page Range.

ISBN 978-3-0365-4968-2 (Hbk)
ISBN 978-3-0365-4967-5 (PDF)

Contents

Editorial

Editorial for the Special Issue on Heat and Mass Transfer in Micro/Nanosystems

Ruijin Wang [1],* and **Junfeng Zhang** [2],*

1 School of Mechanical Engineering, Hangzhou Dianzi University, Hangzhou 310018, China
2 Bharti School of Engineering and Computer Science, Laurentian University, 935 Ramsey Lake Road, Sudbury, ON P3E 2C6, Canada
* Correspondence: wangrj@hdu.edu.cn (R.W.); jzhang@laurentian.ca (J.Z.)

Citation: Wang, R.; Zhang, J. Editorial for the Special Issue on Heat and Mass Transfer in Micro/Nanosystems. *Micromachines* **2022**, *13*, 1151. https://doi.org/10.3390/mi13071151

Received: 20 July 2022
Accepted: 20 July 2022
Published: 21 July 2022

Publisher's Note: MDPI stays neutral with regard to jurisdictional claims in published maps and institutional affiliations.

The miniaturization of components in mechanical and electronic equipment has been the driving force for the fast development of micro/nanosystems. Heat and mass transfer are crucial processes in such systems, and they have attracted great interest in recent years. Tremendous effort, in terms of theoretical analyses, experimental measurements, numerical simulation, and practical applications, has been devoted to improve our understanding of complex heat and mass transfer processes and behaviors in such micro/nanosystems.

This Special Issue is dedicated to showcase recent advances in heat and mass transfer in micro- and nanosystems, with particular focus on the development of new model and theory, the employment of new experimental techniques, the adoption of new computational methods, and the design of novel micro/nanodevices. Thirteen articles have been published after peer-review evaluations, and these articles cover a wide spectrum of active research in the frontiers of micro/nanosystems. For example, Hu et al. [1] studied the satellite droplet generation in piezoelectric methods, and found that there are two key parameters responsible for this phenomenon: the pulse frequency for driving the piezoelectric transducer tube and the fluid flow rate. Optimal operation conditions have also been proposed to eliminate the satellite droplets for deionized water. In the article by Song et al. [2], the authors developed a structural design to visualize the evaporation and condensation processes in the silicon-based ultra-thin loop heat pipe (s-UTLHP), and performed experimental measurements to study the heat transfer mechanism in such devices. To gain a more accurate thermal measurement for microfluidic devices, Meng et al. [3] proposed the use of a liquid metal to fill the gap space between the temperature sensor and the microfluidic substrate, and they also tested this concept on a microchennel chip with gallium. Furthermore, Wang et al. [4] developed a disposable microfluidic chip for the real-time monitoring of sweat rate. This economical and convenient paper-based *sticker* has a diameter of 25 mm and a thickness of 0.3 mm, and it can be applied to the skin at any parts of the body. The chip consists of multiple layers; in particular, the sweat-sensing layer has an impressed wax micro-channel containing chromogenic agent to show sweat absorption amount, which can be read directly from the scale lines on the chip surface. The proposed chip, as a low-cost and convenient wearable device, has potential applications in the real-time monitoring of sweat loss for bodybuilders, athletes, firefighters, etc.

With the rapid advances in computational technologies, numerical modeling and simulations have been proven as a valuable complement for studying complex systems and processes. In this direction, Saghir and Ranman [5] numerically investigated the thermal and hydraulic performances through minichannels with different pin-fin configurations, and their results showed that the wavy pin-fin configuration exhibited the best performance with a high Nusselt number and a low pressure drop. On the other hand, Jbeili and Zhang [6] examined the convective heat transfer performances of flows through porous materials, and found that, in addition to the porosity, the aspect ration of the microscopic porous structure and the possible interfacial thermal resistance can also affect the macroscopic thermal performance of the porous media. For an efficient evaluation of the flow and pressure distributions in a microchannel network, Zhao et al. [7] introduced an electric circuit analogy and applied it to study the effect of microchannel length on the flow behaviors. Another interesting study is presented by Huang et al. [8], where the immersed boundary method has been combined with the lattice Boltzmann method to study the trajectory of a neutrally buoyant circular particle in the pulsatile channel flow. The particle exhibits rich dynamic behaviors which have not been observed in non-pulsatile situations, and the results could be useful for nanoparticle transport in drug delivery applications.

Nanofluids, which are fluid suspensions of nanoparticles, have attracted great attention from scientists for the enhanced thermal performances, and extensive studies have been conducted over the past decades. In this Special Issue, several papers have been devoted to exploring the thermal enhancement mechanisms of nanofluids in various situations. Elsafy and Saghir [9] conducted simulations of the convective heat transfer by considering the nanofluids through straight and wavy microchannels filled with porous materials, and Wu and Zhang [10] studied the effects of the nanoparticle volume fraction of Al_2O_3-water nanofluids and the aspect ratio of rectangular microchannels on heat transfer and pumping power consumption for heat sink applications in electronic devices. Moreover, Rasool et al. [11,12] and Shafig et al. [13] established a mathematical framework to consider the magnetohydrodynamic flows of nanofluids in Darcy–Forchheimer porous media with moving boundaries.

The success of this Special Issue is built on the team work of everyone involved. We would like to thank the authors for contributing their interesting research, and we are also grateful to the anonymous reviewers for their critical comments which are valuable for further improving the article quality. Moreover, we should not forget to mention the help from Drs. Myung-Suk Chun, Khashayar Khoshmanesh, and Kwang-Yong Kim as Academic Editors, and Ms. Violet Cheng and Mr. Toot Jiang at the journal office.

Conflicts of Interest: The authors declare no conflict of interest.

References

1. Hu, Z.; Li, S.; Yang, F.; Lin, X.; Pan, S.; Huang, X.; Xu, J. Formation and Elimination of Satellite Droplets during Monodisperse Droplet Generation by Using Piezoelectric Method. *Micromachines* **2021**, *12*, 921. [CrossRef] [PubMed]
2. Song, W.; Xu, Y.; Xue, L.; Li, H.; Guo, C. Visualization Experimental Study on Silicon-Based Ultra-Thin Loop Heat Pipe Using Deionized Water as Working Fluid. *Micromachines* **2021**, *12*, 1080. [CrossRef] [PubMed]
3. Meng, J.; Yu, C.; Li, S.; Wei, C.; Dai, S.; Li, H.; Li, J. Microfluidics Temperature Compensating and Monitoring Based on Liquid Metal Heat Transfer. *Micromachines* **2022**, *13*, 792. [CrossRef] [PubMed]
4. Wang, H.; Xu, K.; Xu, H.; Huang, A.; Fang, Z.; Zhang, Y.; Wang, Z.; Lu, K.; Wan, F.; Bai, Z.; et al. A One-Dollar, Disposable, Paper-Based Microfluidic Chip for Real-Time Monitoring of Sweat Rate. *Micromachines* **2022**, *13*, 414. [CrossRef] [PubMed]
5. Saghir, M.Z.; Rahman, M.M. Thermo-Hydraulic Performance of Pin-Fins in Wavy and Straight Configurations. *Micromachines* **2022**, *13*, 954. [CrossRef] [PubMed]
6. Jbeili, M.; Zhang, J. Effects of Microscopic Properties on Macroscopic Thermal Conductivity for Convective Heat Transfer in Porous Materials. *Micromachines* **2021**, *12*, 1369. [CrossRef] [PubMed]
7. Zhao, Y.; Zhang, K.; Guo, F.; Yang, M. Dynamic Modeling and Flow Distribution of Complex Micron Scale Pipe Network. *Micromachines* **2021**, *12*, 763. [CrossRef]
8. Huang, L.; Du, J.; Zhu, Z. Neutrally Buoyant Particle Migration in Poiseuille Flow Driven by Pulsatile Velocity. *Micromachines* **2021**, *12*, 1075. [CrossRef] [PubMed]
9. Elsafy, K.M.; Saghir, M.Z. Forced Convection in Wavy Microchannels Porous Media Using TiO_2 and Al_2O_3-Cu Nanoparticles in Water Base Fluids: Numerical Results. *Micromachines* **2021**, *12*, 654. [CrossRef] [PubMed]
10. Wu, H.; Zhang, S. Numerical Study on the Fluid Flow and Heat Transfer Characteristics of Al_2O_3-Water Nanofluids in Microchannels of Different Aspect Ratio. *Micromachines* **2021**, *12*, 868. [CrossRef] [PubMed]
11. Rasool, G.; Shafiq, A.; Alqarni, M.S.; Wakif, A.; Khan, I.; Bhutta, M.S. Numerical Scrutinization of Darcy-Forchheimer Relation in Convective Magnetohydrodynamic Nanofluid Flow Bounded by Nonlinear Stretching Surface in the Perspective of Heat and Mass Transfer. *Micromachines* **2021**, *12*, 374. [CrossRef] [PubMed]
12. Rasool, G.; Shafiq, A.; Hussain, S.; Zaydan, M.; Wakif, A.; Chamkha, A.J.; Bhutta, M.S. Significance of Rosseland's Radiative Process on Reactive Maxwell Nanofluid Flows over an Isothermally Heated Stretching Sheet in the Presence of Darcy-Forchheimer and Lorentz Forces: Towards a New Perspective on Buongiorno's Model. *Micromachines* **2022**, *13*, 368. [CrossRef] [PubMed]
13. Shafiq, A.; Rasool, G.; Alotaibi, H.; Aljohani, H.M.; Wakif, A.; Khan, I.; Akram, S. Thermally Enhanced Darcy-Forchheimer Casson-Water/Glycerine Rotating Nanofluid Flow with Uniform Magnetic Field. *Micromachines* **2021**, *12*, 605. [CrossRef] [PubMed]

 *micromachines*

Article

Numerical Scrutinization of Darcy-Forchheimer Relation in Convective Magnetohydrodynamic Nanofluid Flow Bounded by Nonlinear Stretching Surface in the Perspective of Heat and Mass Transfer

Ghulam Rasool [1,*], Anum Shafiq [2], Marei S. Alqarni [3,4], Abderrahim Wakif [5], Ilyas Khan [6] and Muhammad Shoaib Bhutta [1]

1 Binjiang College, Nanjing University of Information Science and Technology, Wuxi 214105, China; shoaibbhutta@hotmail.com
2 School of Mathematics and Statistics, Nanjing University of Information Science and Technology, Nanjing 210044, China; anumshafiq@ymail.com
3 Department of Mathematics, College of Sciences, King Khalid University, Abha 61413, Saudi Arabia; msalqarni@kku.edu.sa
4 Mathematical Modelling and Applied Computation Research Group (MMAC), Department of Mathematics, King Abdulaziz University, P. O. Box 80203, Jeddah 21589, Saudi Arabia
5 Laboratory of Mechanics, Faculty of Sciences Aïn Chock, Hassan II University, B.P.5366 Mâarif, Casablanca 9167, Morocco; wakif.abderrahim@gmail.com
6 Department of Mathematics, College of Science Al-Zulfi, Majmaah University, Al-Majmaah 11952, Saudi Arabia; i.said@mu.edu.sa
* Correspondence: ghulam46@yahoo.com

Citation: Rasool, G.; Shafiq, A.; Alqarni, M.S.; Wakif, A.; Khan, I.; Bhutta, M.S. Numerical Scrutinization of Darcy-Forchheimer Relation in Convective Magnetohydrodynamic Nanofluid Flow Bounded by Nonlinear Stretching Surface in the Perspective of Heat and Mass Transfer. *Micromachines* **2021**, *12*, 374. https://doi.org/10.3390/mi12040374

Academic Editor: Junfeng Zhang and Ruijin Wang

Received: 9 March 2021
Accepted: 24 March 2021
Published: 1 April 2021

Publisher's Note: MDPI stays neutral with regard to jurisdictional claims in published maps and institutional affiliations.

1. Introduction

A simple base fluid, for example, water, ethylene glycol, and oil, etc., when upgraded with the suspension of nanometric metallic strong conductive particles, is termed as nanofluid. Such a formulation sufficiently intensifies the conduction abilities of the base fluid. Numerous applications have been discovered for the so-called nanofluids in the industrial and engineering aspects as well as in bio-medicine. For example, vehicle cooling, heat exchangers, cooling and transformer cooling, electronic cooling, and many others are typical and widely used applications of nanofluids. These are also applicable in the medical treatments, especially cancer and tumor treatments, resonance imaging, and wound treatment, etc., and they are typically dependent on the conductive nature of nanofluids. Choi [1] introduced the definition of nanofluid in his experimental work where he proved that the suspension of nanoparticles in typical fluids drastically changes the thermo-physical properties of the fluid. Later on, Buongiorno [2] modeled the same concept in the perspective of convective transport of nanofluids. Adding details to the concept of nanofluids, Buongiorno emphasized the fact that the Brownian diffusion and Thermophoresis are two major slip factors in the transport of nanofluids. Afterwards, several interesting attempts have been reported by renowned researchers of fluid mechanics. For instance, Khan et al. [3] reported convection phenomena in nanofluid flow passing a linear stretching surface using the Keller–Box numerical method for the final solutions of the modeled governing problems. An interesting study of Mustafa et al. [4] disclosed an analysis on stagnation spot flow of nanofluids involving linear stretched sheet. For more details on this topic, one can see [5–24] and cross references cited therein.

Flow past a linear as well as nonlinear stretching surface relates the fluid mechanics with several important industrial and engineering setups, such as hot rolling, polymeric extrusion, continuously stretching done in plastic thin films, crystal growth, fiber production, and metallic extrusion, etc. Numerous articles are available in the literature that explain the flow caused by stretching surfaces, whether linear or nonlinear rates. The importance of linear stretching cannot be neglected but the flow caused by non-linear stretching rates have always played a significant role in the above mentioned procedures, especially in polymeric extrusion. In all such scenarios, the work of Cortell [25] is considered to be a pioneer. He considered a viscous fluid for studying heat and mass transfer developments driven by a nonlinear or linear stretching in the sheet. The prescribed wall temperature and constant wall temperature were both discussed in this study. Vajravelu [26] reported a study on exploration of heat transfer developments in a viscous fluid flow past a stretching sheet surface using power law velocity distribution with nonlinear stretching rate. Rana and Bhargava [27] reported a study on the fluid flow analysis and heat transfer aspects involving a nonlinear stretching rate in nanofluids flow over a sheet.

The type of nanofluids governed in the categorical classification of non-Newtonian fluids have gained special attention for numerous engineering and industrial applications based on their extended contributions in nuclear, chemical, metallic, polymeric, and plastic industries. Shampoos, paints, apple sauce, katchup, as well as different type of oils are typical genuine examples of non-Newtonian fluids. Having extended viscosity, the non-Newtonian fluids are best treated by involving the viscoelastic terms (second grade model), which is known as sub-categorical classification of differential non-Newtonian liquids related with the normal stress attribute. For a better understanding of this model, one can read [28–34] and the references cited therein.

The studies mentioned above are either concerned with a linear stretching surface with effective convective heating or nonlinear stretching sheets without the involvement of convective conditions together with the Darcy–Forchheimer model. Here, for the first time, we involve convective conditions to consider a visco-elastic and strictly incompressible nanoliquid (nanofluid) flow bounded by a nonlinear flat stretching surface. Firstly, the model shapes in mathematical form using the famous Navier stokes equations for incompressible non-Newtonian nanofluid. The leading problems are then transformed into highly nonlinear ordinary problems via suitable transformations. A numerical scheme is implemented for finding the final solutions. From now onward, fluid means means incompressible viscoelastic nanofluid. The next section will physically justify the existence of the problem and mathematical expressions with properly defined boundary conditions. The rest of the article comprises of the results and discussion, a graphical display, and concluding remarks.

2. Problem Formulation

In this numerical investigation, we have invoked the convective boundary on the flow of nanofluid passing over a nonlinear flate stretching surface. These conditions are invoked to balance the temperature difference within the system. The system relies on Darcy–Forchheimer medium saturated via nanofluid over the stretching surface. Cartesian coordinates are considered to analyze the fluid flow. The flow direction is assumed along positive $x-$direction, whereas no-movement is allowed towards vertical. An induced magnetic effect is directly invoked to the surface normal direction via MHD; however, a tiny Reynolds number helps to dismiss the magnetic impact. The nonlinear stretching rate is taken into account via n as a positive integer, whereas the stretching velocity is assumed as $u = U_w = mx^n$ at the bottom line. However, the velocity diminishes away from the surface and attains zero value at free surface ($u = 0$). The coefficient of heat and mass flux (h_1 and h_2) are involved in convective boundary conditions. Figure 1 sketches a physical display.

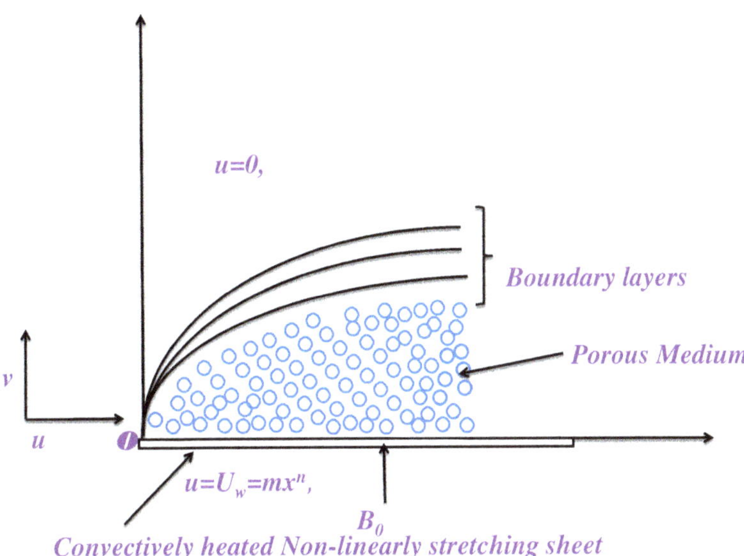

Figure 1. Physical model and coordinate system.

The modified Navier stokes problems are given below:

$$\frac{\partial v}{\partial y} + \frac{\partial u}{\partial x} = 0,\tag{1}$$

$$v\frac{\partial u}{\partial y} + u\frac{\partial u}{\partial x} = v\frac{\partial^2 u}{\partial y^2} - \left[\frac{B_0^2 \sigma}{\rho_f} + \frac{v}{K}\right]u - \frac{C_b}{K^{1/2}}u^2$$
$$- k_1\left(u\frac{\partial^3 u}{\partial x \partial y^2} + \frac{\partial u}{\partial x}\frac{\partial}{\partial y}\left(\frac{\partial u}{\partial y}\right) + v\frac{\partial}{\partial y}\left(\frac{\partial^2 u}{\partial y^2}\right) - \frac{\partial u}{\partial y}\frac{\partial}{\partial x}\left(\frac{\partial u}{\partial y}\right)\right),\tag{2}$$

$$v\frac{\partial T}{\partial y} + u\frac{\partial T}{\partial x} = \alpha\frac{\partial}{\partial y}\left(\frac{\partial T}{\partial y}\right) + \frac{(\rho c)_p}{(\rho c)_f}\left[D_{Br}\left(\frac{\partial C}{\partial y}\frac{\partial T}{\partial y}\right) + \frac{D_{Th}}{T_\infty}\left(\frac{\partial T}{\partial y}\right)^2\right] - \frac{1}{\rho c_f}\frac{\partial q_r}{\partial y},\tag{3}$$

where q_r is known as radiative heat flux. Given by Rosseland's approximation:

$$q_r = -\frac{\partial(T^4)}{\partial y}\left(\frac{4\sigma^*}{3k^*}\right),\tag{4}$$

where $\sigma^* =$ Stefan–Boltzmann constant and $k^* =$ mean absorption, respectively. Here, in,

$$4T_\infty^3 T - 3T_\infty^4 \cong T^4,\tag{5}$$

confirms,

$$\frac{\partial q_r}{\partial y} = -\frac{\partial^2 T}{\partial y^2}\left(\frac{16\sigma^* T_\infty^3}{3k^*}\right).\tag{6}$$

Therefore, the energy equation has the following final form:

$$u\frac{\partial T}{\partial x} + v\frac{\partial T}{\partial y} = \alpha\frac{\partial}{\partial y}\left(\frac{\partial T}{\partial y}\right) + \frac{(\rho c)_p}{(\rho c)_f}\left[\frac{D_{Th}}{T_\infty}\left(\frac{\partial T}{\partial y}\right)^2 + D_{Br}\left(\frac{\partial C}{\partial y}\frac{\partial T}{\partial y}\right)\right] + \frac{\partial^2 T}{\partial y^2}\left(\frac{16\sigma^* T_\infty^3}{3k^*}\right)\frac{1}{(\rho c)_f},\tag{7}$$

$$v\frac{\partial C}{\partial y} + u\frac{\partial C}{\partial x} = D_{Br}\left(\frac{\partial}{\partial y}\frac{\partial C}{\partial y}\right) + \frac{D_{Th}}{T_\infty}\left(\frac{\partial}{\partial y}\frac{\partial T}{\partial y}\right), \tag{8}$$

Equations $(1),(2),(7)$, and (8) are known as the governing equations with the following boundary conditions, as per the present model,

$$u = U_w = mx^n, \; h_1(T_f - T) = -k\frac{\partial T}{\partial y}, \; v = 0, \; h_2(C_f - C) = -D_{Br}\frac{\partial C}{\partial y} \quad \text{at} \quad y = 0, \tag{9}$$

$$u = 0, \; \frac{\partial u}{\partial y} = 0, \; C = C_\infty, \; T = T_\infty \quad \text{as} \quad y \to \infty. \tag{10}$$

In the above governing equations, μ is taken as dynamic viscosity, $v\left(= \mu/\rho_f\right)$ is known as kinematic viscosity, ρ_f is taken as density, σ is involved for electrical conductivity, $\alpha = k/(\rho c)_f$ is well known thermal diffusive force, k is typical thermal conductiveness, $h_1 = h_p x^{\frac{n-1}{2}}$ and $h_2 = h_q x^{\frac{n-1}{2}}$ are known as non-uniform heat and mass (transfer) coefficients, respectively, τ is known as the ratio given between the effective nanoparticles heating capacity versus effective fluid heating capacity, D_{Br} is known for Brownian diffusion, D_{Th} is thermophoretic factor, and B_0 is magnetic effect. Defining,

$$v = -\frac{1}{2}x^{\frac{n-1}{2}}\left[\left(\frac{n-1}{n+1}\right)\frac{\partial f}{\partial \eta}\eta + f(\eta)\right]\sqrt{2mv(n+1)}, \quad u = mx^n\frac{\partial f}{\partial \eta},$$

$$\phi(\eta) = \frac{C - C_\infty}{C_w - C_\infty}, \quad \theta(\eta) = \frac{T - T_\infty}{T_w - T_\infty}, \quad \eta = \frac{1}{2}\sqrt{\frac{2\rho_{fl}m(n+1)}{\mu}}x^{\frac{n-1}{2}}y. \tag{11}$$

Using (15) in $(1), (2), (9)$ and (10) we have

$$f''' + ff'' + k_0\left[\left(\frac{n+1}{2n}\right)f'f^{iv} + (f'')^2 - \left(\frac{n+1}{2n}\right)f'''f'\right]$$
$$- \left(\frac{2n}{1+n}\right)(1 + F_r)(f')^2 - \lambda\left(\frac{2}{n+1}\right)f' - M^2\left(\frac{2}{n+1)}\right)f' = 0, \tag{12}$$

$$(1 + 4/3Rd)\theta'' + Prf\theta' + NbPr\theta'\phi' + NtPr(\theta')^2 = 0, \tag{13}$$

$$\phi'' + LePrf\phi' + \frac{Nt}{Nb}\theta'' = 0, \tag{14}$$

$$f(0) = 0, \quad f' = 1, \quad \phi'(0) = -\gamma_2(1 - \phi(0)), \quad \theta'(0) = -\gamma_1(1 - \theta(0)),$$
$$f'(\infty) = 0, \quad f''(\infty) = 0, \quad \phi(\infty) = 0, \quad \theta(\infty) = 0. \tag{15}$$

Here, γ_i for $i = 1, 2$ are the convective parameters extracted from $h_1 = h_p x^{\frac{n-1}{2}}$ and $h_2 = h_q x^{\frac{n-1}{2}}$ known as non-uniform heat transfer coefficient and mass transfer coefficient, respectively. F_r is used for inertial frame, M is involved for magnetic impact, λ is treated as porosity, Pr is the typical Prandtl, Le is the well known Lewis number, k_0 is taken as viscoelastic parameter, and Nb and Nt Brownian diffusion and thermophoretic factors, respectively. Mathematically,

$$\gamma_1 = \frac{h_p}{k}\sqrt{\frac{2v}{m(n+1)}}, \gamma_2 = \frac{h_q}{D_{Br}}\sqrt{\frac{2v}{m(n+1)}}, \quad F_r = \frac{C_b x}{K^{1/2}}, \quad M^2 = \frac{2\sigma B_0^2}{mx^{n-1}\rho_f},$$

$$\lambda = \frac{2v}{Kmx^{n-1}}, \quad Pr = \frac{v}{\alpha}, \quad Le = \frac{\alpha}{D_{Br}}, \quad k_1 = \frac{k_0 mx^{n-1}}{v}, \tag{16}$$

$$Nb = \frac{\left(\rho_f c\right)_p D_{Br}(C_w - C_\infty)}{\left(\rho_f c\right)_f v}, \quad Nt = \frac{\left(\rho_f c\right)_p D_{Th}(T_w - T_\infty)}{\left(\rho_f c\right)_f v T_\infty}.$$

Furthermore, using $Re_x = mx^{n+1}/v$, the local tiny Reynolds, the non-dimensional forms of physical quantities, are given below:

$$(Re_x)^{1/2}\left(\frac{n+1}{2}\right)^{-1/2}C_f = \left[(1-3k_0)f''(0)\right],$$

$$(Re_x)^{-1/2}\left(\frac{n+1}{2}\right)^{-1/2}Nu = -\left(1+\frac{4}{3}\right)Rd\left[\theta'(0)\right], \tag{17}$$

$$(Re_x)^{-1/2}\left(\frac{n+1}{2}\right)^{-1/2}Sh = -\left[\phi'(0)\right],$$

3. Methodology

RK45 is one of the most frequently used numerical methods for solving IVPs for its accuracy and efficiency. The built-in codes involve basic concept of converting Boundary value problems into the initial value problems and subsequent problems are solved using a parallel scheme with RK45, such as shooting technique, secant method, etc., using an appropriate set of initial guesses to approximate the solutions. Herein, the boundary value problems are converted into initial value problems together with the given boundary conditions and thereafter, careful selection of initial guess for repeated iterations of the numerical scheme are chosen to obtain the solutions.

4. Results and Discussion

In this section, we have described the consequences noticed in flow profiles for all relevant parameters that are involved in leading problems of present nanofluid model. The arguments are comprehended by physical justifications as to why the variation is occurred and what consequence is being witnessed. The results are obtained using the numerical method and the data are graphically plotted in increasing and decreasing curves to witness the difference. The first section of the graphs belongs to the velocity profile, the second to temperature distribution, and third is related with the concentration of nanoparticles. In particular, Figure 2 is plotted to see the variation in velocity profile against augmented trend of visco-elastic parameter k_0. The physical reasoning of the behavior noted for this parameter is connected with the combination of k_1 and kinematic viscosity μ having an inverse relation with fluid viscosity. For larger values of k_0, the fluid viscosity shows a decline, and a consequent increasing trend is witnessed in the relative profile. The magnetic parameter M that is given in Figure 3 with a relevant variation in fluid motion produces a reduction. Physically, the affect of magnetic field is always in anti-directional with the fluid flow because of the surface normal bumps. The certain effect of magnetic field in the direction normal to the fluid flow is, for sure, the reason behind this trend of decline in fluid velocity. The similar trend in velocity profile can be figured out for the Forchheimer number F_r that is given in Figure 4, but this time the decline is related with a higher friction and retardation offered by porous medium. The larger the Forchheimer number, the higher the frictional force that is offered to fluid motion and the consequence is a decline in fluid velocity. The thermal layer receives significant modification in the incremental direction for augmented values of thermal radiation R_d given in Figure 5. The influence of thermal radiation eases the way of heat flux with a more convenient convection and the corresponding profile receives significant enhancement. Physically, thermal radiation is responsible for the enhancement of the thermal state of the fluid. Elevation in thermal Biot number γ_1 apparently results in a significant increase of the thermal layer and associated boundary layer thickness enhances, as shown in Figure 6. The heat flux coefficient that is involved in constitution of the said parameter is a sufficient justification for this behavior of thermal profile. Thermophoresis and Brownian motion are the two important factors of nanofluid flow and boundary layer phenomenon. By definition, Brownian motion is an uncertain movement of some particles (nanoparticles in this case) in the given medium for any time $t > 0$. The parameters are both closed influential to each other as well as on the flow profiles. Here, the thermophoresis enforces a stronger thermophoretic push to the given nanoparticles and the nanoparticles receive certain inpredictive enhanced motion and, therefore, a disturbance appears and thermal profiles increase for both of the given factors shown in Figures 7 and 8, respectively. Figures 9–12 are related with the third important flow profiles, the concentration profiles, which is solely based on the concentration of nanoparticles. In particular, Figure 9 presents the display of variation that is noticed in concentration profile with respect to the elevation in Schmidt number. Physically, the inverse relationship between Brownian diffusion and kinematic viscosity is sufficient to justify the reducing trend noticed in concentration profile for augmented values of Schmidt number. The solute Biot number is an enhancing factor the concentration profile that is given in Figure 10. The coefficient of mass flux involved in the constitutive expression justifies this variation. An enhancement is noticed for elevated values of thermophoresis given in Figure 11. A decline can be seen in concentration profile for augmented Brownian diffusion parameter given in Figure 12. For sure, the decline is based on the in-predictive and uncertain fluctuation of nanoparticles. Figures 13 and 14 are the contour graphs sketched for both the linear and non-linear case, respectively, while Figures 15 and 16 are the density graphs. A minor but significant difference can be seen in both the linear and non-linear case settled at $n = 1$ & 1.5, respectively. Relevant data for skin-friction, heat, and mass flux numbers are given in Tables 1 and 2, respectively. Skin friction jumps for larger porosity and a larger Forchheimer number, therefore, declines the easy movement of fluid. The heat flux and mass flux both receive a reduction for the augmented values of Forchheimer number. Thermal radiation is an enhancing factor for both of the flux rates. An opposite trend is noticed in both Nusselt and Sherwood numbers for thermal Biot number. Heat flux enhances, while mass flux reduces with the passage of time.

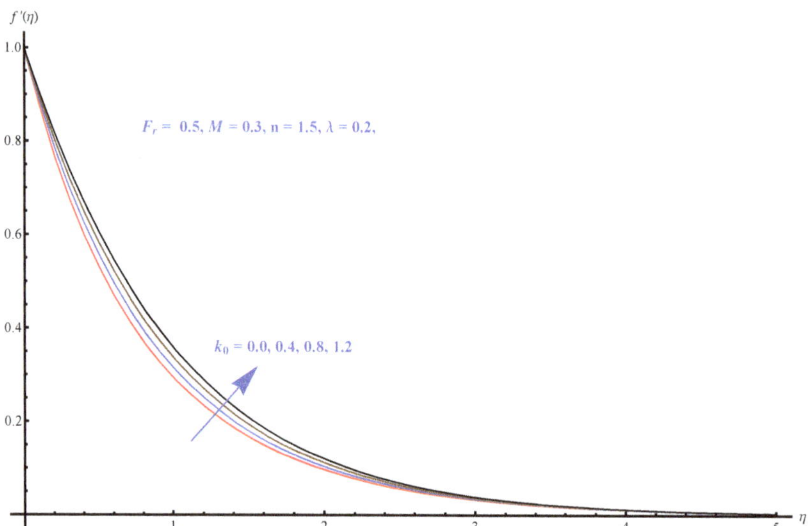

Figure 2. Visco-elastic parameter versus velocity field.

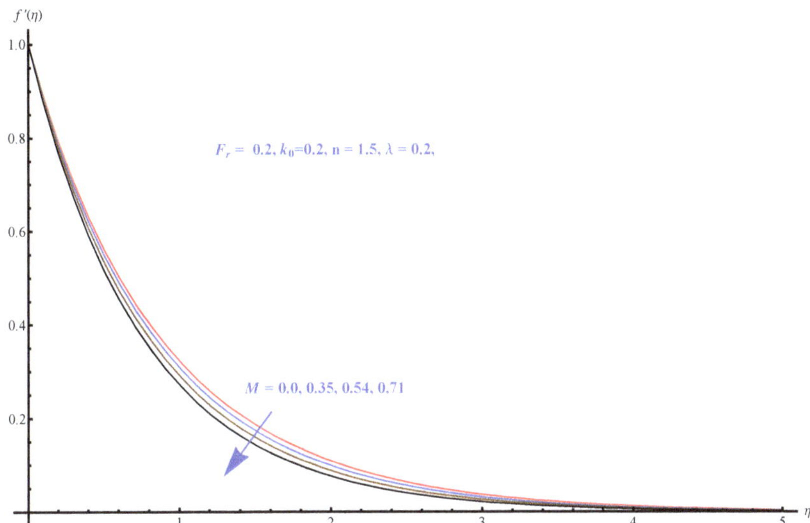

Figure 3. Magnetic number versus velocity field.

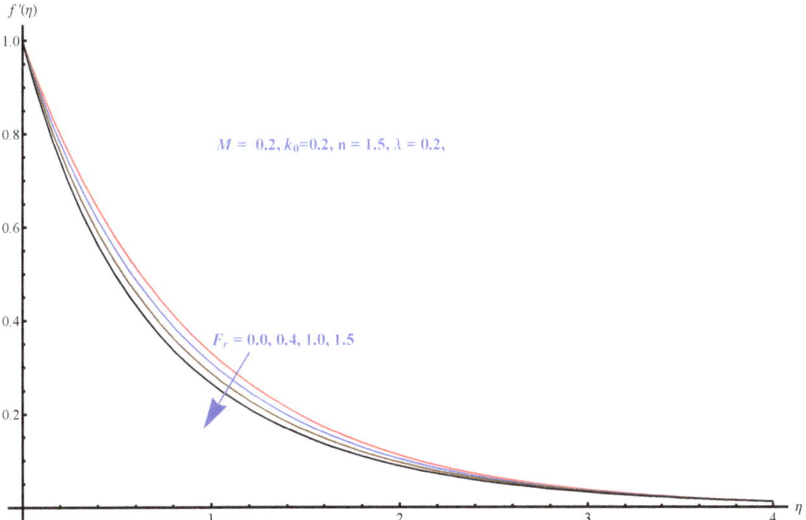

Figure 4. Forchheimer number versus velocity field.

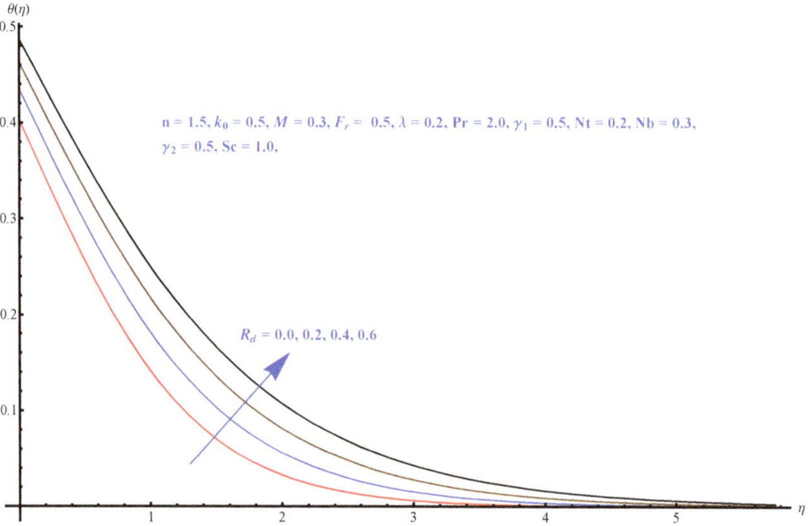

Figure 5. Thermal radiation parameter versus thermal profile.

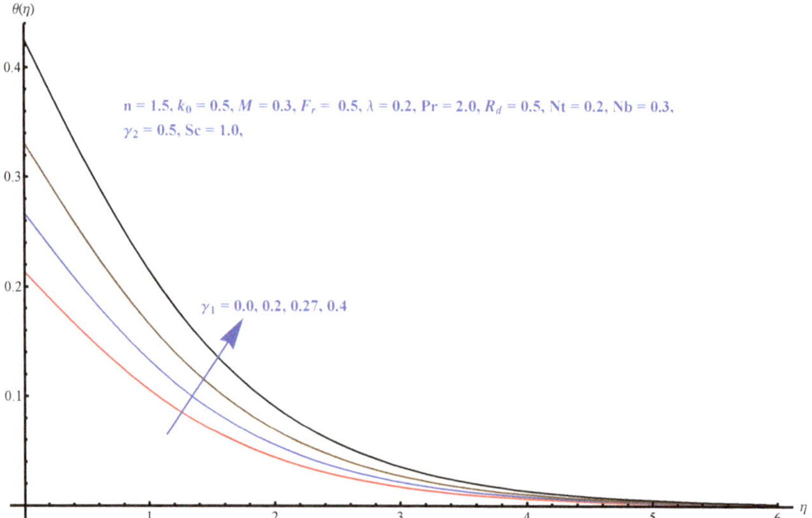

Figure 6. Thermal Biot number versus thermal profile.

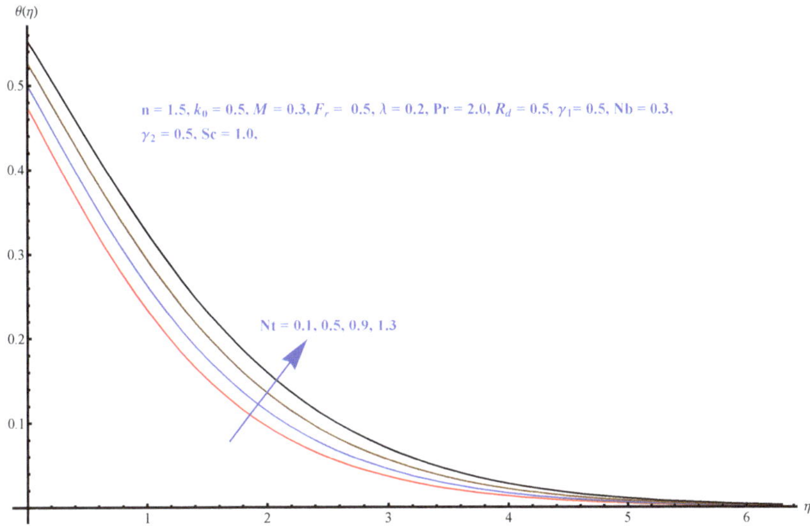

Figure 7. Thermophoresis versus thermal profile.

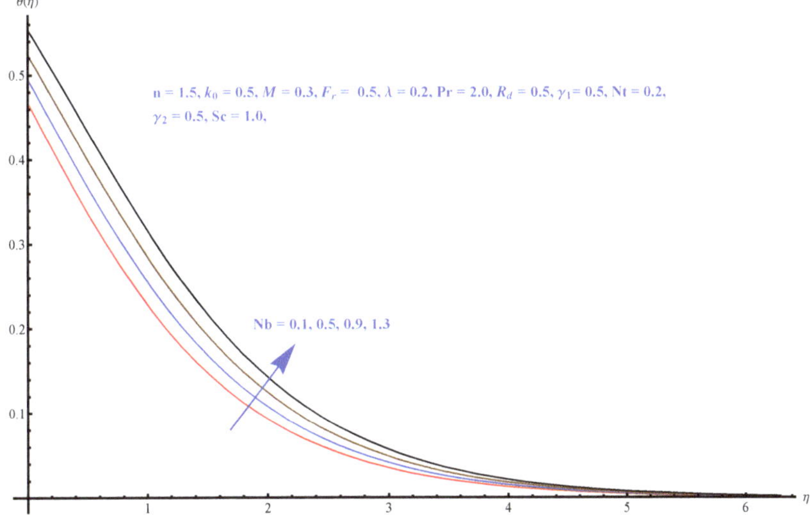

Figure 8. Brownian diffusion versus thermal profile.

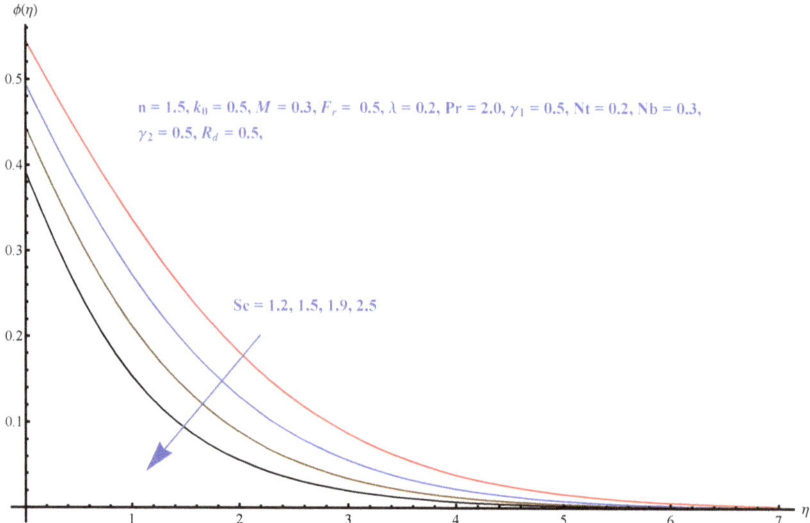

Figure 9. Schmidt number versus concentration profile.

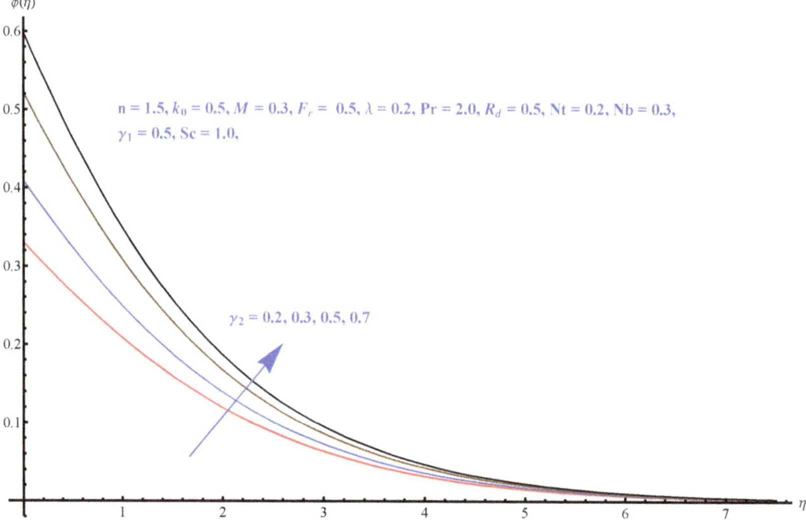

Figure 10. Solute Biot number versus concentration distribution.

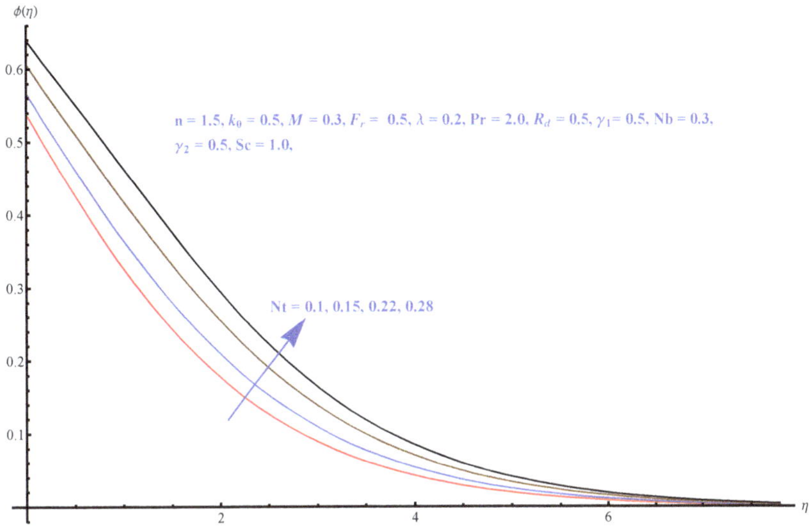

Figure 11. Thermophoresis versus concentration distribution.

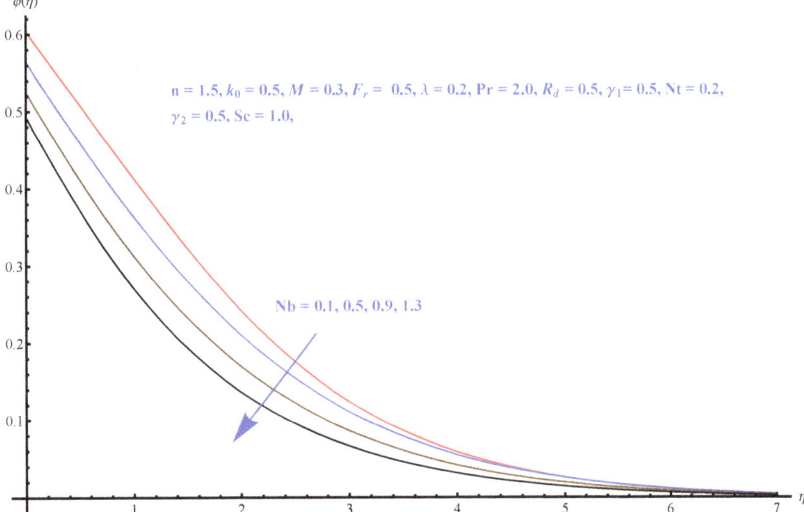

Figure 12. Brownian diffusion versus concentration distribution.

Figure 13. Contour graph for linear case.

Figure 14. Contour graph for nonlinear case.

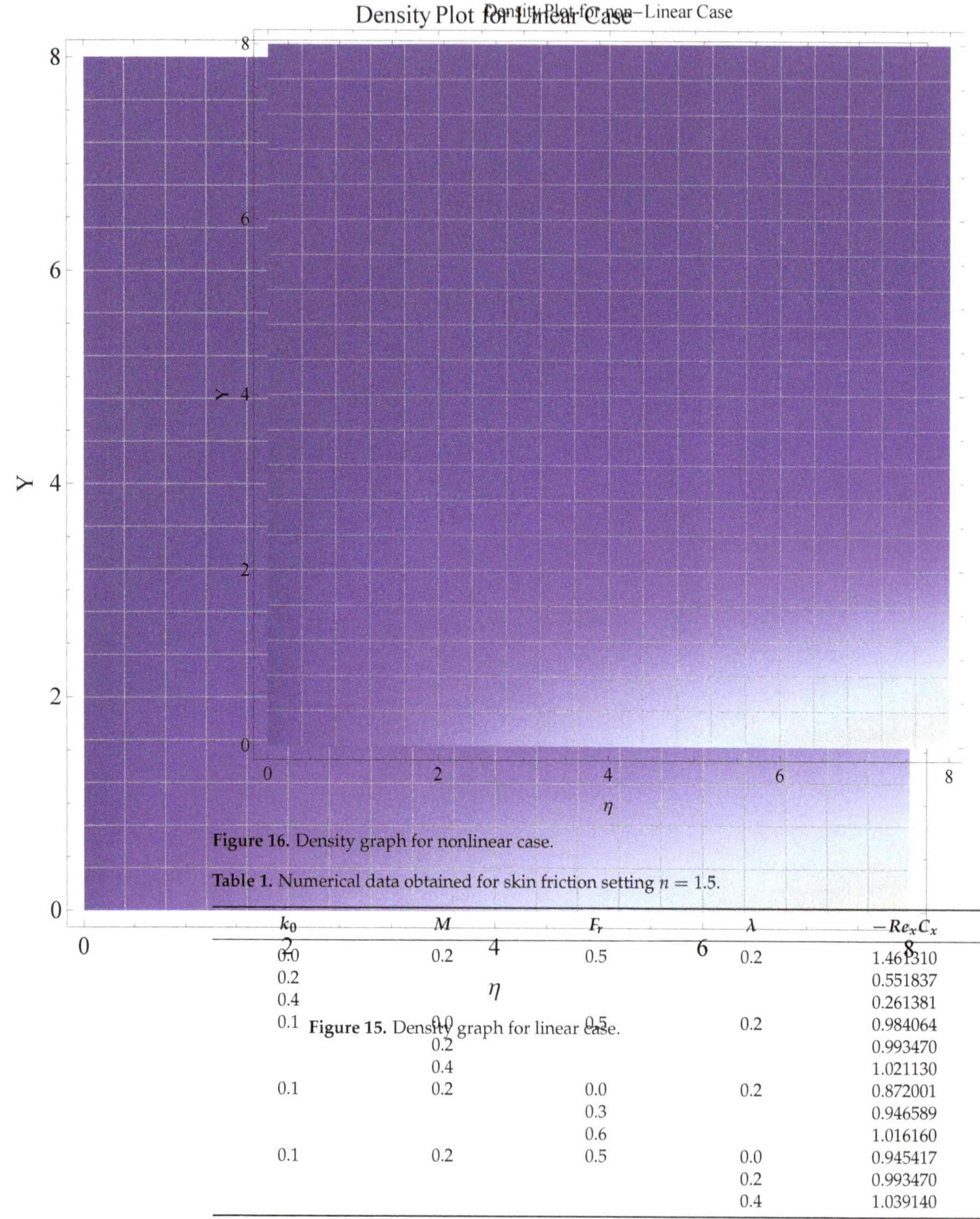

Figure 16. Density graph for nonlinear case.

Figure 15. Density graph for linear case.

Table 1. Numerical data obtained for skin friction setting $n = 1.5$.

k_0	M	F_r	λ	$-Re_x C_x$
0.0	0.2	0.5	0.2	1.461310
0.2				0.551837
0.4				0.261381
0.1	0.0	0.5	0.2	0.984064
	0.2			0.993470
	0.4			1.021130
0.1	0.2	0.0	0.2	0.872001
		0.3		0.946589
		0.6		1.016160
0.1	0.2	0.5	0.0	0.945417
			0.2	0.993470
			0.4	1.039140

5. Conclusions

Here, in this research we have focused on the features of Darcy–Forchheimer relation in nanofluid flow bounded by a convectively heated non-linear stretching plane sheet. A visco-elastic and strictly incompressible nanofluid saturates the designated porous medium under the direct influence of the Darcy–Forchheimer model. The magnetic effect is taken non-uniformly normal to the flow direction. However, the model is bounded to tiny magnetic Reynolds number for practical applications. The salient findings are summarized below:

- A decreasing trend in velocity profile is noted for a stronger effect of Forchheimer number F_r.
- The thermal layer receives significant modification in the incremental direction for augmented values of thermal radiation R_d with a more convenient convection.
- Elevated values of thermal Biot number γ_1 apparently result a in significant increase of the thermal layer.
- The two important factors of nanofluid flow and boundary layer phenomenon, the thermophoresis and the Brownian motion, apparently develop a rising trend in thermal profile.
- The solute Biot number is an enhancing factor for the concentration profile
- Skin friction rises for larger porosity, while both the heat flux and mass flux receive a reduction for augmented values of Forchheimer number.
- A significant part of this study is the contour and density graphs.

Table 2. Numerical results used for Nusselt and Sherwood numbers setting $n = 1.5, k_0 = 0.1$.

M	F_r	λ	R_d	Pr	N_t	N_b	S_c	γ_1	γ_2	$-\theta'(0)$	$-\phi'(0)$
0.0	0.5	0.2	0.5	2.0	0.2	1.8	1.0	0.5	0.5	0.382563	0.284189
0.2										0.381299	0.283494
0.4										0.377592	0.281461
0.3	0.0	0.2								0.388611	0.287151
	0.3									0.385131	0.283876
	0.6									0.378095	0.281800
0.3	0.5	0.0								0.386118	0.286148
		0.3								0.376684	0.280965
		0.6								0.367978	0.276224
0.3	0.5	0.3	0.0	2.0	0.2	1.8	1.0	0.5	0.5	0.243061	0.280613
			0.3							0.326357	0.280770
			0.6							0.410511	0.281073
0.3	0.5	0.3	0.5	1.0	0.2	1.8	1.0	0.5	0.5	0.314527	0.282394
				1.5						0.353191	0.281437
				2.0						0.376684	0.280965
0.3	0.5	0.3	0.5	2.0	0.1	1.8	1.0	0.5	0.5	0.383141	0.284704
					0.3					0.370253	0.277406
					0.5					0.351170	0.267780
0.3	0.5	0.3	0.5	2.0	0.2	1.0	1.0	0.5	0.5	0.453068	0.274166
						1.5				0.390927	0.279511
						2.0				0.348077	0.283071
0.3	0.5	0.3	0.5	2.0	0.2	1.8	1.0	0.5	0.5	0.376684	0.280965
							1.5			0.377924	0.321943
							2.0			0.379042	0.348489
0.3	0.5	0.3	0.5	2.0	0.2	1.8	1.0	0.1	0.5	0.145086	0.285651
								0.3		0.298417	0.282539
								0.6		0.402833	0.280441
0.3	0.5	0.3	0.5	2.0	0.2	1.8	1.0	0.5	0.1	0.458177	0.090822
									0.3	0.408401	0.208285
									0.6	0.364879	0.307814

Author Contributions: Conceptualization, G.R.; methodology, G.R., A.S.; software, G.R., A.S., A.W.; validation, G.R., A.S., A.W., M.S.B.; formal analysis, G.R., A.S., A.W., I.K., M.S.A., M.S.B.; investigation, G.R., A.S., A.W., I.K., M.S.B.; resources, M.S.A.; data curation, G.R., A.S., A.W., M.S.B.; writing—original draft preparation, G.R., A.S., A.W., I.K., M.S.B., M.S.A.; writing—review and editing, G.R., A.S., A.W., I.K., M.S.B., M.S.A.; visualization, G.R., A.S., A.W.,; supervision, M.S.A., I.K.; project administration, M.S.A.; funding acquisition, M.S.A. All authors have read and agreed to the published version of the manuscript.

Data Availability Statement: All data are available within this manuscript.

Acknowledgments: The authors extend their appreciation to the Deanship of Scientific Research at King Khalid University, Abha, Saudi Arabia for funding this work through research groups program under grant number R.G.P-1/128/42.

Conflicts of Interest: The authors declare no conflict of interest.

Nomenclature:

MHD	Magnetohydrodynamics
$RK45$	Runge Kutta 45 method
PDE	Partial Differential Equation
ODE	Ordinary Differential Equation
$u(x,y), v(x,y)$	velocity components in cartesian coordinates/m·s^{-1}
$u = mx^n$	Horizontal velocity /m·s^{-1}
m	constant/s^{-1}
T, C	Local temperature and concentration of nanoparticles
T_∞	Ambient temperature/K
C_b	Drag coefficient
K	Permeability/H · m^{-1}
B_0	Magnetic field/A · m^{-1}
σ	Electric conductivity/(Ohm · m)$^{-1}$
ρ	Density/kg·m^{-3}
k_1	viscoelastic coefficient
α	Thermal diffusivisity/m^2·s-1
q_r, Rd	Radiation
Nu_x	Local Nusselt number
C_f	Skin-friction (wall drag force)
Re_x	Local Reynolds number
Le	Lewis number
D_B	Brownian Diffusion/m^2· s^{-1}
D_T	Thermophoretic diffusion/m^2· s^{-1}
M	Magnetic number
λ	viscoelastic parameter
Pr	Prandtl number
Nt	Thermophoresis parameter
Nb	Brownian diffusion parameter
τ	Ratio of heat capacity of fluid and nanoparticles

References

1. Choi, S.U.S. *Enhancing Thermal Conductivity of Fluids with Nanoparticles*; ASME: New York, NY, USA, 1995; Volume 66, pp. 99–105; FED 231/MD.
2. Buongiorno, J. Convective transport in nanofluids. *ASME J. Heat Transf.* **2006**, *128*, 240–250. [CrossRef]
3. Khan, W.A.; Pop, I. Boundary-layer flow of a nanofluid past a stretching sheet. *Int. J. Heat Mass Transf.* **2010**, *53*, 2477–2483. [CrossRef]
4. Mustafa, M.; Hayat, T.; Pop, I.; Asghar, S.; Obaidat, S. Stagnation-point flow of a nanofluid towards a stretching sheet. *Int. J. Heat Mass Transf.* **2011**, *54*, 5588–5594. [CrossRef]

5.   Khan, N.S.; Gul, T.; Islam, S.; Khan, A.; Shah, Z. Brownian motion and Thermophoresis effects on MHD mixed convective thin film second-grade nanofluid flow with hall effect and heat transfer past a stretching sheet. *J. Nanofluids* **2017**, *6*, 812–829. [CrossRef]

6.   Mebarek-Oudina, F.; Bessaih, R.; Mahanthesh, B.; Chamkha, A.J.; Raza, J. Magneto-Thermal-Convection Stability in an Inclined Cylindrical Annulus filled with a Molten Metal. *Int. J. Numer. Methods Heat Fluid Flow* **2020**. [CrossRef]

7.   Rasool, G.; Shafiq, A. Numerical Exploration of the Features of Thermally Enhanced Chemically Reactive Radiative Powell-Eyring Nanofluid Flow via Darcy Medium over Non-linearly Stretching Surface Affected by a Transverse Magnetic Field and Convective Boundary Conditions. *Appl. Nanosci.* **2020**. [CrossRef]

8.   Dogonchi, A.S.; Selimefendigil, F.; Ganji, D.D. Magneto-hydrodynamic natural convection of CuO-water nanofluid in complex shaped enclosure considering various nanoparticle shapes. *Int. J. Numer. Methods Heat Fluid Flow* **2019**, *29*, 1663–1679. [CrossRef]

9.   Rasool, G.; Wakif, A. Numerical spectral examination of EMHD mixed convective flow of second-grade nanofluid towards a vertical Riga plate using an advanced version of the revised Buongiorno's nanofluid model. *J. Therm. Anal. Calorim.* **2020**. [CrossRef]

10.   Khan, N.S.; Gul, T.; Khan, M.A.; Bonyah, E.; Islam, S. Mixed convection in gravity-driven thin film non-Newtonian nanofluids flow with gyrotactic microorganisms. *Results Phys.* **2017**, *7*, 4033–4049. [CrossRef]

11.   Rasool, G.; Shafiq, A.; Khalique, C.M.; Zhang, T. Magnetohydrodynamic Darcy Forchheimer nanofluid flow over nonlinear stretching sheet. *Phys. Scr.* **2019**, *94*, 105221. [CrossRef]

12.   Zaim, A.; Aissa, A.; Mebarek-Oudina, F.; Mahanthesh, B.; Lorenzini, G.; Sahnoun, M. Galerkin finite element analysis of magneto-hydrodynamic natural convection of Cu-water nanoliquid in a baffled U-shaped enclosure. *Propuls. Power Res.* **2020**, *9*, 383–393. [CrossRef]

13.   Rasool, G.; Zhang, T. Darcy-Forchheimer nanofluidic flow manifested with Cattaneo-Christov theory of heat and mass flux over non-linearly stretching surface. *PLoS ONE* **2019**, *14*, e0221302. [CrossRef] [PubMed]

14.   Swain, K.; Mebarek-Oudina, F.; Abo-Dahab, S.M. Influence of MWCNT/Fe3O4 hybrid-nanoparticles on an exponentially porous shrinking sheet with variable magnetic field and chemical reaction. *J. Therm. Anal. Calorim.* **2021**. [CrossRef]

15.   Abo-Dahab, S.M.; Abdelhafez, M.A.; Mebarek-Oudina, F.; Bilal, S.M. MHD Casson Nanofluid Flow over Nonlinearly Heated Porous Medium in presence of Extending Surface effect with Suction/Injection. *Indian J. Phys.* **2021**. [CrossRef]

16.   Rasool, G.; Shafiq, A.; Khan, I.; Baleanu, D.; Nisar, K.S.; Shahzadi, G. Entropy generation and consequences of MHD in Darcy-Forchheimer nanofluid flow bounded by non-linearly stretching surface. *Symmetry* **2020**, *12*, 652. Available online: https://www.mdpi.com/2073-8994/12/4/652 (accessed on 26 March 2021). [CrossRef]

17.   Marzougui, S.; Bouabid, M.; Mebarek-Oudina, F.; Abu-Hamdeh, N.; Magherbi, M.; Ramesh, K. A computational analysis of heat transport irreversibility phenomenon in a magnetized porous channel. *Int. J. Numer. Methods Heat Fluid Flow* **2020**. [CrossRef]

18.   Parvin, S.; Nasrin, R.; Alim, M.A.; Hossain, N.F.; Chamkha, A.J. Thermal conductivity variation on natural convection flow of water–alumina nanofluid in an annulus. *Int. J. Heat Mass Transf.* **2012**, *55*, 5268–5274. [CrossRef]

19.   Nasrin, R.; Alim, M.A.; Chamkha, A.J. Combined convection flow in triangular wavy chamber filled with water–CuO nanofluid: Effect of viscosity models. *Int. Commun. Heat Mass Transf.* **2012**, *39*, 1226–1236. [CrossRef]

20.   Dogonchi, A.S.; Ismael, M.A.; Chamkha, A.J.; Ganji, D.D. Numerical analysis of natural convection of Cu–water nanofluid filling triangular cavity with semicircular bottom wall. *J. Therm. Anal. Calorim.* **2019**, *135*, 3485–3497. [CrossRef]

21.   Reddy, P.S.; Sreedevi, P.; Chamkha, A.J. MHD boundary layer flow, heat and mass transfer analysis over a rotating disk through porous medium saturated by Cu-water and Ag-water nanofluid with chemical reaction. *Powder Technol.* **2017**, *307*, 46–55. [CrossRef]

22.   Mebarek-Oudina, F. Convective Heat Transfer of Titania Nanofluids of different base fluids in Cylindrical Annulus with discrete Heat Source. *Heat-Transf. Asian Res.* **2019**, *48*, 135–147. [CrossRef]

23.   Lund, L.A.; Omar, Z.; Khan, I.; Raza, J.; Bakouri, M.; Tlili, I. Stability analysis of Darcy-Forchheimer flow of casson type nanofluid over an exponential sheet: Investigation of critical points. *Symmetry* **2019**, *11*, 412. [CrossRef]

24. Swain, K.; Mahanthesh, B.; Mebarek-Oudina, F. Heat transport and stagnation-point flow of magnetized nanoliquid with variable thermal conductivity with Brownian moment and thermophoresis aspects. *Heat Transf.* **2021**, *50*, 754–764. [CrossRef]
25. Cortell, R. Viscous flow and heat transfer over a nonlinearly stretching sheet. *Appl. Math. Comput.* **2007**, *184*, 864–873. [CrossRef]
26. Vajravelu, K. Viscous flow over a nonlinearly stretching sheet. *Appl. Math. Comput.* **2001**, *124*, 281–288. [CrossRef]
27. Rana, P.; Bhargava, R. Flow and heat transfer of a nanofluid over a nonlinearly stretching sheet: A numerical study. *Commun. Nonlinear Sci. Numer. Simul.* **2012**, *17*, 212–226. [CrossRef]
28. Tan, W.C.; Masuoka, T. Stokes first problem for second grade fluid in a porous half space. *Int. J. Non-Linear Mech.* **2005**, *40*, 515–522. [CrossRef]
29. Fetecau, C.; Fetecau, C. Starting solutions for the motion of a second grade fluid due to longitudinal and torsional oscillations of a circular cylinder. *Int. J. Eng. Sci.* **2006**, *44*, 788–796. [CrossRef]
30. Chamkha, A.J. MHD-free convection from a vertical plate embedded in a thermally stratified porous medium with Hall effects. *Appl. Math. Model.* **1997**, *21*, 603–609. [CrossRef]
31. Ramzan, M.; Bilal, M. Time dependent MHD nano-second grade fluid flow induced by permeable vertical sheet with mixed convection and thermal radiation. *PLoS ONE* **2015**, *10*, e0124929. [CrossRef]
32. Hayat, T.; Muhammad, T.; Shehzad, S.A.; Alsaedi, A. Similarity solution to three dimensional boundary layer flow of second grade nanofluid past a stretching surface with thermal radiation and heat source/sink. *AIP Adv.* **2015**, *5*, 017107.[CrossRef]
33. Hayat, T.; Hussain, Z.; Farooq, M.; Alsaedi, A. Effects of homogeneous and heterogeneous reactions and melting heat in the viscoelastic fluid flow. *J. Mol. Liq.* **2016**, *215*, 749–755. [CrossRef]
34. Hayat, T.; Ullah, I.; Muhammad, T.; Alsaedi, A. Magnetohydrodynamic (MHD) three-dimensional flow of second grade nanofluid by a convectively heated exponentially stretching surface. *J. Mol. Liq.* **2016**, *220*, 1004–1012. [CrossRef]

Article

Thermally Enhanced Darcy-Forchheimer Casson-Water/Glycerine Rotating Nanofluid Flow with Uniform Magnetic Field

Anum Shafiq [1], Ghulam Rasool [2,*], Hammad Alotaibi [3,*], Hassan M. Aljohani [3], Abderrahim Wakif [4], Ilyas Khan [5,*] and Shakeel Akram [6]

1 School of Mathematics and Statistics, Nanjing University of Information Science and Technology, Nanjing 210044, China; anumshafiq@ymail.com
2 Binjiang College, Nanjing University of Information Science and Technology, Wuxi 214105, China
3 Department of Mathematics, College of Science, Taif University, P.O. Box 11099, Taif 21944, Saudi Arabia; hmjohani@tu.edu.sa
4 Laboratory of Mechanics, Faculty of Sciences Aïn Chock, Hassan II University, Mâarif, B.P. 5366 Casablanca, Morocco; wakif.abderrahim@gmail.com
5 Department of Mathematics, College of Science Al-Zulfi, Majmaah University, Al-Majmaah 11952, Saudi Arabia
6 College of Electrical Engineering, Sichuan University, Chengdu 610065, China; shakeel.akram@scu.edu.cn
* Correspondence: ghulam46@yahoo.com (G.R.); h-85m@hotmail.com (H.A.); i.said@mu.edu.sa (I.K.)

Citation: Shafiq, A.; Rasool, G.; Alotaibi, H.; Aljohani, H.M.; Wakif, A.; Khan, I.; Akram, S. Thermally Enhanced Darcy-Forchheimer Casson-Water/Glycerine Rotating Nanofluid Flow with Uniform Magnetic Field. *Micromachines* **2021**, *12*, 605. https://doi.org/10.3390/mi12060605

Academic Editor: Kwang-Yong Kim

Received: 8 April 2021
Accepted: 17 May 2021
Published: 23 May 2021

Publisher's Note: MDPI stays neutral with regard to jurisdictional claims in published maps and institutional affiliations.

Abstract: This numerical study aims to interpret the impact of non-linear thermal radiation on magnetohydrodynamic (MHD) Darcy-Forchheimer Casson-Water/Glycerine nanofluid flow due to a rotating disk. Both the single walled, as well as multi walled, Carbon nanotubes (CNT) are invoked. The nanomaterial, thus formulated, is assumed to be more conductive as compared to the simple fluid. The properties of effective carbon nanotubes are specified to tackle the onward governing equations. The boundary layer formulations are considered. The base fluid is assumed to be non-Newtonian. The numerical analysis is carried out by invoking the numerical Runge Kutta 45 (RK45) method based on the shooting technique. The outcomes have been plotted graphically for the three major profiles, namely, the radial velocity profile, the tangential velocity profile, and temperature profile. For skin friction and Nusselt number, the numerical data are plotted graphically. Major outcomes indicate that the enhanced Forchheimer number results in a decline in radial velocity. Higher the porosity parameter, the stronger the resistance offered by the medium to the fluid flow and consequent result is seen as a decline in velocity. The Forchheimer number, permeability parameter, and porosity parameter decrease the tangential velocity field. The convective boundary results in enhancement of temperature facing the disk surface as compared to the ambient part. Skin-friction for larger values of Forchheimer number is found to be increasing. Sufficient literature is provided in the introduction part of the manuscript to justify the novelty of the present work. The research greatly impacts in industrial applications of the nanofluids, especially in geophysical and geothermal systems, storage devices, aerospace engineering, and many others.

Keywords: Darcy-Forchheimer theory; carbon nanotubes; nanofluid; magnetohydrodynamics; thermal radiation

1. Introduction

The contribution of nanomaterials (Nanofluids) in industry and engineering is very diversified. A lot of advantages have been noted by induction of nanomaterials in fluid flow analysis. Pioneered by Choi [1], the term nanofluid is also named as the nanomaterials subject to the type of nanoparticles that are used in the formulation procedure. The chore and basic property of nanomaterials is very important, i.e., enhanced thermal conductivity. The base fluids such as water, ethylene, toluene, and kerosene oil are the most commonly used base fluids in this regard and there are several related works that have been reported in the recent past to practically implement the idea of nanofluids and nanomaterials. For instance, some important studies are mentioned in the following lines. Lin et al. [2] reported discussion on the results obtained for MHD Transient Pseudo-Plastic nanofluid flow giving a highlight of the heat transfer properties and the impact of drag force in this transport. Bai et al. [3] analyzed the Brownian diffusion and the thermophoresis in radiative MHD Maxwell type nanofluid flow. They highlighted both the heat and mass transport phenomena in fluid flow procedures and provided sufficient numerical data to adhere the physical quantities. Madhu et al. [4] analyzed the non-Newtonian fluid behavior using the Maxwell model. The important aspect of this study is the consequence of MHD, as well as thermal radiation, on the flow attributes, especially the drag force component. Sheikholeslami et al. [5] reported the impact of MHD and radiation on Darcy-type flow of nanomaterials using controlled volume based finite element method (CVFEM) scheme. This scheme is more accurate as compared to the conventional analytic methods. Thus, the results were more precise and accurate for implementation in respective industrial applications. Williamson nanofluid flow using bi-directional stretching surface using the Brownian diffusion and thermophoresis are the main features of study reported by Hayat et al. [6] where they discussed three dimensional fluid flow analysis. The three dimensional nanofluid convection in natural flow has been reported by Zadi et al. [7]. The second important aspect of this study is related with studies carried out using disks and, here, this disk is assumed to be permeable under the definition of Darcian medium. Such formulations are highly important in geophysical and geothermal systems, storage devices, aerospace engineering, crystal growing procedures, medical instruments, and many food processing techniques that are based on porous mediums. The purpose behind such formulations is heat and mass transfer analysis by rotating frame. Several studies are available in literature on such formulations. Turkyilmazoglu and Senel [8] reported significant study on viscous nanofluid flow bounded by a porous disk using the usual Van Karman type of transformations. Bödewadt et al. [9] reported boundary layer fluid flow analysis using the rotating frame/disk. This study was used as a base study for investigation of fluid flow analysis using stationary disk by Mustafa et al. [10]. The use of partial slip conditions in nanofluid flow bounded by rotating disk under the impact of MHD is an important study reported by Mustafa [11]. Dogonchi et al. [12] interrogated the impact of heat convection using the magnetic field effect and shape and size factor of the used nanoparticles, using a cavity as the core surface. The importance of Casson material can not be neglected in this research, which is assumed to be a major factor for fluid flow analysis under the present formulation. It has several important applications in industry, such as metallurgy, food processing, bio-engineering and drilling operations, etc. The mixing of Casson material with Water-Glycerine, etc., is the base of Casson nanofluid. This so formulated nanomaterial has shear thinning properties, having infinite values of viscosity even at a zero shearing rate and deformation, below to which there is no flow. Besides, the importance of a porous (Darcy) medium is yet another very important phenomena in fluid flow analysis. It has received utmost attention in the last few decades. The efficiency of the typical energy systems is, therefore, enhanced using such formulations. Law of Darcy is valid for very small Reynolds number and, therefore, the high speed flows cannot be dealt under such theories. Therefore, the improvement was genuinely required which was dealt by Forchheimer [13] for high flow rates. Therefore, the combined Darcy-Forchheimer relation is way more effective to deal with fluid flow analysis in porous medium for moderate flow rates. Several studies have been reported on the significance of Darcy-Forchheimer medium, such as the impact of Cattaneo-Christov model in fluid flow analysis in Darcy-Medium has been reported by Shehzad et al. [14]. Bakar et al. [15] reported boundary layer approximation in stagnation point flow under Darcy-Forchheimer model of heat and mass flux. Hayat et al. [16] reported the significance of Cattaneo-Christov model in Darcy-Forchheimer model using variable thermal conductivity. In their study, Chamkha et al. [17] categorically reported the impact of radiation on the nanofluid flow via wedge using a Darcy type medium. In another article, Chamkha et al. [18] discussed the significance of thermal radiation in the mixed convective nanofluid flow having porous medium. The concept on CNTs was first revealed in early 1991 leading to extensive investigations for its not known properties. The micro-level structure of CNT is usually seen in cylindrical shape rolled from the single sheet called graphite. CNTs are usually divided into two categories as single wall nanotubes and multi walled nanotubes. For sure, the multi wall structure is more complicated as compared to single wall nanotubes. In theoretical fluid mechanics, the nanotubes are analyzed by their properties pre-defined for a particular problem. Several studies related to the structure and the applications of these tubes are available in literature. One can read [19–35] and cross reference therein.

Up till now, the literature survey indicates that there is a gap of study in the context of heat and mass transfer investigation on the Casson-Water/Glycerine nanofluids convection due to radially stretching disk. Therefore, the objective of present investigation is clear and novel, i.e., to explore the variation imparted by Carbon nanotubes, non-linear thermal radiation and Darcy-Forchheimer relation on Casson-Water/Glycerine nanofluid flow due to radially stretching disk. These formulations are highly important in geophysical and geothermal systems, storage devices, aerospace engineering, crystal growing procedures, medical instruments, and many food processing techniques that are based on a porous medium to help understand the fluid flow, heat transfer, especially the drag force intensity at the surface, which is in contact with the fluid. Numerical scheme is implemented for finding the solutions of so-formulated problems. The analysis is carried out via graphical display of the results for various parameters and their impact on the three profile of nanofluids in boundary layer approximations. Furthermore, the variation in skin-friction and Nusselt number is noted via graphical display. The article concludes with physical justifications and major findings of the study.

2. Problem Formulation

In this numerical investigation, we include the influence of non-linear thermal radiation and Carbon nanotubes on viscous incompressible Darcy-Forchheimer nanofluid flow bounded by rotating disk. Thermal convection is analyzed and convective boundary is invoked. The porosity factor appears highly under the implementation of Darcy-Forchheimer model. The formulation is based on two type of materials, i.e., water and glycerine, respectively. The single- and multi-walled carbon nanotubes are considered whose properties are given in the Table 1. The velocity components are taken as (u, v, w), in the direction of (r, ϕ, z), respectively. The rotation of disk is assumed at $z = 0$. One can see the physical scenario in Figure 1.

Table 1. Thermophysical properties of fluid and nanoparticles (for reference see Sheikholeslami et al. [5]).

	Base Fluid					Nanoparticles	
Properties	Water	Kerosene	Glycerin	Ethylene Glycol	Engine Oil	SWCNT	MWCNT
C_p (J/kg K)	4179	2090	2427	2430	1910	427	796
ρ (KG/m^3)	997	783	1259.9	1115	884	2600	1600
k (W/mK)	0.613	0.145	0.286	0.253	0.144	6600	3000
$\beta \times 10^{-5}$	21		48	65	70	27	44
Pr	6.2	21	6.78	203.63	6450	−	−

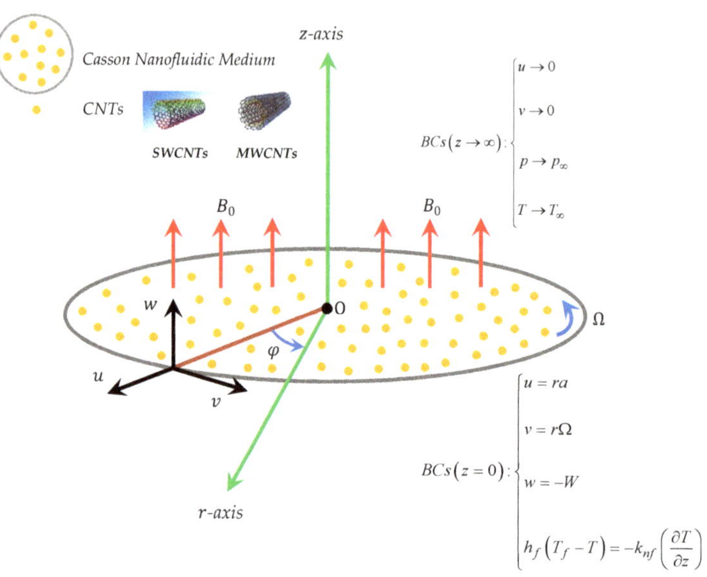

Figure 1. Physical model and coordinate system.

Consider,

$$\frac{\partial u}{\partial r} + \frac{u}{r} + \frac{\partial w}{\partial z} = 0, \tag{1}$$

$$
u\frac{\partial u}{\partial r} - \frac{v^2}{r} + w\frac{\partial u}{\partial z} = \frac{-1}{\rho_{nf}}\frac{\partial p}{\partial r} + \left(1 + \frac{1}{\beta_1}\right)\frac{\mu_{nf}}{\rho_{nf}}\left(\frac{\partial^2 u}{\partial r^2} + \frac{1}{r}\frac{\partial u}{\partial r} - \frac{u}{r^2} + \frac{\partial^2 u}{\partial z^2}\right)
$$
$$
+ \frac{(\rho\beta)_{nf}}{\rho_{nf}}g_1(T - T_\infty) - \frac{\mu_{nf}}{\rho_{nf}}\frac{u}{K^*} - F^* u^2 - \frac{\sigma B_0^2}{\rho}u, \tag{2}
$$

$$
u\frac{\partial v}{\partial r} + \frac{uv}{r} + w\frac{\partial v}{\partial z} = \left(1 + \frac{1}{\beta_1}\right)\frac{\mu_{nf}}{\rho_{nf}}\left(\frac{\partial^2 v}{\partial r^2} + \frac{1}{r}\frac{\partial v}{\partial r} - \frac{v}{r^2} + \frac{\partial^2 v}{\partial z^2}\right)
$$
$$
+ \frac{(\rho\beta)_{nf}}{\rho_{nf}}g_1(T - T_\infty) - \frac{\mu_{nf}}{\rho_{nf}}\frac{v}{K^*} - F^* v^2 - \frac{\sigma B_0^2}{\rho}v, \tag{3}
$$

$$
u\frac{\partial w}{\partial r} + w\frac{\partial w}{\partial z} = -\frac{1}{\rho_{nf}}\frac{\partial p}{\partial z} + \left(1 + \frac{1}{\beta_1}\right)\frac{\mu_{nf}}{\rho_{nf}}\left(\frac{\partial^2 w}{\partial r^2} + \frac{1}{r}\frac{\partial w}{\partial r} + \frac{\partial^2 w}{\partial z^2}\right), \tag{4}
$$

$$
(\rho C_p)_{nf}\left(u\frac{\partial T}{\partial r} + w\frac{\partial T}{\partial z}\right) = \left(k_{nf} + \frac{16\sigma^* T^3}{3k^*}\right)\left(\frac{\partial^2 T}{\partial r^2} + \frac{1}{r}\frac{\partial T}{\partial r} + \frac{\partial^2 T}{\partial z^2}\right)
$$
$$
+ \frac{16\sigma^* T^2}{k^*}\left[\left(\frac{\partial T}{\partial z}\right)^2 + \left(\frac{\partial T}{\partial r}\right)^2\right] + \left(1 + \frac{1}{\beta_1}\right)2\mu_{nf}\left[\left(\frac{\partial u}{\partial r}\right)^2 + \left(\frac{\partial w}{\partial z}\right)^2 + \frac{u^2}{r^2}\right]
$$
$$
+ \mu_{nf}\left(1 + \frac{1}{\beta_1}\right)\left[\left(\frac{\partial v}{\partial z}\right)^2 + \left(\frac{\partial w}{\partial r} + \frac{\partial u}{\partial z}\right)^2 + \left[r\frac{\partial}{\partial r}\left(\frac{v}{r}\right)\right]^2\right]. \tag{5}
$$

The boundary conditions are,

$$
v = r\Omega, \qquad u = ra, \qquad w = -W, \qquad h_f\left(T_f - T\right) = -k_{nf}\left(\frac{\partial T}{\partial z}\right), \qquad \text{at } z=0. \tag{6}
$$

$$
u \to 0, \qquad v \to 0, \qquad p \to p_\infty, \qquad T \to T_\infty, \qquad \text{as} \qquad z \to \infty. \tag{7}
$$

The effective Carbon nanotubes are (see for reference Shaw et al. [36]),

$$
\begin{aligned}
\mu_{nf} &= \frac{\mu_f}{(1-\phi)^{2.5}}, \\[4pt]
\rho_{nf} &= (1-\phi)_{nf} + \phi\rho_{CNT}, \\[4pt]
\alpha_{nf} &= \frac{k_{nf}}{\rho_{nf}}(C_p)_{nf}, \\[4pt]
\nu_{nf} &= \frac{\mu_{nf}}{\rho_{nf}}, \\[4pt]
\frac{K_{nf}}{K_f} &= \frac{(1-\phi) + 2\phi\frac{K_{CNT}}{K_{CNT}-K_f}\log\frac{K_{CNT}+K_f}{2K_f}}{(1-\phi) + 2\phi\frac{K_f}{K_{CNT}-K_f}\log\frac{K_{CNT}+K_f}{2K_f}}, \\[4pt]
(\rho\beta)_{nf} &= (1-\phi)(\rho\beta)_f + \phi(\rho\beta)_{S_1}, \\[4pt]
(\rho C_p)_{nf} &= (1-\phi)(\rho C_p)_f + \phi(\rho C_p)_{CNT},
\end{aligned}
\tag{8}
$$

Using the following transformations,

$$
\begin{aligned}
u &= r\Omega f', \qquad v = r\Omega g, \qquad w = -\sqrt{2\Omega\nu_f}\,f, \\[4pt]
p &= p_\infty - \mu_f\Omega P, \qquad T = T_\infty - (T_\infty - T_f)\theta, \\[4pt]
\eta &= \sqrt{\frac{2\Omega}{\nu_f}}\,z.
\end{aligned}
\tag{9}
$$

the final non-dimensional equations are,

$$
\frac{1}{(1-\phi)^{2.5}\left(1-\phi+\phi\left(\frac{\rho_{CNT}}{\rho_f}\right)\right)}\left(1+\frac{1}{\beta_1}\right)(2f''' - Kf') + 2ff'' - f'^2
$$

$$
+ g^2 + \frac{\left(1-\phi+\frac{(\rho\beta)_{CNT}\phi}{(\rho\beta)_f}\right)}{1-\phi+\frac{\rho_{CNT}}{\rho_f}\phi}\lambda\theta - F_r f'^2 - \frac{M_1^2}{1-\phi+\phi\frac{\rho_{CNT}}{\rho_f}}f = 0,
\tag{10}
$$

$$
\frac{1}{(1-\phi)^{2.5}\left(1-\phi+\phi\left(\frac{\rho_{CNT}}{\rho_f}\right)\right)}\left(1+\frac{1}{\beta_1}\right)(g'' - Kg) + 2fg' - 2f'g + \frac{\left(1-\phi+\phi\frac{(\rho\beta)_{CNT}\phi}{(\rho\beta)_f}\right)}{1-\phi+\frac{\rho_{CNT}\phi}{\rho_f}\phi}\lambda\theta
$$

$$
- F_r g^2 - \frac{M_1^2}{1-\phi+\phi\frac{\rho_{CNT}}{\rho_f}}g = 0,
\tag{11}
$$

$$
\frac{\frac{K_{nf}}{K_f}}{\left(1-\phi+\phi\left(\frac{(\rho C_p)_{CNT}}{(\rho C_p)_f}\right)\right)}\left(1+\frac{4}{3}\frac{K_f}{K_{nf}}R_1\left[1+\left(\theta_f-1\right)\theta\right]^3\right)\theta''
$$

$$
+ \frac{4R_1}{1-\phi+\phi\frac{(\rho C_p)_{CNT}}{(\rho C_p)_f}}\left[\left(1+\left(\theta_f-1\right)\theta\right)^2\left(\theta_f-1\right)\left(\theta'\right)^2\right]
$$

$$
+ Pr\left[f\theta' + \frac{6}{(1-\phi)^{2.5}}\left(1+\frac{1}{\beta_1}\right)\frac{Ec}{Re}(f')^2 + 2\left(1+\frac{1}{\beta_1}\right)Ec\frac{1}{(1-\phi)^{2.5}}\left[(f'')^2 + g'^2\right]\right] = 0.
\tag{12}
$$

Such that,

$$
g = 1, \qquad f' = \delta_1, \qquad f = \delta_1, \qquad -B_i(1-\theta) = \frac{K_{nf}}{K_f}\theta' \qquad \text{at}\ \ \eta = 0,
\tag{13}
$$

$$g \to 0, \qquad f' \to 0, \qquad \theta \to 0, \qquad \text{as } \eta \to \infty. \tag{14}$$

where,

$$F_r = \frac{Cd}{\sqrt{K^*}}, \qquad Pr = \frac{\nu_f}{\alpha_f}, \qquad R_1 = \frac{4\sigma^* T_\infty^3}{K^* K_f},$$

$$S_1 = \frac{W}{\sqrt{2\Omega\nu_f}}, \qquad Ec = \frac{r^2\Omega^2}{(C_p)_f T_\infty \left(\theta_f - 1\right)}, \qquad \delta_1 = \frac{a}{\Omega}, \tag{15}$$

$$Bi = \frac{h_f}{K_f}\sqrt{\frac{\nu_f}{2\Omega}},$$

are Forchheimer number, Prandtl number, nonlinear radiation factor, Suction parameter, Eckert number, stretching strength parameter, and Biot number, respectively. The physical quantities are,

$$Re_r^{1/2} C_f = \frac{1}{(1-\phi)^{2.5}}\left(1 + \frac{1}{\beta}\right)\sqrt{(G'(0))^2 + (f''(0))^2}, \tag{16}$$

$$Re_r^{-1/2} Nu_r = -\left[\frac{K_{nf}}{K_f} + \frac{4}{3}R_1\left[1 + (\theta_f - 1)\theta(0)\right]^3\right]\theta'(0). \tag{17}$$

3. Solution Methodology

The numerical RK45 scheme, together with the shooting technique, is implemented for final solutions of the problems. In order to gain a clear physical insight, firstly, the above Equations (11) and (12), along with the boundary conditions (13) and (14), are converted to an initial value problem and then solved numerically by means of the fourth-order Runge–Kutta method coupled with the shooting technique, with a systematic estimate of $f''(0)$ and $\theta(0)$ according to the corresponding boundary conditions at $f'(\infty)$ and $\theta(\infty)$ with the Newton–Raphson shooting technique. In this method, it is necessary to choose a suitable finite value for $\eta \to \infty$, say η_∞. If the boundary conditions at infinity are not satisfied, then the numerical routine uses the Newton–Raphson method to calculate the corrections to the estimated values of $f''(0)$ and $\theta(0)$. This process is repeated iteratively until convergence is achieved to a specified accuracy, with order 10–5. Assuming the governing parameters are as follows,

$$j_1 = f, j_2 = f', j_3 = f'', j_4 = g, j_5 = g', j_6 = \theta, j_7 = \theta' \tag{18}$$

then, higher order of governing equations can be written as follows:

$$f''' = -\frac{1}{2\frac{1}{(1-\phi)^{2.5}\left(1-\phi+\phi\left(\frac{\rho_{CNT}}{\rho_f}\right)\right)}\left(1+\frac{1}{\beta_1}\right)}$$

$$\left[-K\left(1+\frac{1}{\beta_1}\right)\frac{1}{(1-\phi)^{2.5}\left(1-\phi+\phi\left(\frac{\rho_{CNT}}{\rho_f}\right)\right)}j_2 + 2j_1 j_3 - j_1^2 + j_4^2\right] \tag{19}$$

$$-\frac{1}{2\frac{1}{(1-\phi)^{2.5}\left(1-\phi+\phi\left(\frac{\rho_{CNT}}{\rho_f}\right)\right)}\left(1+\frac{1}{\beta_1}\right)}\left[\frac{\left(1-\phi+\frac{(\rho\beta)_{CNT}\phi}{(\rho\beta)_f}\right)}{1-\phi+\frac{\rho_{CNT}\phi}{\rho_f}}\lambda j_6 - Fr j_2^2 - \frac{M_1^2}{1-\phi+\phi\frac{\rho_{CNT}}{\rho_f}}j_1\right],$$

$$g'' = -\frac{1}{\frac{1}{(1-\phi)^{2.5}\left(1-\phi+\phi\left(\frac{\rho_{CNT}}{\rho_f}\right)\right)}\left(1+\frac{1}{\beta_1}\right)}$$

$$\left[-K\left(1+\frac{1}{\beta_1}\right)\frac{1}{(1-\phi)^{2.5}\left(1-\phi+\phi\left(\frac{\rho_{CNT}}{\rho_f}\right)\right)}j_4 + 2j_1 j_5 - 2j_2 j_4\right] \tag{20}$$

$$-\frac{1}{\frac{1}{(1-\phi)^{2.5}\left(1-\phi+\phi\left(\frac{\rho_{CNT}}{\rho_f}\right)\right)}\left(1+\frac{1}{\beta_1}\right)}\left[\frac{\left(1-\phi+\frac{(\rho\beta)_{CNT}\phi}{(\rho\beta)_f}\right)}{1-\phi+\frac{\rho_{CNT}\phi}{\rho_f}}\lambda j_6 - Fr j_4^2 - \frac{M_1^2}{1-\phi+\phi\frac{\rho_{CNT}}{\rho_f}}j_4\right],$$

$$\theta'' = -\frac{Pr}{\frac{\frac{K_{nf}}{K_f}}{\left(1-\phi+\phi\left(\frac{(\rho C_p)_{CNT}}{(\rho C_p)_f}\right)\right)}\left(1+\frac{4}{3}\frac{K_f}{K_{nf}}R_1[1+(\theta_f-1)j_6]^3\right)}\left(\frac{4R_1}{1-\phi+\phi\frac{(\rho C_p)_{CNT}}{(\rho C_p)_f}}(1+(\theta_f-1)j_6)^2 j_7^2\right)$$

$$\frac{-Pr}{\frac{\frac{K_{nf}}{K_f}}{\left(1-\phi+\phi\left(\frac{(\rho C_p)_{CNT}}{(\rho C_p)_f}\right)\right)}\left(1+\frac{4}{3}\frac{K_f}{K_{nf}}R_1[1+(\theta_f-1)j_6]^3\right)}\left(j_1 j_7 + \frac{6}{(1-\phi)^{2.5}}\left(1+\frac{1}{\beta_1}\right)\frac{Ec}{Re}j_2^2\right) \tag{21}$$

$$\frac{-Pr}{\frac{\frac{K_{nf}}{K_f}}{\left(1-\phi+\phi\left(\frac{(\rho C_p)_{CNT}}{(\rho C_p)_f}\right)\right)}\left(1+\frac{4}{3}\frac{K_f}{K_{nf}}R_1[1+(\theta_f-1)j_6]^3\right)}\left(2\left(1+\frac{1}{\beta_1}\right)Ec\frac{1}{(1-\phi)^{2.5}}(j_3^2+j_5^2)\right)$$

Thus, the above mentioned three equations are used to write down the system of non linear governing ODEs in the form of a matrix subject to the converted boundary conditions according to the new parameters, and solved by using the numerical scheme. The skin friction coefficient and the Nusselt number are also converted accordingly.

4. Results and Discussion

Here in, Casson-water/glycerine MHD Darcy-Forchheimer nanofluid flow analysis subject to a rotating frame is considered. The rate of heat transfer and skin-friction are analyzed. The graphical display of results gives the impact of various parameters involved in the flow model on the main profiles of momentum and energy. The numerical RK45 scheme is invoked to obtain the requisite solutions of the governing non-linear ordinary differential equations. The graphs are sketched from the final solutions to analyze the impact of various parameters on fluid flow profiles. It is pertinent to note that solid lines represent the carbon nanotubes—water dilution, while the dashed lines are used for carbon nanotubes—Glycerine dilute, respectively.

4.1. Radial Velocity

In particular, Figure 2 gives the impact of the Forchheimer number on the momentum boundary layer in the context of radial velocity field. The enhanced Forcheimer number physically relates with more frictional force offered to the fluid in the opposite direction of the movement. Clearly, a decline in both cases, i.e., SWCNTs and MWCNTs can be seen in the figure. Figure 3 represents the behavior of velocity profile subject to variation in Casson parameter. Both, the solid and dashed lines present a declining trend. Physically, the elevated Casson parameter means a reduction in yield stress which in turns correspond to a Newtonian fluid, consequently, the fluid velocity undergoes a restriction. Figure 4 gives the variation in velocity field subject to augmented values of porosity factor. The larger the porosity parameter, the larger the resistance offered by the medium to the fluid flow and consequent result is decline in velocity. Both the cases behave in similar trends. The impact of stretching strength parameter on radial velocity is given in Figure 5. The stronger stretching rate corresponds to declination in the radial component of velocity. Away from disk, the result is significant decline in velocity profile. Figure 6 corresponds to the significance of permeability parameter (K) in the radial velocity field. The velocity profile shows drastic declination in both cases when the values of K are increased. Larger values of K correspond to the dense porous matrix, which in turn offers intensive resistance to the fluid flow, and consequently a stronger retardation is faced by the fluid movement.

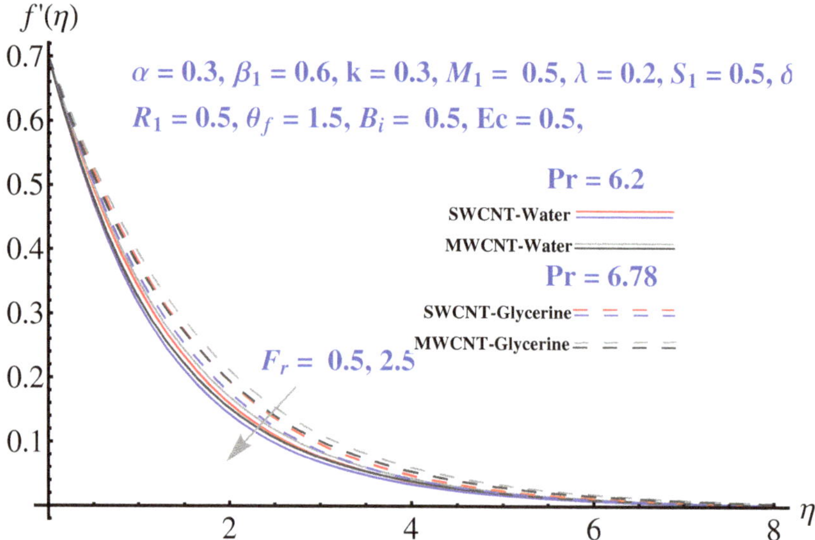

Figure 2. Impact of *Fr* on radial velocity.

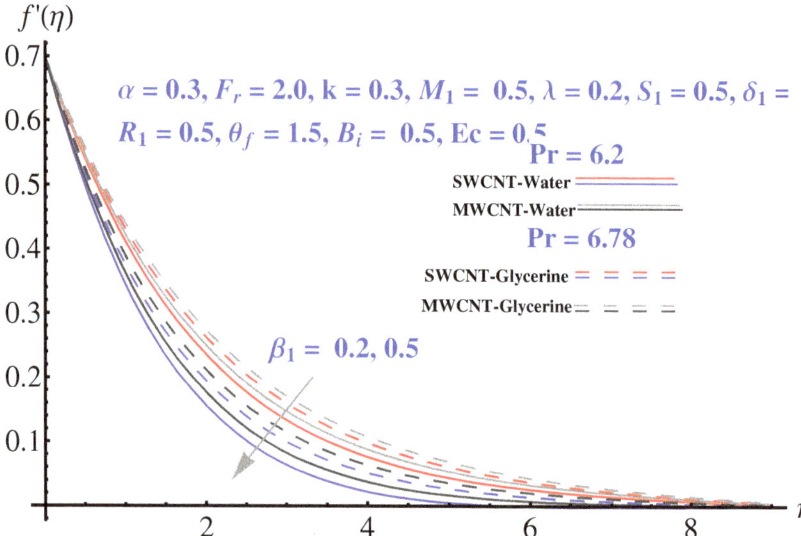

Figure 3. Impact of β_1 on radial velocity.

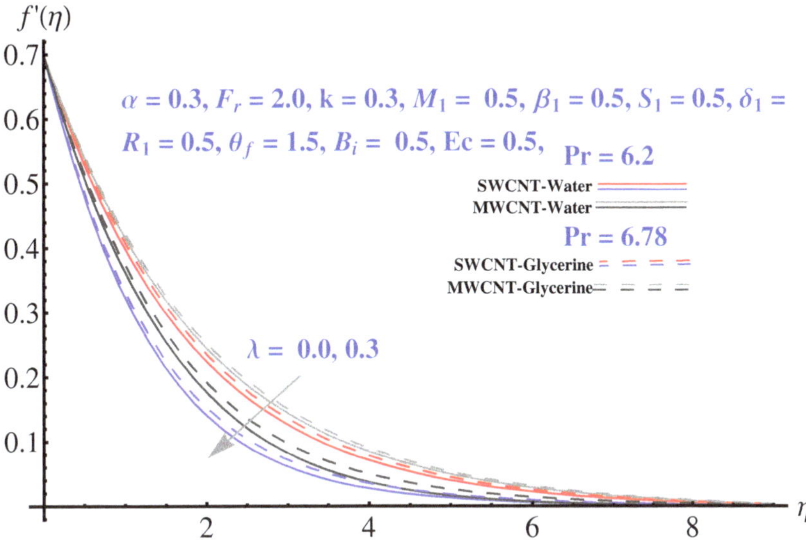

Figure 4. Impact of λ on radial velocity.

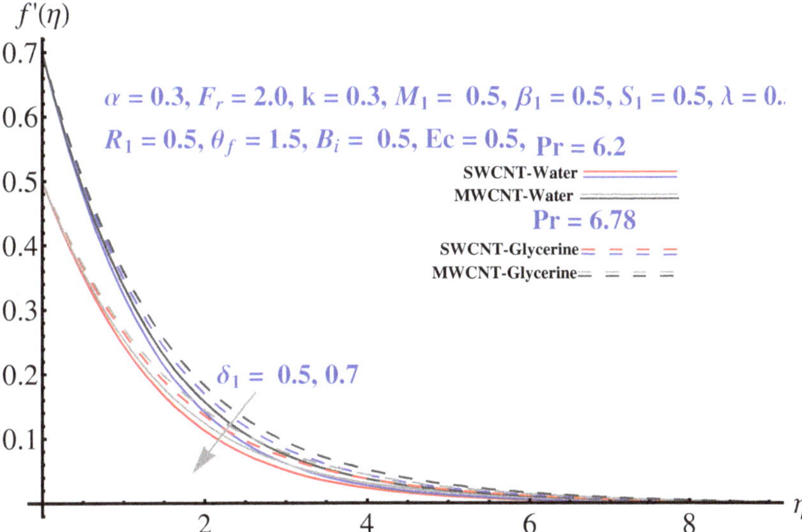

Figure 5. Impact of δ_1 on radial velocity.

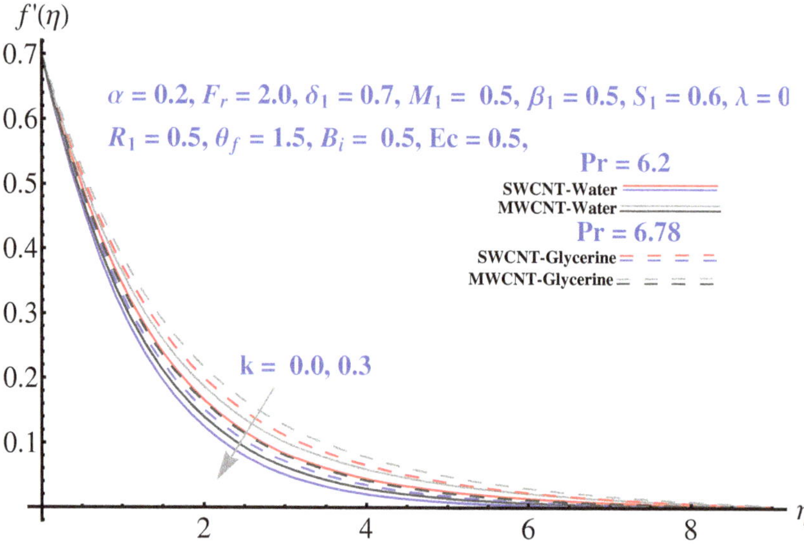

Figure 6. Impact of k on radial velocity.

4.2. Tangential Velocity

Impact of various parameters on Tangential velocity profiles is given in Figures 7–11. In particular, the impact of Casson factor on tangential velocity is given in Figure 7. A larger Casson factor results in decay of the transport rate. Subsequently, a shrinkage appeared in the corresponding boundary layer. Physically, the tensile stress appeared because of the elasticity yields a reduction in fluid movement. The stretching strength parameter results in the decline of the tangential velocity profile, as given in Figure 8. The impact of Forchheimer number, permeability parameter, and porosity parameter on the tangential velocity field is given in Figures 9–11. In both cases, the larger values of corresponding parameters are found to be declining factors for the fluid velocity and the associated boundary layer shrinks up to a significant level. More resistance is offered to the fluid flow that causes disturbance in the smooth movement and, thereby, the velocity profile and associated boundary layer ends up with a reducing trend. The convective condition involved in the governing equations results in scattered diagrams of thermal profile at the boundary.

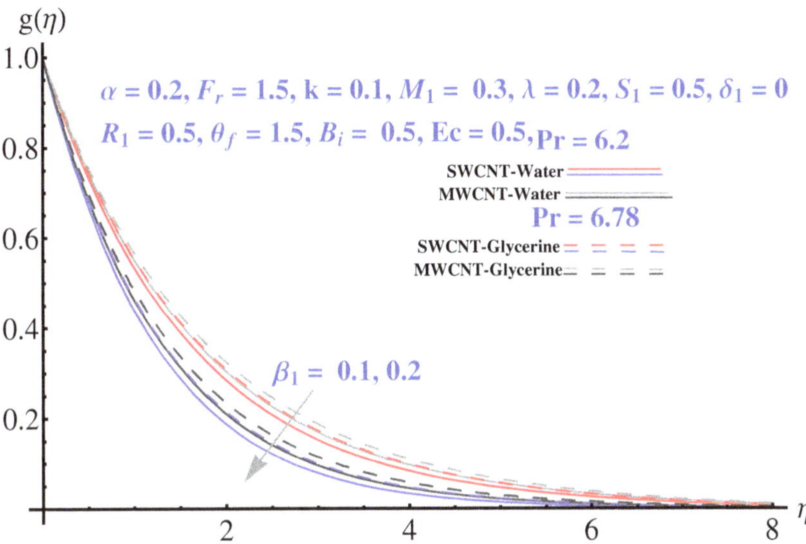

Figure 7. Impact of β_1 on tangential velocity.

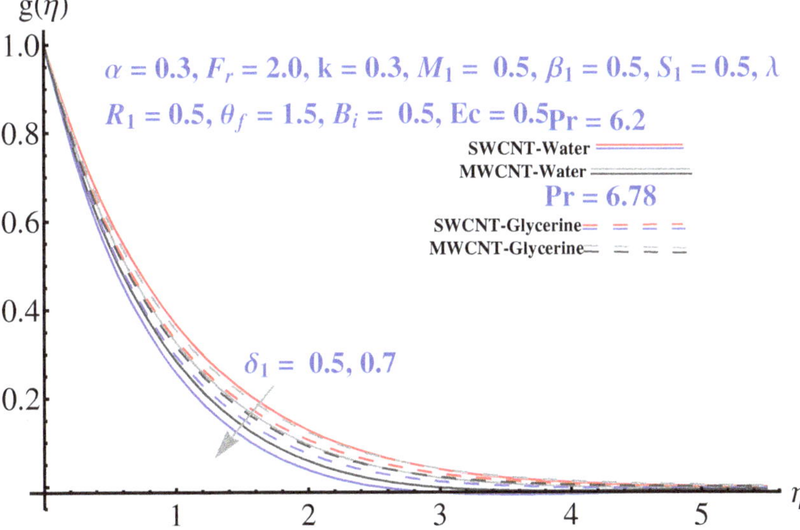

Figure 8. Impact of δ_1 on tangential velocity.

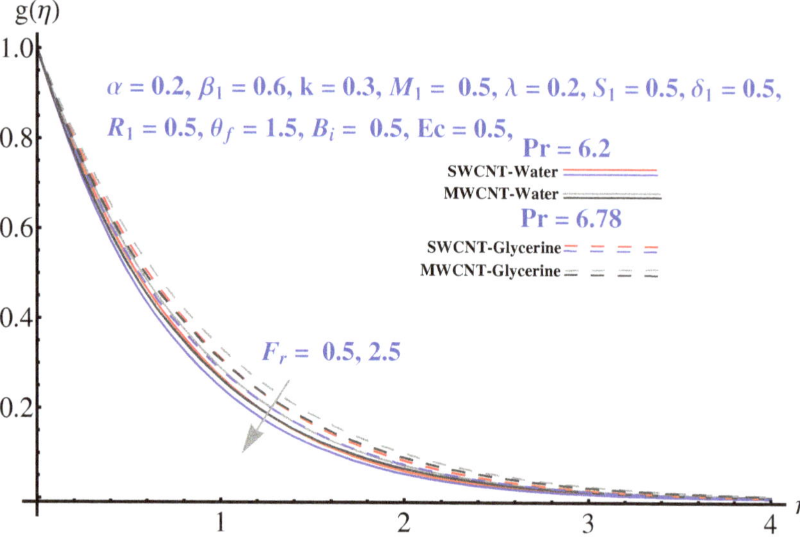

Figure 9. Impact of *Fr* on tangential velocity.

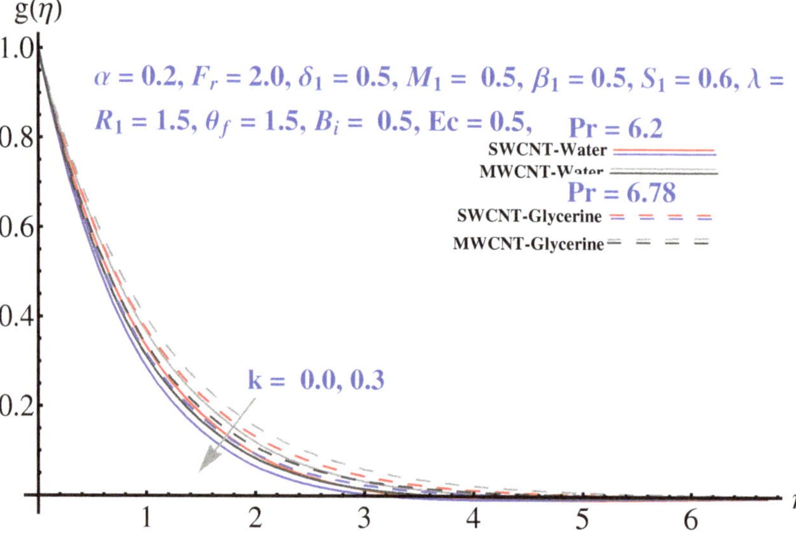

Figure 10. Impact of *k* on tangential velocity.

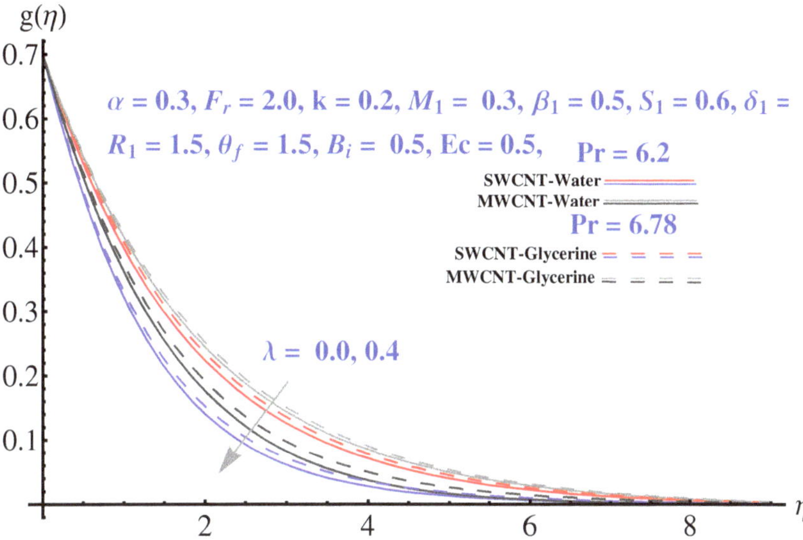

Figure 11. Impact of λ on tangential velocity.

4.3. Temperature Field

The impact of various parameters on thermal profile is given in Figures 12–14. The impact of θ_f on thermal profile is given in Figure 12. In both cases, the profile shows enhancement for elevated values of the corresponding parameter. A significant rise in thermal profile is noted for larger values of Biot number. The convective boundary results in enhancement of temperature facing the disk surface as compared to the ambient fluid. Physically, the trend of justified by the convective boundary. Similarly to Biot number, the enhanced radiation parameter results in more convenience in heat transfer rate and, therefore, the thermal state of the fluid enhances with larger values of radiation factor as given in Figure 14.

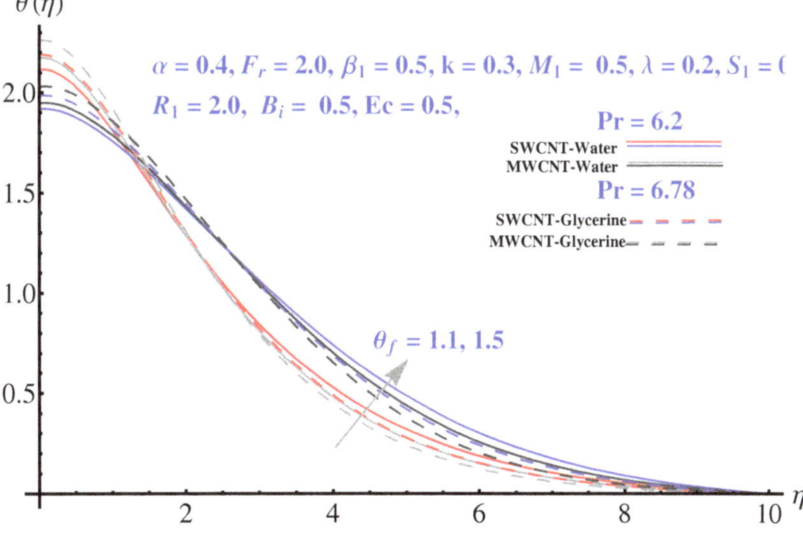

Figure 12. Impact of θ_f on temperature field.

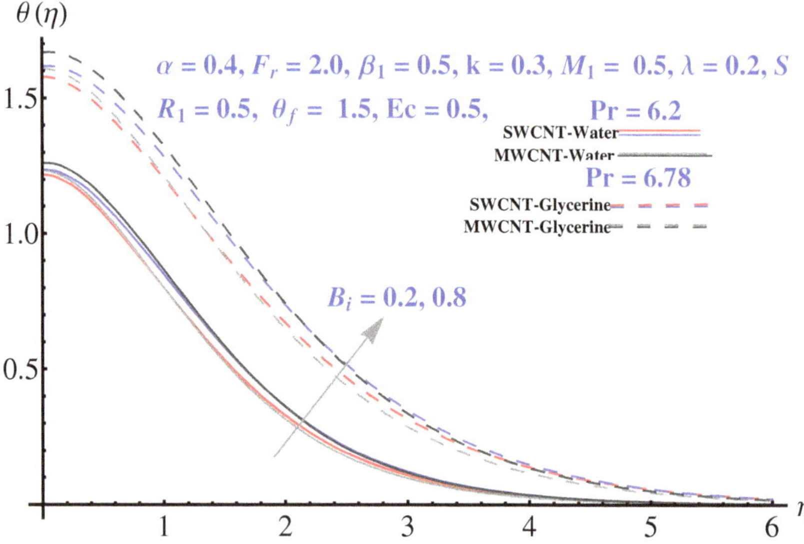

Figure 13. Impact of B_i on temperature field.

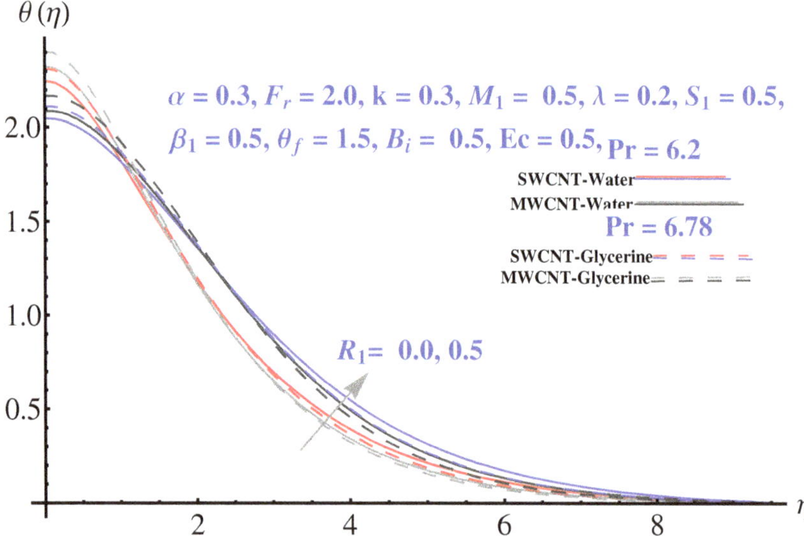

Figure 14. Impact of R_1 on temperature field.

4.4. Contour and Density Graphs

In Figures 15–22, the contour graphs have been sketched for various values of Casson parameter and permeability parameters against the single- and multi-walled Carbon nanotubes—Water/Glycerine dilution. Results are prominent near the surface, as compared to away from the surface. Figures 23 and 24 are the density graphs for both the SWCNT and MWCNT based Water/Glycerine nanofluid.

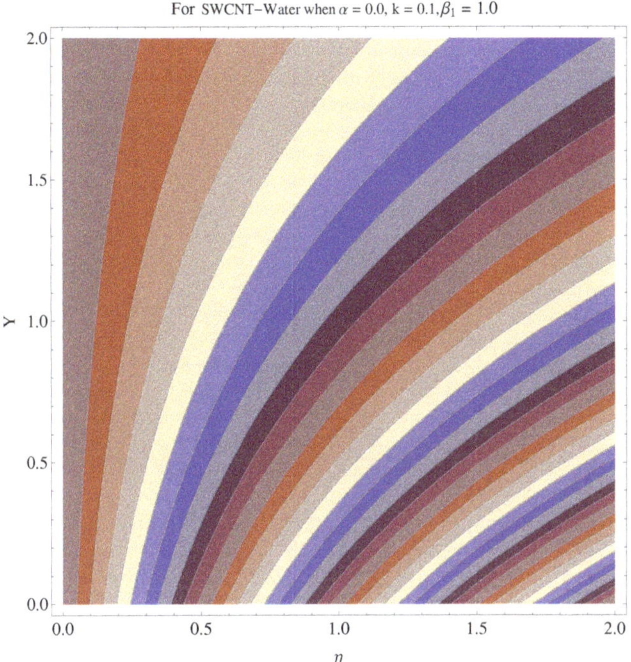

Figure 15. Contour graph at $\beta_1 = 1.0$ for SWCNT.

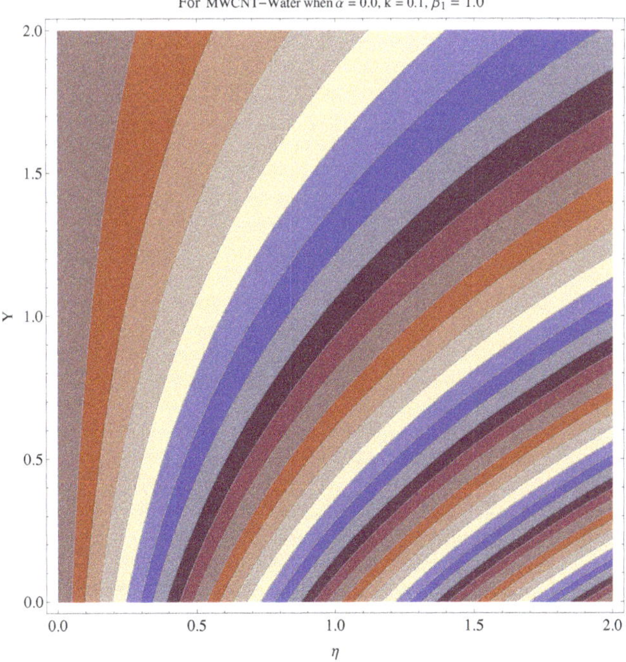

Figure 16. Contour graph at $\beta_1 = 1.0$ for MWCNT.

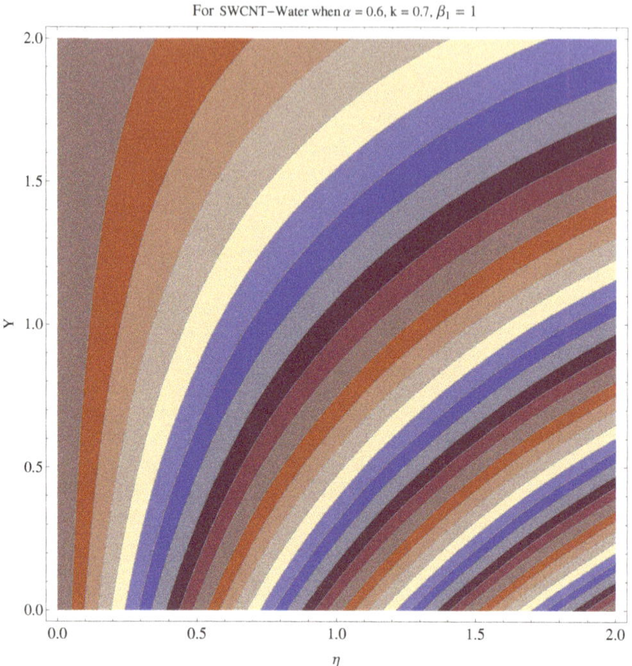

Figure 17. Contour graph at $\beta_1 = 1.0$ for SWCNT at larger k.

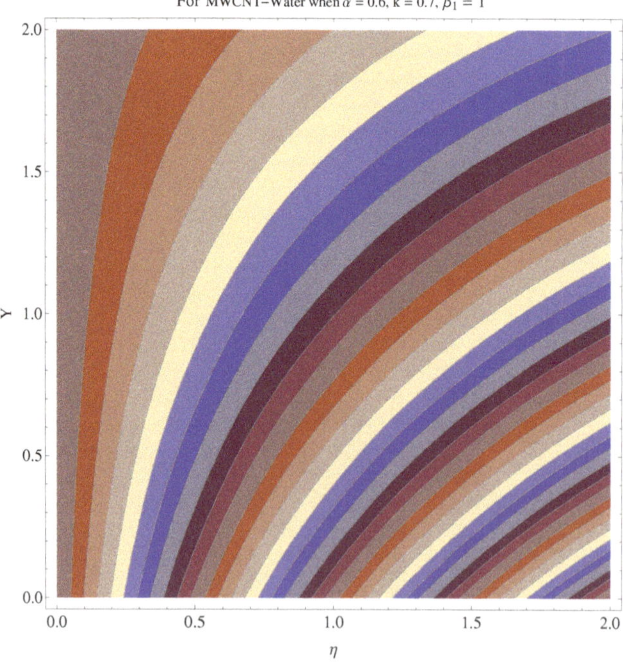

Figure 18. Contour graph at $\beta_1 = 1.0$ for MWCNT at larger k.

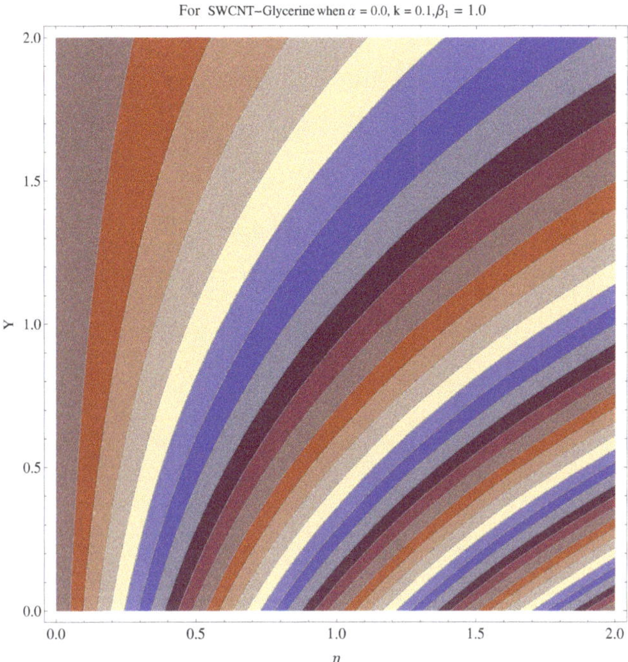

Figure 19. Contour graph at $\beta_1 = 1.0$ for SWCNT-Glycerine.

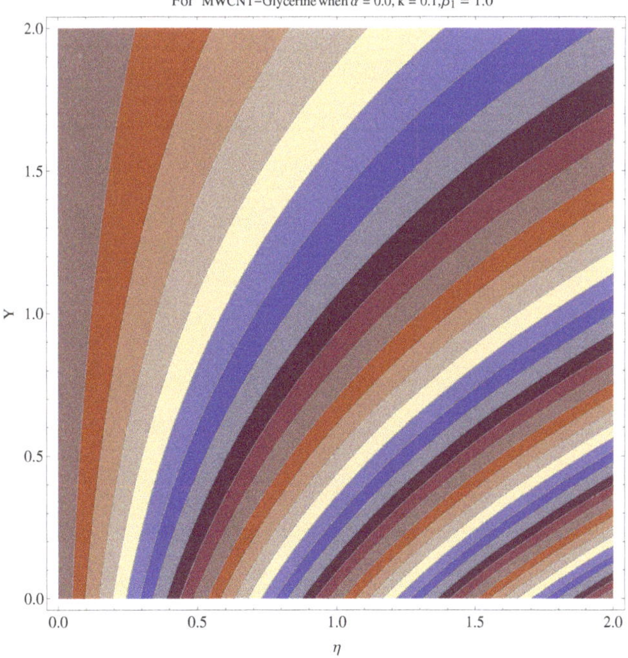

Figure 20. Contour graph at $\beta_1 = 1.0$ for MWCNT-Glycerine.

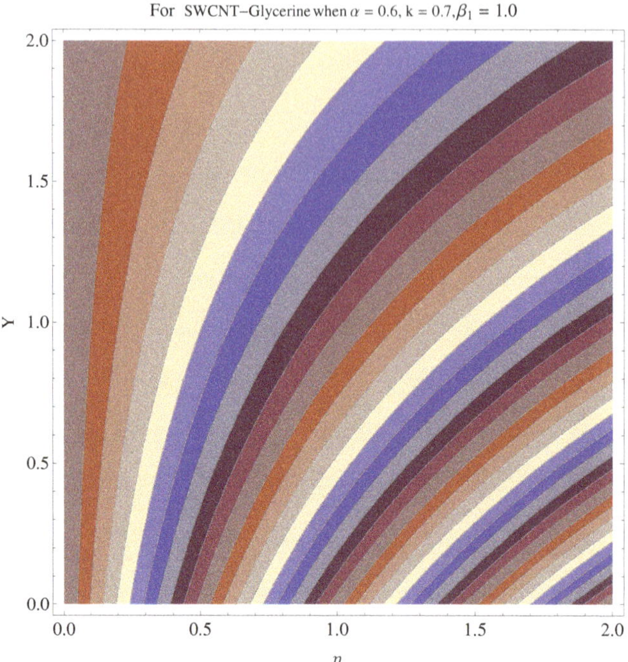

Figure 21. Contour graph at $\beta_1 = 1.0$ for SWCNT-Glycerine at larger k.

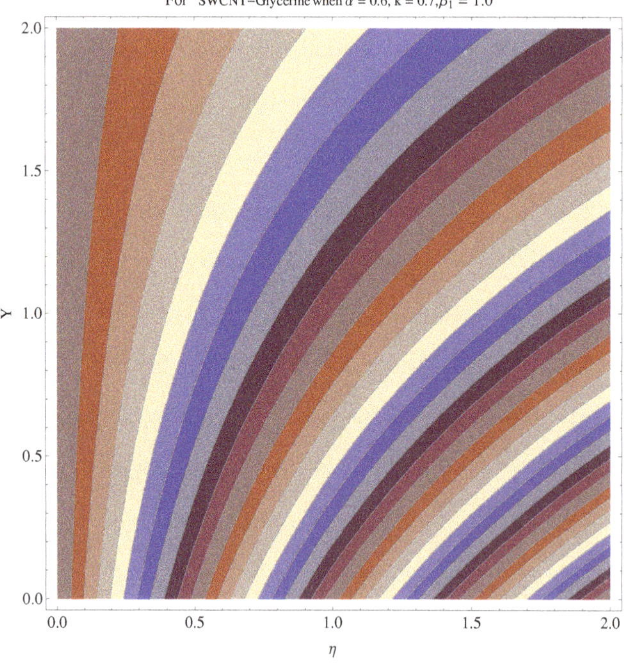

Figure 22. Contour graph at $\beta_1 = 1.0$ for MWCNT-Glycerine at larger k.

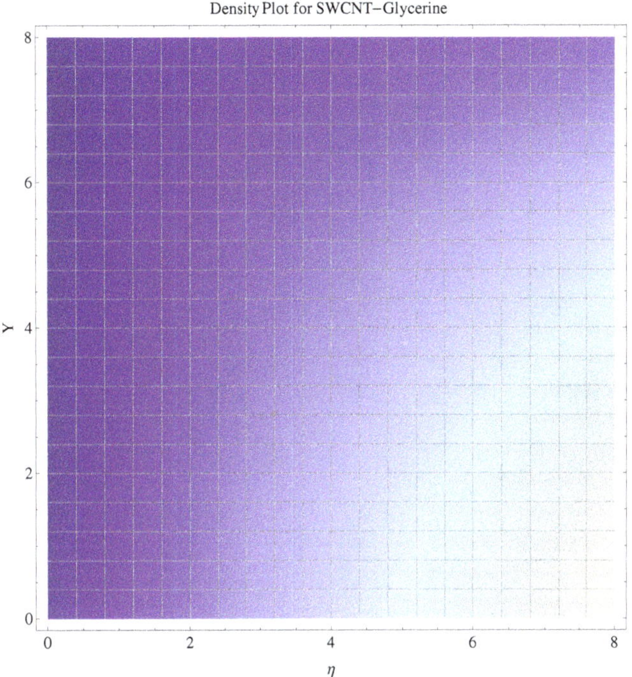

Figure 23. Density graph of single-walled nanotubes—Glycerine dilute.

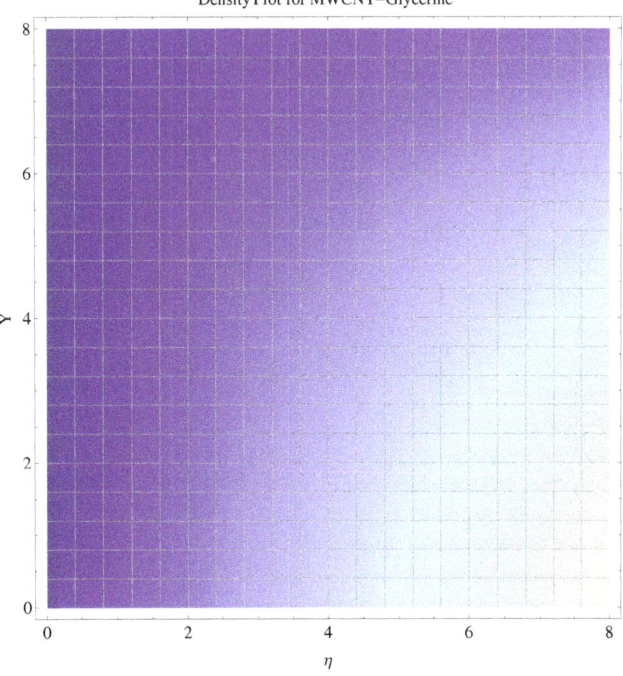

Figure 24. Density graph of multi-walled nanotubes—Glycerine dilute.

4.5. Skin Friction and Nusselt number

The variation of Skin-friction and Nusselt number is given in graphical, as well as tabular date form in Figures 25–32 and in Tables 2 and 3. One can see an enhancement in Skin-friction for larger values of Forchheimer number. Similarly, the non-linear radiation parameter shows an increasing trend in skin-friction. However, the friction faces a decline for enhancement in Casson parameter. The Forchheimer number results in the decline of the nusselt number (heat flux) as compared to the skin-friction. Whereas, the non-linear radiation parameter significantly increases the heat flux rate. Similar to skin-friction, Casson parameter results in decline of heat flux.

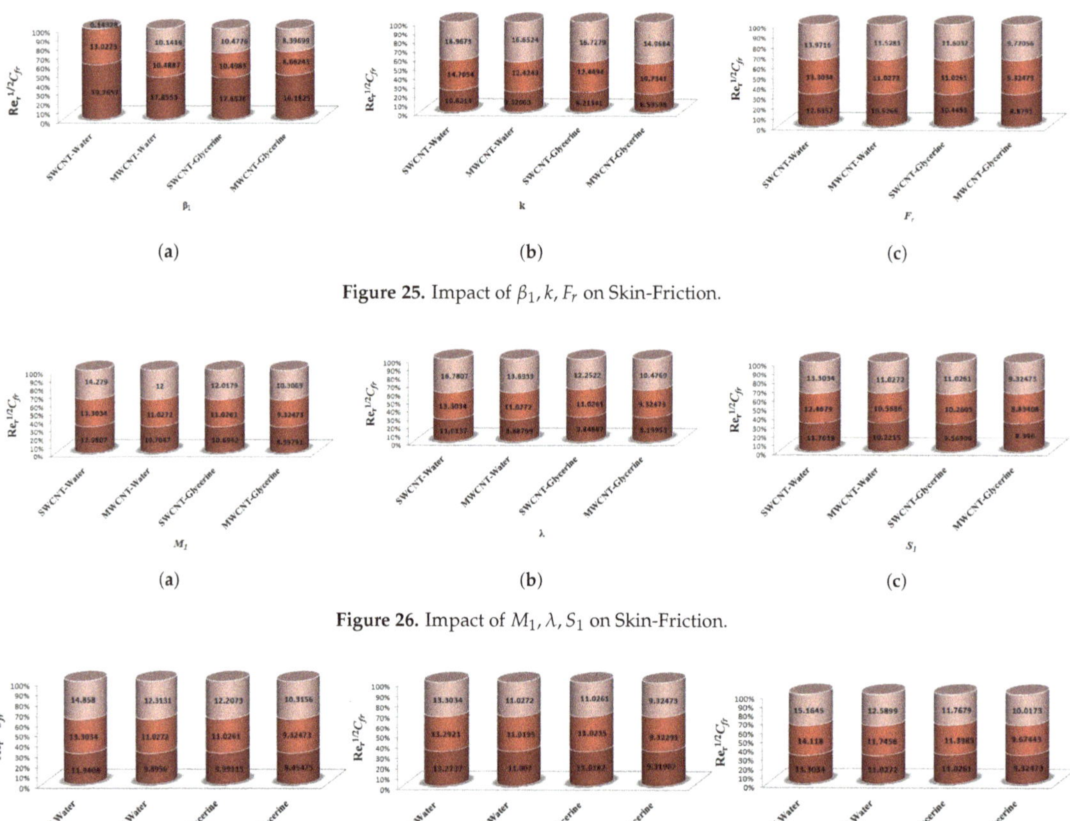

Figure 25. Impact of β_1, k, F_r on Skin-Friction.

Figure 26. Impact of M_1, λ, S_1 on Skin-Friction.

Figure 27. Impact of δ_1, R_1, Ec on Skin-Friction.

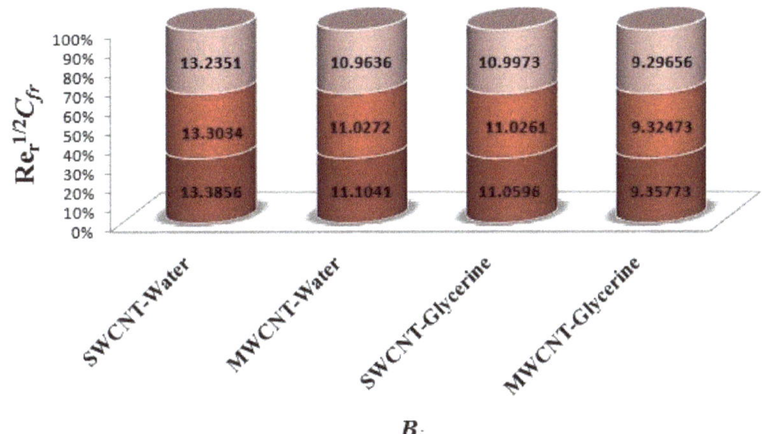

Figure 28. Impact of B_i on Skin-friction.

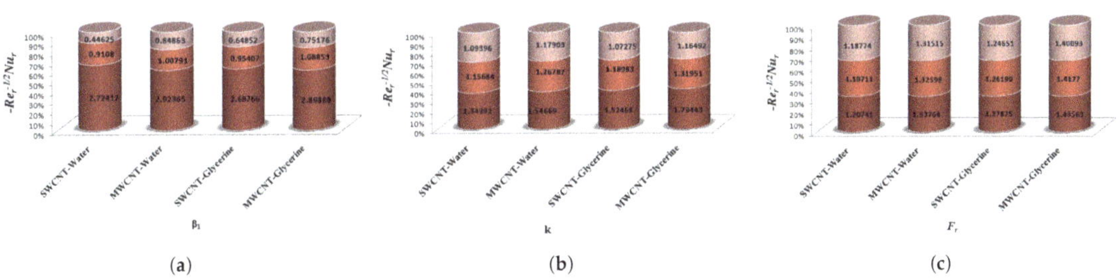

(a) **(b)** **(c)**

Figure 29. Impact of β_1, k, F_r on Nusselt number.

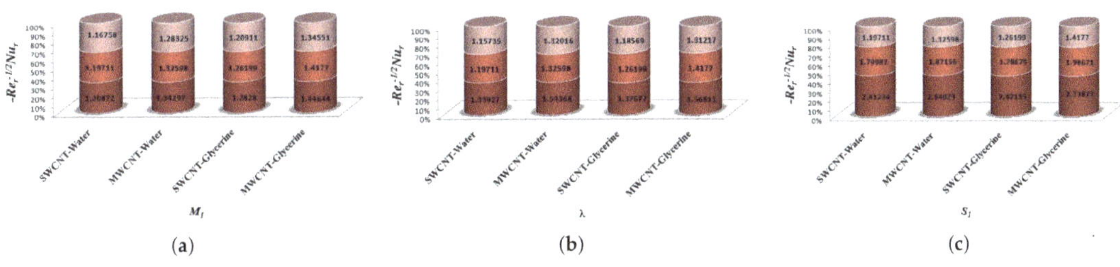

(a) **(b)** **(c)**

Figure 30. Impact of M_1, λ, S_1 on Nusselt number.

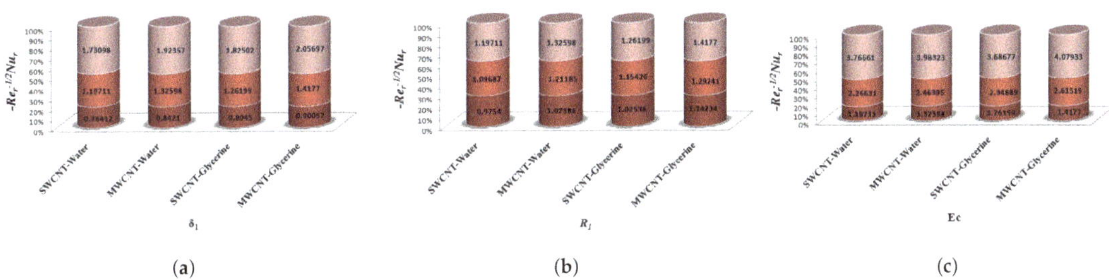

(a) **(b)** **(c)**

Figure 31. Impact of δ_1, R_1, Ec on Nusselt number.

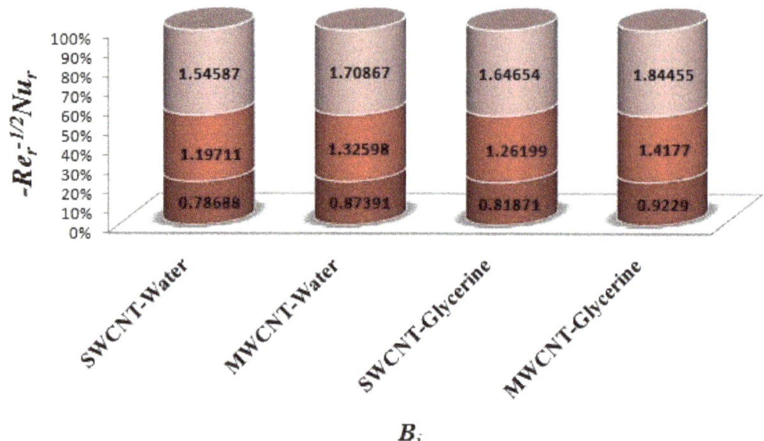

Figure 32. Impact of B_i on Nusselt number.

Table 2. Skin friction for water and Glycerine $\theta_f = 1.5$.

| | | | | | | | | | | $Re_r^{1/2}C_{fr}$ | | | |
| | | | | | | | | | | Water | | Glycerine | |
β_1	k	F_r	M_1	λ	S_1	δ_1	R_1	Ec	B_i	SWCNT	MWCNT	SWCNT	MWCNT
0.1	0.2	2.5	0.5	0.2	0.6	0.7	2.0	0.3	0.5	19.7657	17.8553	17.6526	16.1825
1.0										13.0225	10.4887	10.4983	8.66245
4.5										0.14328	10.1416	10.4776	8.39699
0.5	0.0	2.5	0.5	0.2	0.6	0.7	2.0	0.3	0.5	10.6214	8.32063	8.21341	6.53598
	0.3									14.7054	12.4243	12.4494	10.7341
	0.6									18.9673	16.6524	16.7279	14.9684
0.5	0.2	2.0	0.5	0.2	0.6	0.7	2.0	0.3	0.5	12.6357	10.5266	10.4493	8.87910
		2.5								13.3034	11.0272	11.0261	9.32473
		3.0								13.9716	11.5281	11.6032	9.77056
0.5	0.2	2.5	0.0	0.2	0.6	0.7	2.0	0.3	0.5	12.9807	10.7047	10.6962	8.99791
			0.5							13.3034	11.0272	11.0261	9.32473
			1.0							14.2790	12.0000	12.0179	10.3069
0.5	0.2	2.5	0.5	0.0	0.6	0.7	2.0	0.3	0.5	11.0137	8.88799	9.84887	8.19953
				0.2						13.3034	11.0272	11.0261	9.32473
				0.4						16.7807	13.6939	12.2522	10.4769
0.5	0.2	2.5	0.5	0.2	$0.0S_1$	0.7	2.0	0.3	0.5	11.7618	10.2215	9.56906	8.39600
					0.3					12.4679	10.5686	10.2605	8.83408
					0.6					13.3034	11.0272	11.0261	9.32473
0.5	0.2	2.5	0.5	0.2	0.6	0.6	2.0	0.3	0.5	11.9408	9.8956	9.99115	8.45475
						0.7				13.3034	11.0272	11.0261	9.32473
						0.8				14.858	12.3131	12.2073	10.3156
0.5	0.2	2.5	0.5	0.2	0.6	0.7	1.0	0.3	0.5	13.2737	11.007	11.0182	9.31907
							1.5			13.2921	11.0195	11.0235	9.32291
							2.0			13.3034	11.0272	11.0261	9.32473
0.5	0.2	2.5	0.5	0.2	0.6	0.7	2.0	0.3	0.5	13.3034	11.0272	11.0261	9.32473
								0.6		14.1180	11.7456	11.3985	9.67443
								0.9		15.1645	12.5899	11.7679	10.0173
0.5	0.2	2.5	0.5	0.2	0.6	0.7	2.0	0.3	0.1	13.3856	11.1041	11.0596	9.35773
									0.3	13.3034	11.0272	11.0261	9.32473
									0.6	13.2351	10.9636	10.9973	9.29656

Table 3. Nusselt number for water and Glycerine $\theta_f = 1.5$.

| | | | | | | | | | | $-Re_r^{-1/2}Nu_r$ | | | |
| | | | | | | | | | | Water | | Glycerine | |
β_1	k	F_r	M_1	λ	S_1	δ_1	R_1	Ec	B_i	SWCNT	MWCNT	SWCNT	MWCNT
0.1	0.2	2.5	0.5	0.2	0.6	0.7	2.0	0.3	0.5	2.72417	2.92365	2.68766	2.89888
1.0										0.91080	1.00791	0.95407	1.08853
4.5										0.44625	0.84863	0.64852	0.75176
0.5	0.0	2.5	0.5	0.2	0.6	0.7	2.0	0.3	0.5	1.34392	1.54669	1.52461	1.79443
	0.3									1.15684	1.26787	1.18983	1.31951
	0.6									1.09396	1.17903	1.07275	1.16492
0.5	0.5	2.0	0.5	0.2	0.6	0.7	2.0	0.3	0.5	1.20741	1.33764	1.27875	1.43563
		2.5								1.19711	1.32598	1.26199	1.41770
		3.0								1.18774	1.31515	1.24651	1.40093
0.5	0.2	2.5	0.0	0.2	0.6	0.7	2.0	0.3	0.5	1.20872	1.34297	1.28280	1.44644
			0.5							1.19711	1.32598	1.26199	1.41770
			0.6							1.16758	1.28325	1.20911	1.34551
0.5	0.2	2.5	0.5	0.0	0.6	0.7	2.0	0.3	0.5	1.33927	1.53368	1.37677	1.56811
				0.2						1.19711	1.32598	1.26199	1.41770
				0.4						1.15735	1.32016	1.18569	1.31217
0.5	0.2	2.5	0.5	0.2	0.0	0.7	2.0	0.3	0.5	2.41234	2.64023	2.47135	2.73877
					0.3					1.70987	1.87156	1.78679	1.98671
					0.6					1.19711	1.32598	1.26199	1.41770
0.5	0.2	2.5	0.5	0.2	0.6	$\delta_1$0.6	2.0	0.3	0.5	0.76412	0.84210	0.80450	0.90057
						0.7				1.19711	1.32598	1.26199	1.41770
						0.8				1.73098	1.92357	1.82502	2.05697
0.5	0.2	2.5	0.5	0.2	0.6	0.7	1.0	0.3	0.5	0.97540	1.07386	1.02538	1.14234
							1.5			1.09687	1.21185	1.15426	1.29241
							2.0			1.19711	1.32598	1.26199	1.41770
0.5	0.2	2.5	0.5	0.2	0.6	0.7	2.0	0.3	0.5	1.19711	1.32598	1.26199	1.41770
								0.6		2.26631	2.46395	2.34889	2.61519
								0.9		3.76661	3.98323	3.68677	4.07933
0.5	0.2	2.5	0.5	0.2	0.6	0.7	2.0	0.3	0.1	0.78688	0.87391	0.81871	0.92290
									0.3	1.19711	1.32598	1.26199	1.41770
									0.6	1.54587	1.70867	1.64654	1.84455

5. Conclusions

The present investigation aims to reveal the significance of non-linear thermal radiation on Casson-water/glycerine MHD Darcy-Forchheimer fluid flow analysis subject to a rotating frame. The rate of heat transfer and skin-friction are analyzed. The graphical display of results gives the impact of various parameters involved in the flow model on the main profiles of momentum and energy. The numerical RK45 scheme is invoked to obtain the requisite solutions of the governing non-linear ordinary differential equations. Salient features are listed below:

- The enhanced Forcheimer number results in a decline in radial velocity;
- Casson parameter restricts the fluid velocity at larger values;
- The larger the porosity parameter, the larger the resistance offered by the medium to the fluid flow and the consequent result is the decline in velocity;
- The larger permeability parameter (K) results a drastic declination in both cases (SWCNT and MWCNT) when the values of K are increased;
- A larger Casson factor results in the shrinkage in the corresponding boundary layer of thermal field;
- Forchheimer number, permeability parameter and porosity parameter on tangential decrease the velocity field;

- The convective boundary results in enhancement of temperature facing the disk surface as compared to the ambient fluid;
- The contour graphs have been sketched for various values of Casson parameter and permeability parameter against the single- and multi-walled Carbon nanotubes—Water/Glycerine nanofluid.
- Skin-friction for larger values of Forchheimer number.
- The non-linear radiation parameter significantly increases the heat flux rate.

Author Contributions: All authors contributed equally to this manuscript. All authors have read and agreed to the published version of the manuscript.

Funding: Taif University Researches Supporting Project number (TURSP-2020/279), Taif University, Taif, Saudi Arabia.

Acknowledgments: Taif University Researches Supporting Project number (TURSP-2020/279), Taif University, Taif, Saudi Arabia.

Conflicts of Interest: The authors have no conflicts of interest.

Nomenclature

MHD	Magnetohydrodynamic
ODEs	Ordinary Differential Equations
PDEs	Partial Differential Equation
CNTs	Carbon Nanotubes
ρ_{nf}	density of given nanofluid
ρ_{CNTs}	density of carbon nanotubes
u, v, w	Velocity components/m·s^{-1}
P	Pressure /Pa
μ_f	dynamic viscosity of considered base fluid /$\text{kgm}^{-1}\,\text{s}^{-1}$
μ_{nf}	dynamic viscosity of the given nanofluid /$\text{kgm}^{-1}\,\text{s}^{-1}$
T	Local temperature/K
T_∞	Ambient temperature/K
f, g	Nondimensional velocity components
g_1	Gravitational force
Gr	Grashof number
K	Permeability parameter
Fr	Forchheimer number (inertia)
β_1	Casson parameter
ϕ	Volume fraction (solid)
Ω	Angular Velocity (angular)
C_f	Skin-friction (wall drag force)
Nu_x	Local Nusselt number
Re_x	Local Reynolds number
M_1	magnetic number
Pr	Prandtl number
Ec	Eckert parameter
λ	Ratio of Grashof and Reynolds square
$S = \dfrac{W}{\sqrt{2\Omega v_f}}$	Suction parameter

References

1. Choi, S.U.S. Enhancing Thermal Conductivity of Fluids with Nanoparticles. In *Developments and Applications of Non-Newtonian Flows*; Siginer, D.A., Wang, H.P., Eds.; FED-Vol. 231/MD-Vol. 66; ASME: New York, NY, USA, 1995; pp. 99–105.
2. Lin, Y.; Zheng, L.; Zhang, X.; Ma, L.; Chen, G. MHD pseudoplastic nanofluid unsteady flow and heat transfer in a finite thin film over stretching surface with internal heat generation. *Int. J. Heat Mass Transf.* **2015**, *84*, 903–911. [CrossRef]
3. Bai, Y.; Liu, X.; Zhang, Y.; Zhang, M. Stagnation-point heat and mass transfer of MHD Maxwell nanofluids over a stretching surface in the presence of thermophoresis. *J. Mol. Liq.* **2016**, *224*, 1172–1180. [CrossRef]
4. Madhu, M.; Kishan, N.; Chamkha, A.J. Unsteady flow of a Maxwell nanofluid over a stretching surface in the presence of magnetohydrodynamic and thermal radiation effects. *Propuls. Power Res.* **2017**, *6*, 31–40. [CrossRef]

5. Sheikholeslami, M.; Sajjadi, H.; Amri Delouei, A.; Atashafrooz, M.; Li, Z. Magnetic force and radiation influences on nanofluid transportation through a permeable media considering Al_2O_3 nanoparticles, mixed convection of a nanofluid in a threedimensional channel. *J. Therm. Anal. Calorim.* **2019**, *136*, 2477–2485. [CrossRef]
6. Hayat, T.; Kiyani, M.Z.; Alsaedi, A.; Khan, M.I.; Ahmad, I. Mixed convective three dimensional flow of Williamson nanofluid subject to chemical reaction. *Int. J. Heat Mass Transf.* **2018**, *127*, 422–429. [CrossRef]
7. Zadi, M.; Hashemi Pour, S.M.R.; Karimdoost Yasuri, A.; Chamkha, A.J. Mixed convection of a nanofluid in a three-dimensional channel. *J. Therm. Anal. Calorim.* **2019**, *136*, 2461–2475.
8. Turkyilmazoglu, M.; Senel, P. Heat and mass transfer of the flow due to a rotating rough and porous disk. *Int. J. Therm. Sci.* **2013**, *63*, 146–158. [CrossRef]
9. Bödewadt, U.T. Die Drehströmungüberfestem Grund. *Z. Angew. Math. Mech.* **1940**, *20*, 241–252. [CrossRef]
10. Mustafa, M.; Khan, J.A.; Hayat, T.; Alsaedi, A. On Bödewadt flow and heat transfer of nanofluids over a stretching stationary disk. *J. Mol. Liq.* **2015**, *211*, 119–125. [CrossRef]
11. Mustafa, M. MHD nanofluid flow over a rotating disk with partial slip effects: Buongiorno model. *Int. J. Heat Mass Transf.* **2017**, *108*, 1910–1916. [CrossRef]
12. Dogonchi, A.S.; Armaghani, T.; Chamkha, A.J.; Ganji, D.D. Natural convection analysis in a cavity with an inclined elliptical heater subject to shape factor of nanoparticles and magnetic field. *Arab. J. Sci. Eng.* **2019**. [CrossRef]
13. Forchheimer, P. Wasserbewegungdurchboden. *Z. Ver. Dtsch. Ing.* **1901**, *45*, 1782–1788.
14. Shehzad, S.A.; Abbasi, F.M.; Hayat, T.; Alsaedi, A. Cattaneo—Christov heat flux model for Darcy–Forchheimer flow of an Oldroyd-B fluid with variable conductivity and non-linear convection. *J. Mol. Liq.* **2016**, *224*, 274–278. [CrossRef]
15. Bakar, S.A.; Arifin, N.M.; Nazar, R.; Ali, F.M.; Pop, I. Forced convection boundary layer stagnation-point flow in Darcy-Forchheimer porous medium past a shrinking sheet. *Front. Heat Mass Transf.* **2016**, *7*, 38.
16. Hayat, T.; Muhammad, T.; Al-Mezal, S.; Liao, S.J. Darcy-Forchheimer flow with variable thermal conductivity and Cattaneo-Christov heat flux. *Int. J. Numer. Methods Heat Fluid Flow* **2016**, *26*, 2355–2369. [CrossRef]
17. Chamkha, A.J.; Abbasbandy, S.; Rashad, A.M.; Vajravelu, K. Radiation effects on mixed convection over a wedge embedded in a porous medium filled with a nanofluid. *Transp. Porous Media* **2012**, *91*, 261–279. [CrossRef]
18. Chamkha, A.J.; Abbasbandy, S.; Rashad, A.M.; Vajravelu, K. Radiation effects on mixed convection about a cone embedded in a porous medium filled with a nanofluid. *Meccanica* **2013**, *48*, 275–285. [CrossRef]
19. Donaldson, K.; Aitken, R.; Tran, L.; Stone, V.; Duffin, R.; Forrest, G.; Alexander, A. Carbon nanotubes: A review of their properties in relation to pulmonary toxicology and workplace safety. *Toxicol. Sci.* **2006**, *92*, 5–22. [CrossRef] [PubMed]
20. Coleman, J.N.; Khan, U.; Blau, W.J.; Gunko, Y.K. Small but strong: A review of the mechanical properties of carbon nanotube–polymer composites. *Carbon* **2006**, *44*, 1624–1652. [CrossRef]
21. Kim, G.M.; Nam, I.W.; Yang, B.; Yoon, H.N. ; Lee, H.K.; Park, S. Carbon nanotube (CNT) incorporated cementitious composites for functional construction materials: The state of the art. *Compos. Struct.* **2019**, *227*, 111244. [CrossRef]
22. Janas, D.; Koziol, K.K. A review of production methods of carbon nanotube and graphene thin films for electrothermal applications. *Nanoscale* **2014**, *6*, 3037–3045. [CrossRef] [PubMed]
23. Keidar, M.; Waas, A.M. On the conditions of carbon nanotube growth in the arc discharge. *Nanotechnology* **2004**, *15*, 1571–1575. [CrossRef]
24. Mukul, K.; Yoshinori, A. Chemical vapor deposition of carbon nanotubes: A review on growth mechanism and mass production. *J. Nanosci. Nanotechnol.* **2010**, *10*, 3739–3758.
25. Dai, H.J. Carbon nanotubes: Synthesis, integration, and properties. *Acc. Chem. Res.* **2002**, *35*, 1035–1044. [CrossRef] [PubMed]
26. Ahmadian, A.; Bilal, M.; Khan, M.A.; Asjad, M.I. The non-Newtonian maxwell nanofluid flow between two parallel rotating disks under the effects of magnetic field. *Sci. Rep.* **2020**, *10*, 17088. [CrossRef] [PubMed]
27. Karen, C.; Fredy, O.; Camilo, Z.; Bernardo, H.; Elizabeth, P.; Robison, B.S. Surfactant concentration and pH effects on the zeta potential values of alumina nanofluids to inspect stability. *Colloids Surf. A Physicochem. Eng. Asp.* **2019**, *583*, 123960.
28. Choudhary, R.; Khurana, D.; Kumar, A.; Subudhi, S. Stability analysis of Al_2O_3/water nanofluids. *J. Exp. Nanosci.* **2017**, *12*, 140–151. [CrossRef]
29. Lim, C.Y.; Lim, A.E.; Lam, Y.C. pH change in electroosmotic flow hysteresis. *Anal. Chem.* **2017**, *89*, 9394–9399. [CrossRef]
30. Lim, A.E.; Lim, C.Y.; Lam, Y.C. Electroosmotic flow hysteresis for dissimilar anionic solutions. *Anal. Chem.* **2016**, *88*, 8064–8073. [CrossRef]
31. Alotaibi, H.; Althubiti, S.; Eid, M.R.; Mahny, K.L. Numerical treatment of mhd flow of casson nanofluid via convectively heated non-linear extending surface with viscous dissipation and suction/injection effects. *Comput. Mater. Contin.* **2020**, *66*, 229–245. [CrossRef]
32. Rafique, K.; Alotaibi, H.; Nofal, T.A.; Anwar, M.I.; Misiran, M.; Khan, I. Numerical Solutions of Micropolar Nanofluid over an Inclined Surface Using Keller Box Analysis. *J. Math.* **2020**, *6617652*. [CrossRef]
33. Rasool, G.; Shafiq, A.; Alqarni, M.S.; Wakif, A.; Khan, I.; Bhutta, M.S. Numerical Scrutinization of Darcy-Forchheimer Relation in Convective Magnetohydrodynamic Nanofluid Flow Bounded by Nonlinear Stretching Surface in the Perspective of Heat and Mass Transfer. *Micromachines* **2021**, *12*, 374. [CrossRef]
34. Rasool, G.; Khan, W.A.; Bilal, S.M.; Khan, I. MHD Squeezed Darcy Forchheimer Nanofluid flow between two h-distance apart horizontal plates. *Open Phys.* **2020**, *18*, 1–8. [CrossRef]

35. Rasool, G.; Shafiq, A. Numerical Exploration of the Features of Thermally Enhanced Chemically Reactive Radiative Powell-Eyring Nanofluid Flow via Darcy Medium over Non-linearly Stretching Surface Affected by a Transverse Magnetic Field and Convective Boundary Conditions. *Appl. Nanosci.* **2020**. [CrossRef]
36. Shaw, S.; Dogonchi, A.S.; Nayak, M.K.; Makinde, O.D. Impact of Entropy Generation and Nonlinear Thermal Radiation on Darcy–Forchheimer Flow of $MnFe_2O_4$-Casson/Water Nanofluid due to a Rotating Disk: Application to Brain Dynamics. *Arab. J. Sci. Eng.* **2020**, *45*, 5471–5490. [CrossRef]

 *micromachines*

Article

Forced Convection in Wavy Microchannels Porous Media Using TiO₂ and Al₂O₃–Cu Nanoparticles in Water Base Fluids: Numerical Results

Kholoud Maher Elsafy and Mohamad Ziad Saghir *

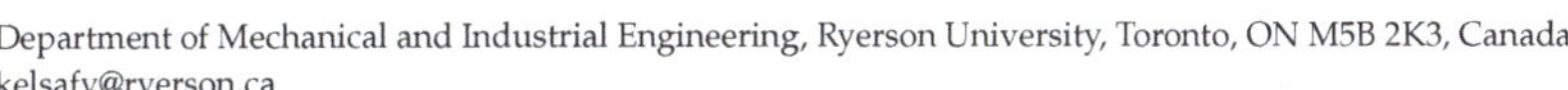

Department of Mechanical and Industrial Engineering, Ryerson University, Toronto, ON M5B 2K3, Canada; kelsafy@ryerson.ca
* Correspondence: zsaghir@ryerson.ca

Abstract: In the present work, an attempt is made to investigate the performance of three fluids with forced convection in a wavy channel. The fluids are water, a nanofluid of 1% TiO_2 in a water solution and a hybrid fluid which consists of 1% Al_2O_3–Cu nanoparticles in a water solution. The wavy channel has a porous insert with a permeability of 10 PPI, 20 PPI and 40 PPI, respectively. Since Reynolds number is less than 1000, the flow is assumed laminar, Newtonian and steady state. Results revealed that wavy channel provides a better heat enhancement than a straight channel of the same dimension. Porous material increases heat extraction at the expenses of the pressure drop. The nanofluid of 1% TiO_2 in water provided the highest performance evaluation criteria.

Keywords: porous cavity; wavy channels; nanofluids; forced convection; heat enhancement; pressure drop; mesh model

Citation: Elsafy, K.M.; Saghir, M.Z. Forced Convection in Wavy Microchannels Porous Media Using TiO_2 and Al_2O_3–Cu Nanoparticles in Water Base Fluids: Numerical Results. *Micromachines* **2021**, *12*, 654. https://doi.org/10.3390/mi12060654

Academic Editors: Junfeng Zhang and Ruijin Wang

Received: 5 May 2021
Accepted: 31 May 2021
Published: 2 June 2021

Publisher's Note: MDPI stays neutral with regard to jurisdictional claims in published maps and institutional affiliations.

1. Introduction

Thermal engineering is one of the major fields in which energy conservation and sustainable development demands are increasing requiring more research efforts towards more energy efficient equipment and processes. The petroleum and process industries helped in many ways through the years to enhance the efficiency and find promising solutions [1]. A new era of microelectronic applications and devices has started requiring more thermal management efficiency specially with the obvious increase of the heat flux of chips. Microchannels heat sinks have been investigated firstly by Tuckerman and Pease [2] and a lot after because of the excellent heat transfer efficiency and capacity plus the small dimensions that are needed nowadays in most new technological applications.

Flow and heat transfer from irregular surfaces are often encountered in many engineering applications to enhance heat transfer such as micro-electronic devices, flat-plate solar collectors and flat-plate condensers in refrigerators, geophysical applications, underground cable systems, electric machinery, cooling system of micro-electronic devices, etc. In addition, roughened surfaces could be used in the cooling of electrical and nuclear components where the wall heat flux is known [3]. An extensive amount of research has been performed in the last decades to obtain a better understanding of flow mixing and heat transfer enhancement in channels with geometrical inhomogeneities, such as serpentine channels, asymmetric and symmetric wavy walls, natural convection heat transfer in wavy porous enclosures, wavy microchannels, etc. [3–6]. In recent years, it has been shown that nanofluids can be applied in heat exchangers to enhance the heat transfer, leading to higher heat exchanger efficiency [7].

The advantages of wavy channels are the ease of manufacturing and the significant enhancement of heat transfer as it was operated [8,9]. Direct liquid cooling incorporating microchannels is considered one of the promising solutions to the problem [10,11]. Secondary flow (Dean vortices) is generated when liquid coolant flows through the wavy microchannels. It is found that along the flow direction, the quantity and the location of the vortices may change leading to chaotic advection, which can greatly enhance the convective fluid mixing.

Choi [12] was the first to mention the term nanofluid which refers to any liquid that contains solid metallic particles in submicron scale/nano scale (e.g., Ag, TiO_2, Cu or Al_2O_3). These nanoparticles then are added to water to form a nanofluid which was found to increase the thermal conductivity and enhance the heat transfer performance of the base working fluids [13]. Widely speaking, it was found from different results that the type of nanoparticles used in the base fluid is the main factor to enhance the performance of heat transfer of traditional work fluids such as Ag nanofluid which gave experimentally the higher performance value due to the inherent properties of this metal oxide that overcomes the thermal diffusion that happens due to the small sizes of its particles [14,15]. Lee et al. [16] used the base fluids such as water and ethylene glycol and added nanoparticles of Al_2O_3 and CuO to measure the thermal conductivity. The results obtained by them showed that ethylene glycol based nanofluid achieved a higher value of thermal conductivity than the water-based type. Xuan and Roetzel [17] were also studying the performance of heat transfer using nanofluids and had some interesting results. They were able to derive many equations using single and double-phase flow techniques for the analysis of convective heat transfer in nanofluid. The heat transfer enhancement of nanofluids and the laminar flow were investigated by Heris et al. [18,19] in a circle tube, and the results were typical compared to the other research. Flow and heat transfer characteristic of copper-water nanofluid in a two-dimensional channel was studied by Santra et al. [20] using Cu water nanofluid. The results were notable as with the increase of volume fraction of the solid nanoparticles and Reynolds number, the heat transfer rate was found to be higher.

Saghir research team [21–32] have experimentally and numerically investigated the forced convection of Al_2O_3–Cu hybrid nanofluid, Al_2O_3/water nanofluid in porous media at different flow rates and heating conditions in a straight channel. Results revealed that fluid such as water with metallic nanoparticle can enhance the heat extraction by no more than 6% when compared to water circulation. They concentrated their effort in straight channel with identical dimension.

In the present paper, attempt is made to investigate the performance of wavy channel using the same geometrical parameters and heating condition as the straight channel's cases done by Saghir et al. The novelty of this research is to be able to enhance the heat removal by allowing the flow to circulate a longer pathway when compared to rectangular case. Pressure drop is a main concern for engineering application thus the need to evaluate the Performance Enhancement Criteria. Section 2 presents the problem description, followed by the finite element formulation in Section 3. Section 4 is allocated for the case studied and the conclusion is in Section 5.

2. Problem Description

In this work, we are investigating a numerical approach of forced convection of nanofluids and hybrid fluid in three-wavy porous channels configuration as shown in Figure 1. The objective is to investigate the heat transfer performance of these fluids that could act as coolants. The three fluids used in the current analysis are water, titanium dioxide/water based nanofluid (1% TiO_2/water) and aluminum oxide/copper nanoparticle/water-based hybrid fluid (1% Al_2O_3–Cu/water). For the hybrid fluid, the single nanoparticles are composed of 90% Al_2O_3 and 10% copper [27].

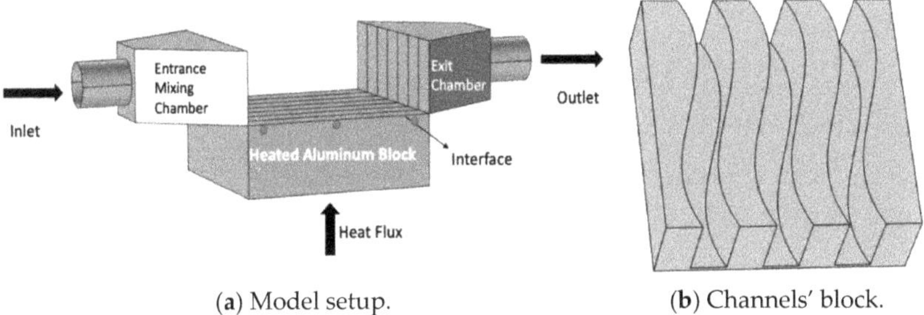

(**a**) Model setup.   (**b**) Channels' block.

Figure 1. (**a**) Model setup. (**b**) Channels' block.

The numerical setup model consists of an inlet tube, a mixing chamber, three wavy porous channels insert, an exit chamber and an outlet tube. The described setup is placed over a heated aluminum block that represents the hot surface. The heated surface of the aluminum block is in direct contact with the bottom of the three channels block. The channel dimensions have a width of 0.00535 m, a height of 0.0127 m and a length of 0.0375 m. The heated aluminum block dimensions are 0.0375 m × 0.0375 m × 0.0127 m. The fluid enters the system with a specific velocity u_{in} and temperature T_{in}. The temperature was calculated numerically along the fluid path 1mm below the interface of the aluminum block at the centre as shown in Figure 1a. Different flow rates were applied corresponding to 0.05 US gallon per minutes (USGPM), 0.1 USGPM, 0.15 USGPM and 0.2 USGPM. This corresponds to a flow rate of 3.15×10^{-6} m^3/s, 6.3×10^{-6} m^3/s, 9.45×10^{-6} m^3/s and 1.26×10^{-5} m^3/s, respectively. The wavy channels are assumed porous, and the permeabilities used were 10 pore per inches (10 PPI), 20 PPI and 40 PPI. This corresponds to a permeability of 9.557×10^{-7} m^2, 2.38×10^{-7} m^2 and 3.38×10^{-8} m^2, respectively. The porosity is maintained constant at 0.91. The inlet pipe diameter is set equal to 0.01 m, and the heat flux applied to the model, as shown in Figure 1a, has an intensity of 75,000 W/m^2. Table 1 presents the physical properties of the fluid used in our simulation. The main reason for selecting these fluids is that we have conducted experimental measurement with these fluids in a straight channel configuration, and no sedimentation of the nanoparticles has been observed.

Table 1. Thermo-physical properties of the fluid used in the analysis [27–29].

Fluid	μ_{nf} (kg/m·s)	ρ_{nf} (Kg/m^3)	Cp_{nf} (J/Kg·K)	k_{nf} (W/m·K)	Pr (Prandtl Number)
Water	0.001002	998.2	4182	0.613	6.83
1% TiO$_2$-0.99 Water	0.001019	1030	4040	0.835	4.93
1% (Al$_2$O$_3$–Cu)-0.99 Water	0.0016025	1024	4067	0.657	11.11

3. Governing Equation and Boundary Conditions

In the present work, we attempt to solve the Navier-Stokes equation for the fluid in the entrance and exit chamber combined with the Brinkman formulation for the flow in the porous channels and the energy equation for the fluid in the setup. In addition, the heat conduction equation is solved for the solid surface. The problem is assumed steady state and the flow is in laminar regime. The set of equations used in our model is as follows:

Momentum equations along x, y and z directions, respectively,

$$\rho_{nf}\left(u\frac{\partial u}{\partial x} + v\frac{\partial u}{\partial y} + w\frac{\partial u}{\partial z}\right) = -\frac{\partial p}{\partial x} + \mu_{nf}(\frac{\partial^2 u}{\partial x^2} + \frac{\partial^2 u}{\partial y^2} + \frac{\partial^2 u}{\partial z^2}) \tag{1}$$

$$\rho_{nf}\left(u\frac{\partial v}{\partial x} + v\frac{\partial v}{\partial y} + w\frac{\partial v}{\partial z}\right) = -\frac{\partial p}{\partial y} + \mu_{nf}(\frac{\partial^2 v}{\partial x^2} + \frac{\partial^2 v}{\partial y^2} + \frac{\partial^2 v}{\partial z^2}) \tag{2}$$

$$\rho_{nf}\left(u\frac{\partial w}{\partial x}+v\frac{\partial w}{\partial y}+w\frac{\partial w}{\partial z}\right)=-\frac{\partial p}{\partial z}+\mu_{nf}\left(\frac{\partial^2 w}{\partial x^2}+\frac{\partial^2 w}{\partial y^2}+\frac{\partial^2 w}{\partial z^2}\right) \tag{3}$$

Continuity equation,

$$(\frac{\partial u}{\partial x}+\frac{\partial v}{\partial y}+\frac{\partial w}{\partial z})=0 \tag{4}$$

Energy conservation equation,

$$(\rho\,Cp)_{nf}(u\frac{\partial T}{\partial x}+v\frac{\partial T}{\partial y}+w\frac{\partial T}{\partial z})=k_{nf}(u\frac{\partial^2 T}{\partial x^2}+v\frac{\partial^2 T}{\partial y^2}+w\frac{\partial^2 T}{\partial z^2}) \tag{5}$$

For the porous flow, the following formulation are used. In particular:

$$\frac{\mu_{nf}}{\kappa}u=-\frac{\partial p}{\partial x}+\mu_{nf}(\frac{\partial^2 u}{\partial x^2}+\frac{\partial^2 u}{\partial y^2}+\frac{\partial^2 u}{\partial z^2}) \tag{6}$$

$$\frac{\mu_{nf}}{\kappa}v=-\frac{\partial p}{\partial y}+\mu_{nf}\left(\frac{\partial^2 v}{\partial x^2}+\frac{\partial^2 v}{\partial y^2}+\frac{\partial^2 v}{\partial z^2}\right) \tag{7}$$

$$\frac{\mu_{nf}}{\kappa}w=-\frac{\partial p}{\partial z}+\mu_{nf}(\frac{\partial^2 w}{\partial x^2}+\frac{\partial^2 w}{\partial y^2}+\frac{\partial^2 w}{\partial z^2}) \tag{8}$$

Energy formulation for the porous flow,

$$(\rho_{nf}Cp_{nf})_{eff}(u\frac{\partial T}{\partial x}+v\frac{\partial T}{\partial y}+w\frac{\partial T}{\partial z})=(k_{nf})_{eff}(u\frac{\partial^2 T}{\partial x^2}+v\frac{\partial^2 T}{\partial y^2}+w\frac{\partial^2 T}{\partial z^2}) \tag{9}$$

The effective conductivity and the heat capacity combine the porous material and the flow. For further information, readers should consult the reference by Bayomy and Saghir [29]. For the purpose of studying the performance of the wavy channel, two important parameters have been investigated. The first is the local Nusselt number, and the second is the Performance Enhancement criterion. The Nusselt number is defined as the ratio of the convective heat coefficient multiplied by the inlet pipe diameter over the water conductivity (i.e., $\frac{hD}{k_w}$). The heat convection coefficient is known as the ratio of the heat flux over the temperature θ which is the temperature calculated at 1 mm below the interface minus the inlet temperature T_{in} (i.e., $\theta = T - T_{in}$). The blue dots shown in Figure 1a indicate the location where the temperature was calculated.

The performance evaluation criterion is an important parameter which combines the average Nusselt number and the friction factor. Nanofluid and hybrid fluid exhibit a larger pressure drop when compared to water. In our analysis, the performance evaluation criteria is shown in Equation (10).

$$PEC=\frac{\overline{Nu}}{(f)^{0.333}} \tag{10}$$

Here $\overline{Nu}$ is the Nusselt number defined earlier, and f is the friction coefficient in the channels.

The fanning friction coefficient is known to be represented by Equation (11).

$$f=\frac{4(\Delta p)}{(\frac{L}{D})(\rho_{nf})(u_{in}^2)} \tag{11}$$

The pressure difference shown in Equation (11) is the pressure taken at the middle of the inlet mixing chamber to the one at the middle of the exit chamber. Here L is the channel Length equal to 0.0375 m without waviness.

Boundary Conditions and Solution Approach

The boundary conditions used in the model consist of applying an inlet velocity u_{in} and an inlet temperature T_{in}. The heat flux is applied at the bottom of the heated block, and the remaining external surface is insulated. Figure 1a shows the graphical location of the boundary condition. Porous material is used as an insert inside the channel. At the exit of the flow, a free flow boundary condition is applied. Different approaches exist in COMSOL to tackle the convergence criteria. In this particular model, the default solver used was the segregated method. Details about this approach could be found in any finite element's textbook. The convergence criterion is clearly explained in COMSOL manual. In a short summary, the convergence criteria were set as follows: at every iteration, the average relative error of u, v, w, p and T were computed. These were obtained using the following relation:

$$R_c = \frac{1}{n \cdot m} \sum_{i=1}^{i=m} \sum_{j=1}^{j=n} \left| \frac{(F_{i,j}^{s+1} - F_{i,j}^{s})}{F_{i,j}^{s+1}} \right| \tag{12}$$

where F represents one of the unknowns, viz., u, v, w, p, or T; s is the iteration number; and (i, j) represents the coordinates on the grid. Convergence is reached if R_c for all the unknowns is below 1×10^{-6} in two successive iterations. For further information on detailed solution method, the reader is referred for COMSOL software manual [33].

4. Mesh Sensitivity Analysis, Convergence Criteria and Model Verification

The mesh sensitivity is examined in purpose of determining the optimal mesh required for the analysis. In the table below, we demonstrated different mesh sensitivity which were investigated following the terminology used by COMSOL software.

The mesh levels that COMSOL supports and the elements number for each mesh level are shown in Table 2. The average Nusselt number was evaluated at 1 mm below the interface in the aluminum block, and the results are represented in the Figure 2a. Here, the heat flux applied is equal to 75,000 W/m^2, and the conductivity of the water was used to evaluate the Nusselt number. It is evident that a coarse or normal mesh level will be suitable to be used in the COMSOL model. Figure 2b, presents the finite element mesh used in our simulation with normal mesh level.

Table 2. Mesh information for different level of meshing [33].

COMSOL Mesh.	Details on Number of Elements at the Boundary and in the Domain
Coarser	71,301 domain elements, 9686 boundary elements, 908 edge elements
Coarse	157,028 domain elements, 17,902 boundary elements, 1276 edge elements
Normal	320,985 domain elements, 29,642 boundary elements, 1674 edge elements
Fine	643,293 domain elements, 47,756 boundary elements, 2140 edge elements

The model has been tested against experimental data to demonstrate the model accuracy. Welsford, Saghir et al. [31], Plant and Saghir [24,27] and Delisle, Saghir et al. [32] conducted experimental measurement of heat enhancement in straight porous channel. They used the same proposed numerical model except the channels were straight channels. Identical boundary conditions are used as well. Results revealed a good agreement between the experimental measurement and the numerical code. Thus, the accuracy of the current numerical model.

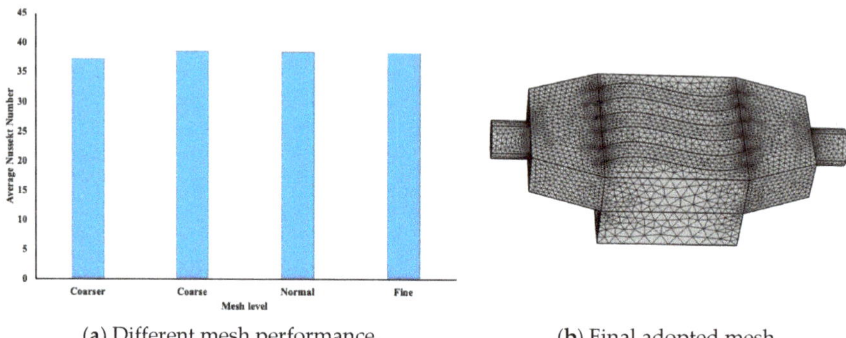

<table>
<tr><td>(**a**) Different mesh performance.</td><td>(**b**) Final adopted mesh.</td></tr>
</table>

Figure 2. Mesh sensitivity analysis. (**a**) Different mesh performance. (**b**) Final adopted mesh.

5. Results and Discussion

In the present study, an attempt is made to investigate the effectiveness of wavy channel in improving heat enhancement. Different flow rates were applied corresponding to 0.05 US gallon per minutes (USGPM), 0.1 USGPM, 0.15 USGPM and 0.2 USGPM. This corresponds to a flow rate of $Q_1 = 3.15 \times 10^{-6}$ m^3/s, $Q_2 = 6.3 \times 10^{-6}$ m^3/s $Q_3 = 9.45 \times 10^{-6}$ m^3/s and $Q_4 = 1.26 \times 10^{-5}$ m^3/s, respectively. The question the authors raised is whether a wavy channel leads to a better performance enhancement criterion when compared to a straight channel? Different fluids are used in the current analysis with water, then a 1% TiO$_2$ nanoparticles diluted in 99% water, and finally, a hybrid fluid which consists of 1% nanoparticles containing 90% Al$_2$O$_3$ and 10% copper diluted in 99% water. The differences between these fluids are their conductivity, density and viscosity. Thermal conductivity may affect the Nusselt number whereas the viscosity, specific heat and density can affect the friction coefficient and thus the pressure drop. Different flow rates will be applied, but the heating condition remains the same with a heat flux of 75,000 W/m^2. The model was studied for three different permeabilities by maintaining the porosity constant at 0.9.

5.1. Heat Enhancement Using Water as Working Fluid

Figure 3 presents the temperature distribution 1 mm below the interface for the three porous material types and for four different flow rates as stated earlier. The temperature profile shows that at the beginning of the flow entrance, the boundary layer is very small, thus allowing the heat to pass from the solid block to the fluid. However, as the boundary layer starts developing, it appears that it reduces the amount of heat extracted from block. This temperature profile is identical regardless of the permeability of the material. By carefully examining the temperature magnitude, it is noticeable that as the permeability varies from 10 PPI to 40 PPI, the heat extraction is improving. This is shown as the temperature in the heated block drop in magnitude. It is evident that having porous material helps to absorb more heat. To study further the heat extraction, Figure 4 presents the local Nusselt number variation for the case or a permeability of 40 PPI.

A negative slope for the local Nusselt number variation is an indication that as the boundary layer developed, the heat extraction decreased accordingly. It is evident for this case that as the flow rate increased, the Nusselt number increased. This increase is noticeable around 3% for the highest flow rate of 0.2 USGPM. For a constant flow rate of 0.2 USGPM, one may notice that the average Nusselt number increases by 3% when compared to the case of 10 PPI and by an additional 2% when compared to 20 PPI case. This is another indication that as the permeability increases the heat extraction improves. Similar observations are noted for three remaining flow rates.

5.2. Heat Enhancement Using 1% TiO$_2$ in Water as Working Fluid

The previous model is repeated by using Titanites as the working fluid. As indicated earlier, a 1% Titanium oxide in water solution is used. The nanoparticles diameter is found to be around 31 nm. Figure 5 presents the local Nusselt number variation along the flow for the four flow rates used and for a permeability of 20 PPI.

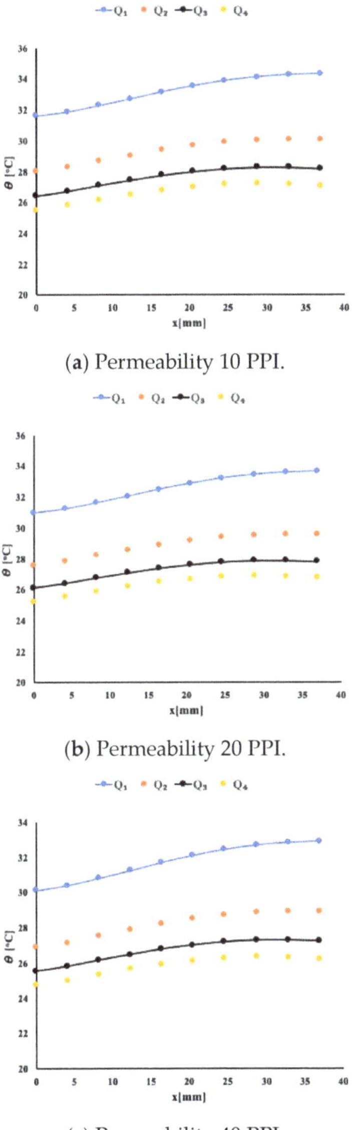

(**a**) Permeability 10 PPI.

(**b**) Permeability 20 PPI.

(**c**) Permeability 40 PPI.

Figure 3. Temperature distribution with water as working fluid.

Since the thermal conductivity of the Titanite is higher, one expects a heat enhancement improvement when compared to water solution. By carefully examining the average Nusselt number ad for the same permeability, the 1% TiO$_2$/water nanofluid exhibits a higher average Nusselt number when compared to water. This improvement is merely

0.5% for a flow rate of 0.2 USGPM. As the permeability increases to 40 PPI, the additional increase in average Nusselt number is found to be around 0.5% as well. Then, one may conclude that nanofluid increases the amount of heat removal.

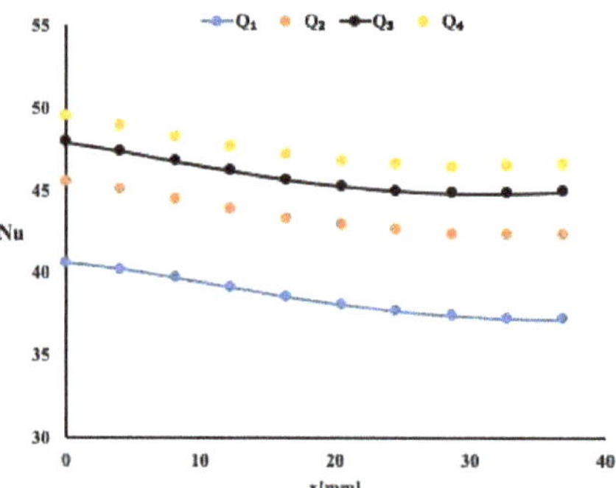

Figure 4. Local Nusselt number variation for different flow rates.

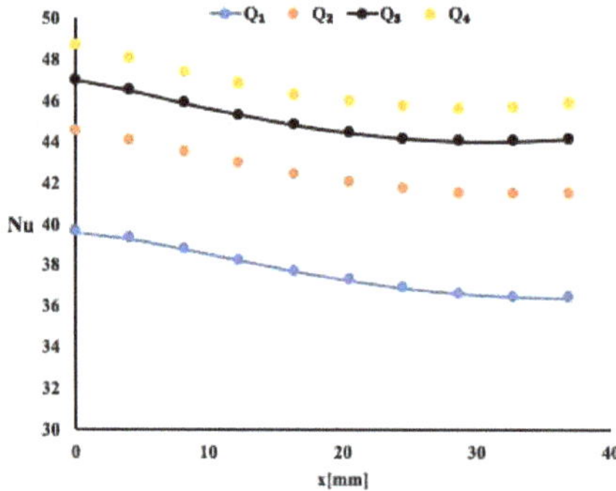

Figure 5. Local Nusselt number variation for a permeability of 20 PPI.

5.3. Heat Enhancement Using 1% (Al_2O_3–Cu) in Water as Working Fluid

The model was repeated with a hybrid fluid consisting of a single nanoparticle containing Al_2O_3 and Copper. Figure 6 displays the local Nusselt number variation for the case of a 10 PPI permeability and for the four flow rates. The Nusselt number variation profile is identical to the other cases, but its magnitude is different.

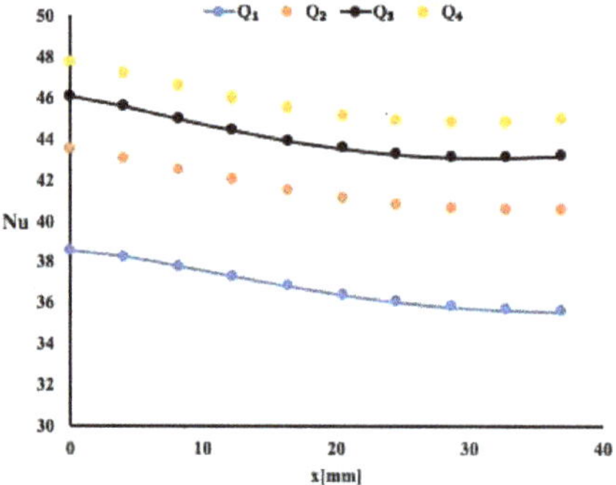

Figure 6. Local Nusselt number variation for a permeability of 10 PPI. (hybrid fluid is the working fluid).

A similar observation was made regarding the average Nusselt number when compared to water. It is evident that there was not a large change when compared to water. This may be due to the fact that water can provide a good cooling fluid if it is given the chance to circulate more in a hot surface. It is worth noting that using the thermal conductivity of the water in the Nusselt number calculation offers an opportunity to make a correct comparison between the three fluids.

5.4. Performance Evaluation Criteria for All Fluids

In the previous sections, we have investigated the importance of heat removal based on different flow rates and porous medium permeabilities. However, an important parameter worth investigation is the pressure drop. One may find a suitable fluid for heat removal but at the expense of higher pressure drop. To overcome this issue, it is important to combine the heat effect and the fluid effect by using the performance evaluation criteria. As shown in Equation (10), it is defined as the ratio of the average Nusselt number and the friction factor to the power one third. The friction factor is defined in Equation (11), which combines the physical properties of the fluid, the inlet velocity and the geometrical dimension of the channel. Figure 7 displays the average Nusselt number, the pressure drops and the performance evaluation criteria for all cases when the permeability is set at 40 PPI. Figure 7c displays the PEC and shows that the TiO_2/water nanofluid exhibits a slightly better performance than water, and the hybrid is the worst candidate between the three studied fluids. In order to determine the reason, Figure 7a displays the average Nusselt number for the three fluids. Almost all of them have a close to identical heat enhancement which may indicate that the waviness of the channel enhances heat removal regardless the fluid used. However, in Figure 7b, a large pressure drop is found for the hybrid fluid contrary to the other remaining two fluids. Thus, there is justification for a higher PEC for the water and the TiO_2 nanofluid.

As the permeability changes to 20 PPI, the fluid circulates with less obstruction, thus less pressure drop is observed. Figure 8 presents the PEC for all cases at a permeability of 20PPI. It is evident from this figure that the nanofluid slightly outperformed the water and more evident that the hybrid performance is very weak. It is obvious that as the flow rates increase, the PEC increases accordingly.

Finally Figure 9 presents the cases when the permeability is set at 10 PPI. Larger pore provides less pressure drop and thus higher PEC. Nevertheless, the same observation is justified here which indicates that the TiO_2 nanofluid exhibits the highest performance evaluation criteria followed by the water. The hybrid fluid due to its high pressure drop exhibits the lowest PEC.

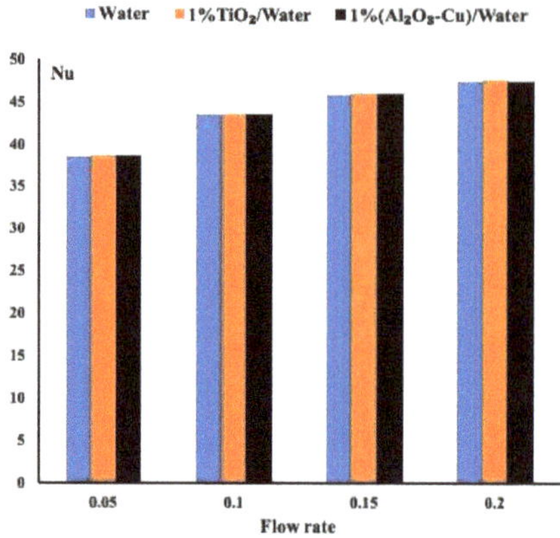

(**a**) Average Nusselt number for all cases for 40 PPI.

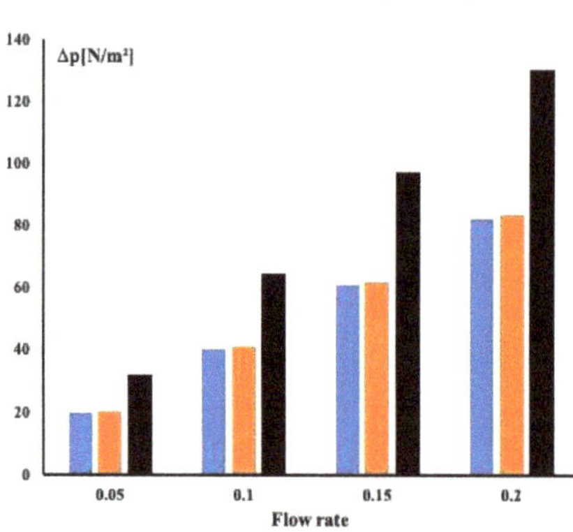

(**b**) Pressure drop for all cases for 40 PPI.

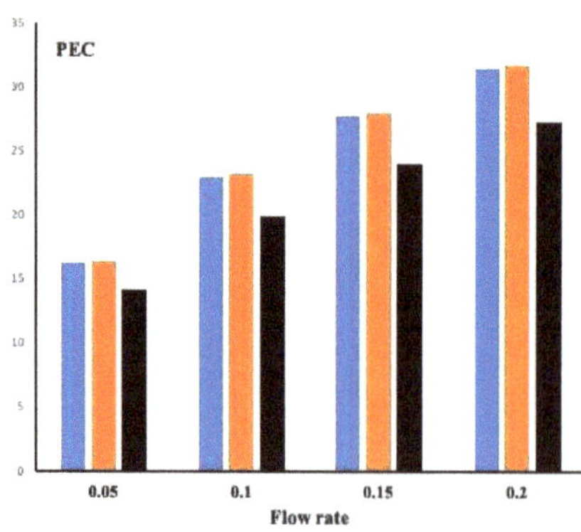

(**c**) PEC for all cases for 40 PPI.

Figure 7. Average Nusselt number, pressure drop and PEC for all cases at a permeability of 40 PPI.

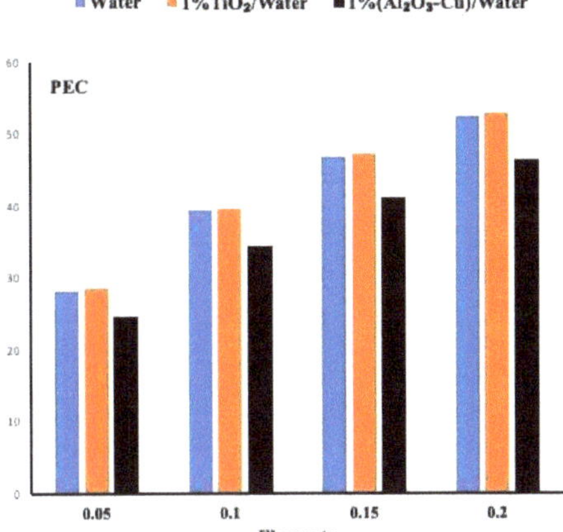

Figure 8. PEC for all cases when the permeability is set at 20 PPI.

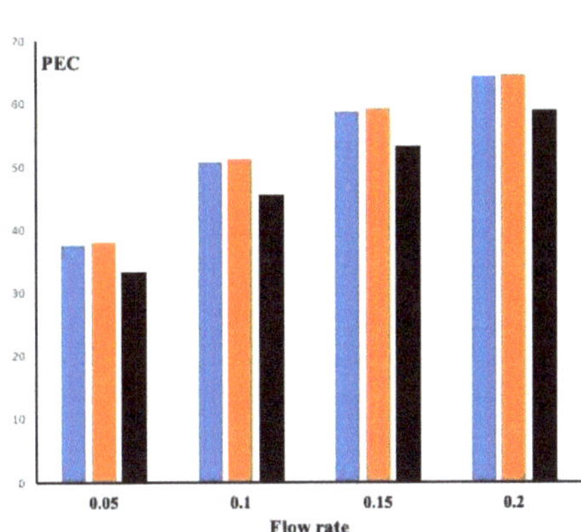

Figure 9. PEC for all cases and for a permeability of 10 PPI.

6. Conclusions

This paper presented a numerical study of the heat performance of three different fluids mainly a single fluid water, a nanofluid 1% TiO$_2$/water and a hybrid fluid 1% (Al$_2$O$_3$–Cu)/water. Three wavy channels with a porous insert, having the three different permeabilities of 10 PPI, 20 PPI and 40 PPI, were investigated numerically. For each case, four different flow rates were implemented. Results reveal the following:

1. Channel waviness allows the flow to circulate longer than a straight channel, thus providing higher heat extraction.
2. By increasing the flow rate, heat enhancement is improved.
3. The presence of porous material helps in heat removal, and as the permeability increases, the pressure drop in the channel decreases accordingly.
4. Amongst the three fluids, the TiO$_2$ nanofluid exhibits slightly better performance than the water based on the Performance Evaluation Criteria coefficient. The hybrid fluid provided less performance due to the large pressure drop it exhibited.

Author Contributions: Conceptualization, K.M.E. and M.Z.S.; methodology, K.M.E.; software, K.M.E.; validation, K.M.E. and M.Z.S.; formal analysis, K.M.E.; investigation, K.M.E.; resources, M.Z.S.; data curation, K.M.E.; writing—original draft preparation, K.M.E.; writing—review and editing, M.Z.S.; visualization, K.M.E.; supervision, M.Z.S.; project administration, M.Z.S.; funding acquisition, M.Z.S. All authors have read and agreed to the published version of the manuscript.

Funding: This research was funded by Qatar Foundation, grant number NPRP12S-0123-190011.

Institutional Review Board Statement: Not Applicable.

Informed Consent Statement: Not Applicable.

Data Availability Statement: Not Applicable.

Acknowledgments: This research was funded by the National Science and Engineering Research Council Canada (NSERC), the Faculty of Engineering and Architectural Science at Ryerson University, and the Qatar Foundation, grant number NPRP12S-0123-190011.

Conflicts of Interest: The authors declare no conflict of interest.

Nomenclature

ρ_{nf}	Density
μ_{nf}	Viscosity
Cp_{nf}	Specific heat
k_{nf}	Conductivity
κ	Permeability
$(\rho_{nf}Cp_{nf})_{eff}$	Effective heat capacity
$(k_{nf})_{eff}$	Effective conductivity
Δp	Pressure difference
u,v,w	Velocity in x, y and z
p	Pressure
L	Block Length
$\overline{Nu}$	Average Nusselt number
Nu	Local Nusselt number
f	Fanning friction factor
PEC	Performance Evaluation Criterion
D	Hydraulic diameter
u_{in}	Inlet velocity

References

1. Webb, R.L.; Kim, N.H. *Principles of Enhanced Heat Transfer*, 2nd ed.; Taylor & Francis: New York, NY, USA, 2005.
2. Tuckerman, D.; Pease, R. High-performance heat sinking for VLSI. *IEEE Electron Device Lett.* **1981**, *2*, 126–129. [CrossRef]
3. Snyder, B.; Li, K.; Wirtz, R. Heat transfer enhancement in a serpentine channel. *Int. J. Heat Mass Transf.* **1993**, *36*, 2965–2976. [CrossRef]
4. Guzmán, A.M.; Cárdenas, M.J.; Urzúa, F.A.; Araya, P.E. Heat transfer enhancement by flow bifurcations in asymmetric wavy wall channels. *Int. J. Heat Mass Transf.* **2009**, *52*, 3778–3789. [CrossRef]
5. Khanafer, K.; Al-Azmi, B.; Marafie, A.; Pop, I. Non-Darcian effects on natural convection heat transfer in a wavy porous enclosure. *Int. J. Heat Mass Transf.* **2009**, *52*, 1887–1896. [CrossRef]
6. Castellões, F.V.; Quaresma, J.N.; Cotta, R.M. Convective heat transfer enhancement in low Reynolds number flows with wavy walls. *Int. J. Heat Mass Transf.* **2010**, *53*, 2022–2034. [CrossRef]
7. Yener, Y.; Kakaç, S.; Avelino, M.; Okutucu, T. Single-phase forced convection in micro-channels—A state-of-the-art review. In *Microscale Heat Transfer—Fundamentals and Applications*; NATO ASI Series; Kakaç, S., Vasiliev, L.L., Bayazitoglu, Y., Yener, Y., Eds.; Kluwer Academic: Amsterdam, The Netherlands, 2005; pp. 1–24.
8. Wang, G.; Vanka, S.P. Convective heat transfer in periodic wavy passages. *Int. J. Heat Mass Transf.* **1995**, *38*, 3219–3230. [CrossRef]
9. Fabbri, G. Heat transfer optimization in corrugated wall channels. *Int. J. Heat Mass Transf.* **2000**, *43*, 4299–4310. [CrossRef]
10. Wang, C.-C.; Chen, C.-K. Forced convection in a wavy-wall channel. *Int. J. Heat Mass Transf.* **2002**, *45*, 2587–2595. [CrossRef]
11. Xia, G.; Chai, L.; Zhou, M.; Wang, H. Effects of structural parameters on fluid flow and heat transfer in a microchannel with aligned fan-shaped re-entrant cavities. *Int. J. Therm. Sci.* **2011**, *50*, 411–419. [CrossRef]
12. Choi, S. Enhancing thermal conductivity of fluids with nanoparticles. In *Development and Applications of Non-Newtonian Flows*; Siginer, D.A., Wang, H.P., Eds.; ASME FED-231/MD; ASME: New York, NY, USA, 1995; Volume 66, pp. 99–105.

13. Chiam, Z.L.; Lee, P.S.; Singh, P.K.; Mou, N. Investigation of fluid flow and heat transfer in wavy micro-channels with alternating secondary branches. *Int. J. Heat Mass Transf.* **2016**, *101*, 1316–1330. [CrossRef]
14. Mohammed, H.; Gunnasegaran, P.; Shuaib, N. Influence of channel shape on the thermal and hydraulic performance of microchannel heat sink. *Int. Commun. Heat Mass Transf.* **2011**, *38*, 474–480. [CrossRef]
15. Zhou, J.; Hatami, M.; Song, D.; Jing, D. Design of microchannel heat sink with wavy channel and its time-efficient optimization with combined RSM and FVM methods. *Int. J. Heat Mass Transf.* **2016**, *103*, 715–724. [CrossRef]
16. Lee, S.; Choi, S.U.-S.; Li, S.; Eastman, J. Measuring Thermal Conductivity of Fluids Containing Oxide Nanoparticles. *J. Heat Transf.* **1999**, *121*, 280–289. [CrossRef]
17. Xuan, Y.; Roetzel, W. Conceptions for heat transfer correlation of nanofluids. *Int. J. Heat Mass Transf.* **2000**, *43*, 3701–3707. [CrossRef]
18. Heris, S.Z.; Esfahany, M.N.; Etemad, G. Numerical investigation of nanofluid laminar convective heat transfer through a circular tube. *Numer. Heat Transf.* **2007**, *52*, 1043–1058. [CrossRef]
19. Heris, S.Z.; Etemad, S.G.; Esfahany, M.N. Convective Heat Transfer of a Cu/Water Nanofluid Flowing Through a Circular Tube. *Exp. Heat Transf.* **2009**, *22*, 217–227. [CrossRef]
20. Santra, A.K.; Sen, S.; Charaborty, N. Study of heat transfer due to laminar flow of copper-water nanofluid through two isothermally heated parallel plates. *Int. J. Therm. Sci.* **2009**, *48*, 391–400. [CrossRef]
21. Saghir, M.Z.; Rahman, M.M. Forced convection of Al_2O_3, Fe_3O_4, ND-Fe_3O_4, and (MWCNT-Fe_3O_4) mixtures in rectangular channels: Experimental and numerical results. *Int. J. Energy Res.* **2020**. [CrossRef]
22. Saghir, M.Z.; Rahman, M.M. Forced Convection of Al_2O_3-Cu, TiO_2-SiO_2, FWCNT-Fe_3O_4, and ND-Fe_3O_4 Hybrid Nanofluid in Porous Media. *Energies* **2020**, *13*, 2902. [CrossRef]
23. Saghir, M.Z.; Welsford, C. Forced Convection in Porous Media Using Al_2O_3 and TiO_2 Nanofluids in Differing Base Fluids. *Energies* **2020**, *13*, 2665. [CrossRef]
24. Plant, R.D.; Saghir, M.Z. Numerical and experimental investigation of high concentration aqueous alumina nanofluids in a two and three channel heat exchanger. *Int. J. Thermofluids* **2021**, *9*, 100055. [CrossRef]
25. Saghir, M.Z.; Rahman, M.M. Toward the effectiveness of mono-nanofluid and hybrid nanofluid for heat enhancement: Experimental and numerical results. *Int. J. Energy Res.* **2021**, in press.
26. Alhajaj, Z.; Bayomy, A.; Saghir, M. A comparative study on best configuration for heat enhancement using nanofluid. *Int. J. Thermofluids* **2020**, *7–8*, 100041. [CrossRef]
27. Plant, R.D.; Hodgson, G.K.; Impellizzeri, S.; Saghir, M.Z. Experimental and Numerical Investigation of Heat Enhancement using a Hybrid Nanofluid of Copper Oxide/Alumina Nanoparticles in Water. *J. Therm. Anal. Calorim.* **2020**, *141*, 1951–1968. [CrossRef]
28. Welsford, C.; Thanapathy, P.; Bayomy, A.M.; Ren, M.; Saghir, M.Z. Heat Enhancement using Aluminum Metal Foam: Experimental and Numerical Approach. *J. Porous Media* **2020**, *23*, 249–266. [CrossRef]
29. Bayomy, A.M.; Saghir, M.Z. Thermal Performance of Finned Aluminum Heat Sink Filled with ERG Aluminum Foam: Experimental and Numerical Approach. *Int. J. Energy Res.* **2020**, *44*, 4411–4425. [CrossRef]
30. Alhajaj, Z.; Bayomy, A.; Saghir, M.Z.; Rahman, M. Flow of nanofluid and hybrid fluid in porous channels: Experimental and numerical approach. *Int. J. Thermofluids* **2020**, *1*, 100016. [CrossRef]
31. Welsford, C.; Delisle, C.; Plant, R.D.; Saghir, M.Z. Effects of Nanofluid Concentration and Channeling on the Thermal Effectiveness of Highly Porous Open-Cell Foam Metals: A Numerical and Experimental Study. *J. Therm. Anal. Calorim.* **2020**, *140*, 1507–1517. [CrossRef]
32. Delisle, C.; Welsford, C.; Saghir, M.Z. Forced convection study with micro-porous channels and nanofluid: Experimental and numerical. *J. Therm. Anal. Calorim.* **2020**, *140*, 1205–1214. [CrossRef]
33. *COMSOL User Manual, Version 5.6*; COMSOL: Boston, MA, USA, 2021.

 *micromachines*

Article

Dynamic Modeling and Flow Distribution of Complex Micron Scale Pipe Network

Yao Zhao, Kai Zhang *, Fengbei Guo and Mingyue Yang

College of Metrology & Measurement Engineering, China Jiliang University, Hangzhou 310018, China; s1802080449@cjlu.edu.cn (Y.Z.); s20020804019@cjlu.edu.cn (F.G.); s1902080445@cjlu.edu.cn (M.Y.)
* Correspondence: zkzb3026@cjlu.edu.cn; Tel.: +86-137-57-135-320

Abstract: A fluid simulation calculation method of the microfluidic network is proposed as a means to achieve the flow distribution of the microfluidic network. This paper quantitatively analyzes the influence of flow distribution in microfluidic devices impacted by pressure variation in the pressure source and channel length. The flow distribution in microfluidic devices with three types of channel lengths under three different pressure conditions is studied and shows that the results obtained by the simulation calculation method on the basis of the fluid network are close to those given by the calculation method of the conventional electrical method. The simulation calculation method on the basis of the fluid network studied in this paper has computational reliability and can respond to the influence of microfluidic network length changes to the fluid system, which plays an active role in Lab-on-a-chip design and microchannel flow prediction.

Keywords: microfluidic; flow distributions; fluid network

Citation: Zhao, Y.; Zhang, K.; Guo, F.; Yang, M. Dynamic Modeling and Flow Distribution of Complex Micron Scale Pipe Network. *Micromachines* **2021**, *12*, 763. https://doi.org/10.3390/mi12070763

Academic Editors: Junfeng Zhang and Ruijin Wang

Received: 28 April 2021
Accepted: 22 June 2021
Published: 28 June 2021

Publisher's Note: MDPI stays neutral with regard to jurisdictional claims in published maps and institutional affiliations.

1. Introduction

Nowadays, the rapid development of science and technology constantly drives the upgrading of scientific and technological products. The portability, integration and intelligentization of scientific and technological products are the key points and difficulties in current science and technology product development. Meanwhile, with the improvement of living standards, health issues have become the focus of people's attention so that people put forward higher standards for the accuracy of test results and detection methods. However, the real-time detection and rapid diagnosis technology in medical and health fields are currently less developed and need to be improved urgently. Therefore, the microfluidic chip technology arises at the right moment.

Microfluidic chip technology refers to building biological or chemical laboratories on chips that are only a few square centimeters in size. Different types of fluid channels, whose diameters are in the dozens to hundreds of micrometers range, are built on the microfluidic chip based on various functions. These channels have many utilities, such as biological or chemical reactions and extraction or detection of reagents. Meanwhile, the microfluidic transmission process can be controlled through the design of the microchannel network [1].

The main functions of the microfluidic system are transferring and mixing, reaction separation, and flow control of microfluidics in the microchannels. Because of the small size of the fluid channel, the fluid flowing in the microchannel is mostly in the laminar flow state, such as particles, droplets or bubbles, which generally belong to the field of low Reynolds number fluid theory in microchannels [2].

In order to calculate and control the flow of the microfluidic network more accurately, the study method that is commonly used by researchers is the electrical equivalent method [3]. By using the electrical equivalent method to liken the micropipe to an electrical device, the microfluidic network can be integrated and analyzed. However, it cannot simulate the issues of the local resistance well in the microfluidic pipe, and the diverting resistance in the microfluidic network needs to be calculated by mathematic methods to be more accurate [4]. This manner aforementioned is the calculation method of a fluid network which is on the basis of the three conservation equations of basic fluid mechanics, including the conversation of mass, momentum and energy [5]. Pressure value and flow rate can be obtained by simultaneous calculations of mass and momentum and heat is solved by the energy equation. However, this study method is rarely reported at present.

Recently, the flow characteristics of microfluidic networks have been studied by many scholars. However, most of them focus on computational fluid dynamics (CFD) and electrical analogy methods [3,6]. For example, Ali Y. Alharbi's group studied the flow process of fractal-like branching networks in micropipes and the pressure drop characteristics and force mode in branching microchannels using three-dimensional computational fluid dynamics approaches [7]. Dalei Jing and Yongping Chen et al. also studied the heat and mass transfer in branched microfluidic channels, but the equivalent methods of microfluidic resistance are all hypotheses of using circuits [8,9]. Bassiouny's group [10,11] developed analytical equations to predict the fluid flow distribution in rectangular, U-shaped and Z-shaped manifolds. Kim's group [12] analyzed the influence of a Z-manifold on channel flow distribution and studied three different shapes, including rectangle, trapezoid and triangle. Delsman's group [6] performed a numerical analysis of the flow distribution in a multi-channel microfluidic device with Z-shape and inline manifolds. They obtained the results of multiple manifold designs by changing the shape and flow direction of the device's inlet and outlet. Wang's group [13] took friction and momentum effects into account in the manifold and modified the U-shaped manifold model established by Bassiouny and Martine; they found that the effects of friction and momentum could increase and decrease the pressure drop in the manifold, respectively. Therefore, it is observed that a uniform distribution of fluid flow can be achieved by balancing these two opposite effects properly.

In this paper, we use the fluid network calculation method to analyze the flow characteristics of the microfluidic network and the conservation equation of mass and momentum to solve the pressure and flow in the microfluidic network. The method that we used was on the basis of the structural characteristics and flow process of the actual network. The flow distribution in a microfluidic device with five microchannels is analyzed and compared with the conventional electrical method. The fluid network method takes the formula of resistance term in the calculation program into account, and its calculation results are closer to the actual flow process than those using the conventional electrical method.

2. Model Development

In this study, it is assumed that the fluid flowing in the microfluidic network is water and that there is no phase change during transportation. Therefore, it can be considered a single-phase incompressible fluid. The hypotheses made to establish the model in this study can be expressed as follows:

(a)   No power source in the fluid network model.
(b)   The fluid state in the nodes is uniform (the internal pressure is equal).
(c)   The flow resistance only takes the equivalent frictional resistance along the pipeline into account and keeps the flow resistance coefficient constant.
(d)   The cross-sectional area in the same branch pipe remains unchanged, and the working medium parameters are represented by the weighted average of the connected node parameters.

The construction of the microfluidic pipe network model mainly includes the following two parts:

(a)   Nodes, including pipe transitions and other essential components in the microfluidic network.
(b)   Branches, a connecting component between two nodes.

The directed topology model of the microfluidic network is shown in Figure 1. In the microfluidic network, both the micropump, as a power source, and the network outlet are regarded as boundary nodes. Moreover, the intermediate branches include various resistance components, such as a micropipe and microvalve.

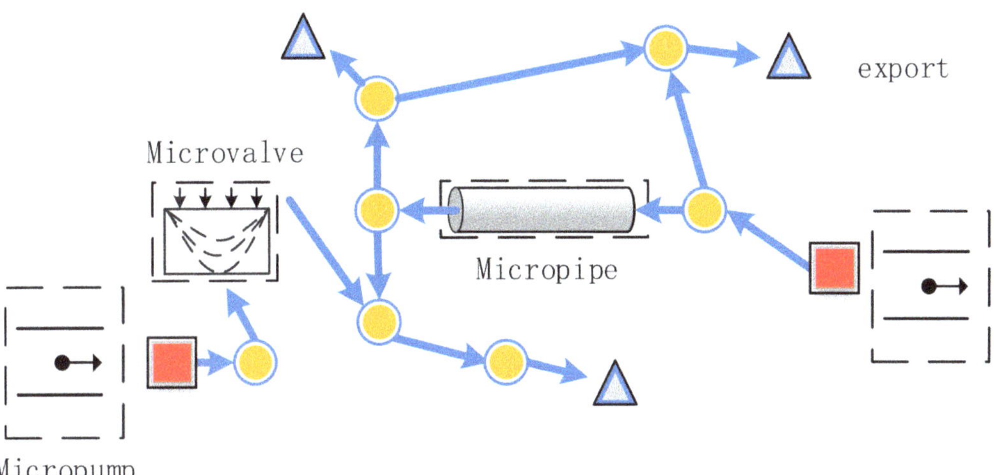

Figure 1. Directed topology model of the microfluidic network.

For a stable microfluidic network, the continuity equation (continuous mass) in the flow process can be expressed as follows:

$$V_i \frac{d\rho_i}{dt} = \sum_{j=1}^{N} D_{ij} G_{ij} \tag{1}$$

Without considering the influence of heat from $\frac{d\rho}{dt} = \frac{\partial \rho}{\partial p}\frac{dp}{dt} + \frac{\partial \rho}{\partial H}\frac{dH}{dt} \approx \frac{\partial \rho}{\partial p}\frac{dp}{dt}$, the following equation can be obtained:

$$C_i \frac{dp_i}{dt} = \sum_{j=1}^{N} D_{ij} G_{ij} \tag{2}$$

where, $C_i = V_i \frac{\partial \rho_i}{\partial p_i}$ is the compressibility (kg/MPa) of the working fluid; p is the pressure of the working fluid; ρ is the density of the working fluid; t is the time; N is the total number of nodes; G_{ij} is the mass flow between nodes i and j; V_i is the volume of node i; D_{ij} is the connection mode between nodes i and j ($i = 1, 2, \ldots, N; j = 1, 2, \ldots, N$) and the specific meaning is:

$$D_{ij} \begin{cases} 1, \text{there is a connection between nodes } i \text{ and } j, \text{ and the flow direction is from } j \text{ to } i \\ 0, \text{there is no connection between nodes } i \text{ and } j \\ -1, \text{there is a connection between nodes i and j, and the flow direction is from } i \text{ to } j \end{cases}$$

The continuity equation is the flow conservation relationship at a node, and the flow state of the node is affected by the nodes connected to it. Figure 2 shows a schematic diagram of the volume elements of all nodes which are connected to i. It is important to note that a node can be connected to N nodes logically. However, a node to connect the number of other nodes is limited in the actual model.

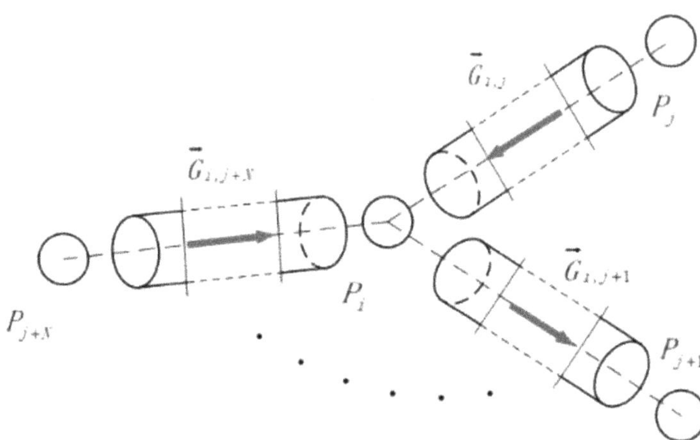

Figure 2. Schematic diagram of the topological structure of the *i*-th node of the pipeline. *P* is the pressure at node *i*. G_{ij} is the flow rate between node *i* and node *j*.

The momentum conservation equation in the microfluidic network can be expressed as follows:

$$\rho_{ij} L_{ij} \frac{dU_{ij}}{dt} = D_{ij} \times \left(P_j - P_i + D_{ij} \times H_{ij} \right) - h_w \tag{3}$$

where, H_{ij} is the pressure generated by macro kinetic energy, potential energy and power source between node *i* and *j*; U_{ij} is the fluid velocity between node *i* and *j*; D_{ij} has the same meaning as D_{ij} in Equations (1) and (2).

In the formula, $h_w = \sum \left[0.5\lambda \frac{L}{d} + \xi \right]$ is the on-way and local resistance loss [14]; λ is the on-way resistance coefficient; L is the length of pipe; d is the diameter of the pipe; ξ is the local pressure loss.

According to the pipeline inertia coefficient $I_{ij} = \frac{L_{ij}}{A_{ij}}$, the friction resistance coefficient on the pipeline $R_{ij} = \frac{h_w}{\left(A_{ij} u_{ij} \rho_{ij} \right)^2}$, and no power supply in the pipeline, $H_{ij} = 0$ is introduced into the Equation (3):

$$I_{ij} \frac{dG_{ij}}{dt} = D_{ij} \times \left(P_j - P_i \right) - R_{ij} \times G_{ij}^2 \tag{4}$$

R_{ij} is the resistance characteristic term between node *i* and *j*, which can be expanded as follows:

$$R_f = \left(\frac{\lambda L}{d} + \xi \right) \frac{1}{2\rho A^2} \tag{5}$$

where, λ is the equivalent resistance coefficient on the way of the microchannel; ξ is the local resistance coefficient of the microchannel; ρ is the working medium density; A is the cross-sectional area of the microchannel.

Equations (2) and (4) are nonlinear differential and algebraic equations. For dynamic simulation, the priority is to ensure real-time performance, which requires high robustness of the calculation process. Using an implicit Euler integration algorithm to calculate, the equation can be obtained by equations (2) and (4) as follows:

$$C_i^{r+1} \frac{P_i^{r+1} - P_i^r}{\Delta t} = \sum_{j=1}^{N} D_{ij}^r \sqrt{\frac{D_{ij}^r \left(P_j^{r+1} - P_i^{r+1} \right)}{R_{ij}}} \tag{6}$$

where, r is the discrete-time variable, time is $t = t_0 + r\Delta t, r = 0, 1, 2, \dots$ Equation (6) is completely decoupled. The pressure of each node can be obtained by solving the N-order nonlinear algebraic equations composed by equation (6) in the process of dynamic simulation. Under the slight perturbation condition, C_i^{t+1} can be substituted by C_i^t, which has little influence on precision and can reduce the calculation amount. After solving the pressure of each node, the momentum equation is introduced to obtain the branch flow. The calculation process is shown in Figure 3.

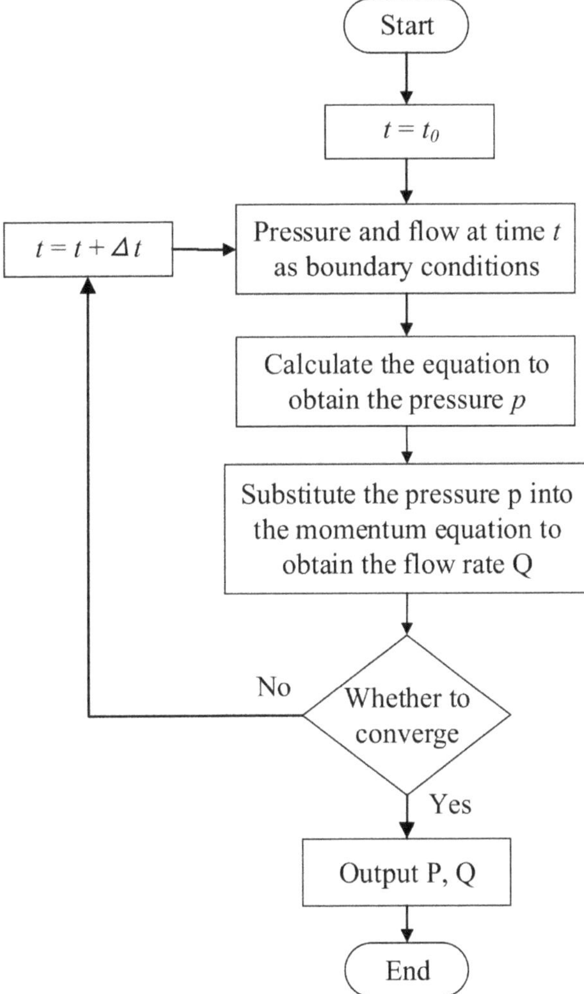

Figure 3. Schematic diagram of calculation process using fluid network method.

3. Results and Discussions

3.1. Model Design and Electrical Equivalent

The basic structure of the microfluidic system model includes micropump, micromixer, microvalve, and microchannel with appropriate size. The flow characteristics of the fluid in the chip can be analyzed at the micron or even nanometer level through these unique miniaturized structures and control devices.

In this case, taking the microscopic scale issues of fluid flow into account, a fluid flow model of an incompressible fluid with a low Reynolds number is adopted to study pipeline pressure distribution under microscale conditions and flow distribution characteristics in the process of fluid flow. Therefore, we designed a microfluidic flow distribution model that has a single entry with three exports, such as Figure 4, as a simulation tool. The variation of fluid flow value in different pipe segments under different inlet pressures was observed and compared with the results obtained by using the electrical method. Since there is no mixing of the two fluids, in this case, the micromixer is not considered in the design of the structure, and the position of the microvalve does not affect the final equivalence study of the fluid flow distribution. Therefore, the electrical analogy method can treat microchannels as resistors.

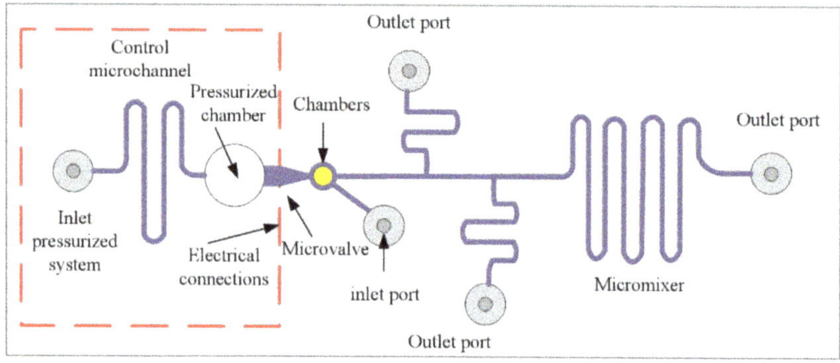

Figure 4. Structure of the microfluidic platform.

Figure 4 shows the specific device design of the microfluidic model. The first is the drive system, which consists of a pressurized chamber, a microchannel controller, and a microvalve that activates the pulse. According to the actual characteristics of our pipe network, we choose the pressure pump as the driving device, whose main role is to provide driving pressure for the microfluidic network. The size of the pressure chamber can be determined on the basis of the size of the microvalve and microchannel used. In the electrical model, this is a power supply that provides voltage. The pressurized chamber is one of the ways to provide the driving pressure. Through the combination of the pressure chamber and the micro valve as a controllable pressure source, the driving pressure can be adjusted according to the system demand.

In order to verify the accuracy of our algorithm, we have mentioned previously the conventional electrical equivalence method in which the microfluidic network is equivalent to a circuit model and the fluid flow parameters are equivalent to electrical parameters. According to the electrical abstraction of our model, Figure 4, the microfluidic network with circuit characteristics can be obtained, as shown in Figure 5. Where, R_{F1} is the microchannel resistance of the microvalve output, R_{F2} is the main channel resistance of fluid flow, and R_{F3}, R_{F4}, R_{F5} is the branch channel resistance of fluid flow, respectively. Corresponding to the resistance of the storage tank where the fluid is located and taking the processing technology of the microfluidic chip into account, the cross-section of the microfluidic tube is generally rectangular, and its resistance is [15]:

$$R_f = \frac{12uL}{(1 - 0.63(h/w))h^3w} \tag{7}$$

where, L is the length of the microchannel, h is its height, w is its width, and u is the viscosity of the fluid. Because of the greater sizes of reservoir and outlet, their resistances to remaining resistors are ignored. Excluding the influence of structure and considering only the logical model of the circuit, the equivalent circuit can be obtained, as shown in Figure 6.

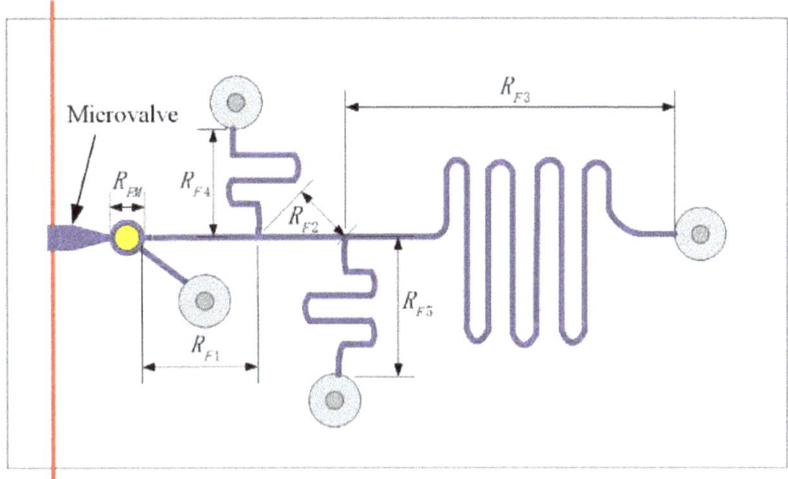

Figure 5. Fluidic resistors of the microfluidic circuit.

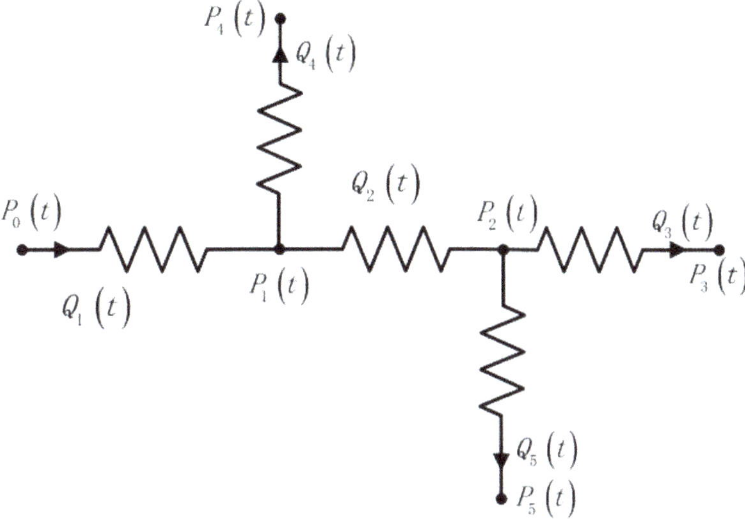

Figure 6. Electrical model of microfluidic platform.

$P_0(t)$ represents the pressure in the supercharging chamber, that is, namely the pressure value at the inlet of the channel; $P_1(t)$ and $P_2(t)$ corresponding to the pressure value at the node of the channel; $P_3(t)$, $P_4(t)$, and $P_5(t)$ corresponding to the pressure value at the outlet of the channel and atmospheric pressure is considered. $Q_1(t)$ and $Q_2(t)$ are the fluid flows in the main channel, and $Q_3(t)$, $Q_4(t)$, and $Q_5(t)$ are the fluid flows in each branching channel.

In this paper, we use Simulink to build the model because of using the circuit analogy method to analyze the parameters of the fluid network in the microchannel. Simulink is a graphical modeling tool in MATLAB software, which can build and calculate various circuit models. The circuit analogy built by Simulink satisfies basic electrical laws, such as Ehrhoff's and Ohm's law. According to the structure in Figure 6, the relevant power parameter relationship can be obtained, and the specific expression is as follows:

$$P_0(t) - P_1(t) = Q_1(t)R_{F1} \tag{8}$$

$$P_1(t) - P_2(t) = Q_2(t)R_{F2} \tag{9}$$

$$P_1(t) - P_4(t) = Q_4(t)R_{F4} \tag{10}$$

$$P_2(t) - P_3(t) = Q_3(t)R_{F3} \tag{11}$$

$$P_2(t) - P_5(t) = Q_5(t)R_{F5} \tag{12}$$

$$Q_1(t) = Q_3(t) + Q_4(t) + Q_5(t) \tag{13}$$

$$Q_2(t) = Q_3(t) + Q_5(t) \tag{14}$$

$$P_0(t) = \frac{P_i V_i}{(V(t) + V_i)} \tag{15}$$

$$V(t) = \int q(t)dt \tag{16}$$

3.2. Model Calculation Results

According to the electrical method, the equivalent resistance and the dimensions of each branching channel in this example are shown in Table 1. The main local drag coefficient in this model is brought about by the micro three-way pipe. For the time being, we do not consider the local resistance coefficient ξ, but only consider the resistance of the micropipe, which is compared with the circuit model of the electrical method.

Table 1. Dimensions of the device and values of some parameters of the setup.

Parameter	R_{F1}	R_{F2}	R_{F3}	R_{F4}	R_{F5}
Width (μm)	500	500	350	350	350
Height (μm)	500	500	350	350	350
Length (mm)	10	8	50	30	30
Equivalent resistance $\left(10^9\,Pa \cdot s/m^3\right)$	4.64	3.71	96.6	57.9	57.9

In order to reduce the activation energy required for fluid flow, the microvalve needs to have a higher length-width ratio. A PDMS film deformation valve can be selected, and both the height of the valve and the main channel are designed to be 500 μm, whose main role is to control the inlet pressure of the fluid network. The driving pressure here is the network pressure, so the micro-valve resistance does not need to be considered. Meanwhile, considering the uniformity of the fluid flow in the microchannel, the distance of the fluid movement should be as long as possible, and the cross-sectional area design of the fluid flow channel should be narrow. Therefore, the length percentages of channels 3, 4, and 5 are set to 5:3:3 and the height is set to 350 μm.

Above all, we keep other parameters constant and change the driving pressure, $P_0(t)$, in the microfluidic network, set as 15 kPa, 18 kPa, and 20 kPa, respectively. In the outlet boundary condition of the microchannel, the pressure values, $P_3(t)$, $P_4(t)$ and $P_5(t)$, are set as 11 kPa, 12 kPa and 11 kPa, respectively. The results of the fluid network algorithm and electrical simulation of Simulink are shown in Table 2. Both the results indicate that the flow in the microfluidic network satisfies the relationship of flow distribution. The flow rate of Q_2 in the main microchannel is the sum of Q_3 and Q_5. The flow rate of Q_1 is the sum of Q_2 and Q_4. Moreover, there is a certain nonlinear relationship among the microchannels when the input pressure of the pressure pump increases. The electrical method is very similar to the results of the fluid network method we used, which proves the correctness of the fluid network method and can be used to calculate the microfluidic network.

Table 2. Model Simulation and Simulink calculation results under different pressures.

Entrance Pressure (KPa)		15	18	20
Q_1 (mL/min)	Simulink Data	7.623	13.990	18.235
	Simulation Data	7.767	13.812	17.945
	Relative Error (%)	1.883	1.272	1.585
Q_2 (mL/min)	Simulink Data	5.126	8.895	11.407
	Simulation Data	5.217	8.802	11.246
	Relative Error (%)	1.772	1.054	1.414
Q_3 (mL/min)	Simulink Data	1.921	3.333	4.275
	Simulation Data	2.000	3.211	4.086
	Relative Error (%)	4.100	3.671	4.421
Q_4 (mL/min)	Simulink Data	2.497	5.095	6.827
	Simulation Data	2.550	5.011	6.700
	Relative Error (%)	2.111	1.653	1.871
Q_5 (mL/min)	Simulink Data	3.205	5.561	7.132
	Simulation Data	3.217	5.590	7.160
	Relative Error (%)	0.378	0.515	0.388

The first example shows that the total flow rate of the system, $Q_1(t)$, increases with the increase of the input pressure. Microchannel 3 is still the outlet with a smaller flow rate, due to the longer length of microchannel 3, so its equivalent resistance value is greater than that of microchannel 4 and 5. Afterward, the increase of pressure leads to the flow rate rise, which is mainly reflected in the shunting of micro-channel 4 and 5. While the flow rate increase in micro-channel 3 is relatively lower. Both calculation results show this point, as shown in Figure 7, and can prove that the computational results of the fluid network method are closer to those of the electrical method, which has certain computational accuracy.

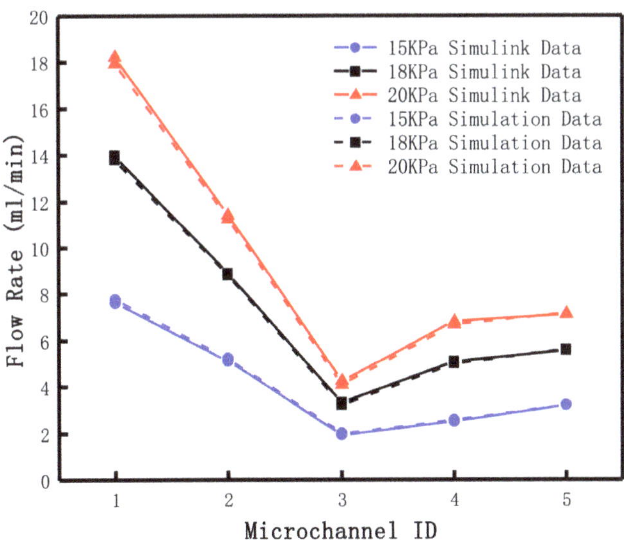

Figure 7. Schematic diagram of different driving pressures.

In the second example, in order to study the influence of microchannel structure on flow characteristics, the width and depth of microchannel 4 were kept at 350 μm. The flow distributions at different channel lengths (40 mm, 50 mm and 60 mm) were studied. The boundary condition at the inlet was set as 15 kPa and the three outlet pressures, $P_3(t)$, $P_4(t)$, and $P_5(t)$, are set as 11 kPa, 12 kPa, and 11 kPa, respectively. The calculation results of the two methods are shown in Table 3 and Figure 8. The diagram shows that the flow rate in microchannel 4 decreases with the increase of its length, which is because the increase of the length of the microchannel leads to the resistance value rise. On the other hand, comparing microchannel 3 with microchannel 5, it can be found that the flow rate in these two channels is relatively stable. Under certain pressure conditions, the flow rate change caused by changing the structure of microchannel 4 does not affect the flow rate in microchannel 3 and 5.

Table 3. Simulation results under different lengths of microchannel 4.

Microchannel 4 Length (mm)		40	50	60
Microchannel 4 Resistance $(10^{10} Pa \cdot s/m^3)$		7.73	9.66	11.6
Q_1 (mL/min)	Simulink Data	7.090	6.764	6.541
	Simulation Data	6.855	6.488	6.680
	Relative Error (%)	3.305	4.078	2.104
Q_2 (mL/min)	Simulink Data	5.188	5.226	5.252
	Simulation Data	5.047	5.016	5.342
	Relative Error (%)	2.733	4.014	1.721
Q_3 (mL/min)	Simulink Data	1.944	1.958	1.968
	Simulation Data	1.894	1.912	2.026
	Relative Error (%)	2.580	2.369	2.943
Q_4 (mL/min)	Simulink Data	1.903	1.538	1.290
	Simulation Data	1.81	1.472	1.337
	Relative Error (%)	4.863	4.297	3.663
Q_5 (mL/min)	Simulink Data	3.244	3.267	3.284
	Simulation Data	3.152	3.104	3.316
	Relative Error (%)	2.825	5.000	0.989

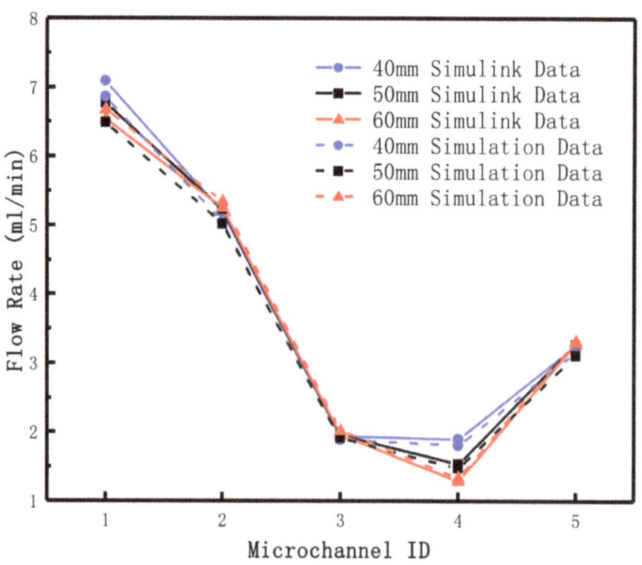

Figure 8. Schematic diagram of calculation results under different lengths of microchannel 4.

In the first example in this section, we have verified the calculation accuracy of the fluid network algorithm and electrical methods. Compared the results, which are similar. However, the resistance items of the fluid network algorithm can be more detailed, such as three or more pipe diversion coefficients of local resistance, and is a good way to add items in resistance, R_{ij}, in the model. Electrical methods can treat micro-channels or micro-valves as resistance items, but it is difficult to deal with local resistance items such as three-way or multi-way, and some approximate formulas are generally adopted to calculate it [15]. According to Figures 7 and 8, we know that the feedback results of the fluid network method and electrical method to the pressure and structure changes in the microfluidic network are similar, which can be used for the simulation of the fluid flow process in the microfluid network and set up test-rig in the follow-up work. According to the actual measurement parameters, the calculation model can be modified to improve its accuracy.

4. Conclusions

In this paper, a computational method of microfluidic pipeline network on the basis of fluid network is studied and compared with the conventional electrical equivalent method. It is found that the computational results obtained by using the fluid network method are closer to those given by using the electrical method. Therefore, we believe that the computational accuracy of the fluid network method is the same as that of the electrical method. From the research results of the simulation examples established in this paper, it can be concluded that the fluid network method can reflect the fluid flow process under different pressures, and the effects caused by structural changes in the microfluidic network are consistent with the results obtained by the electrical equivalent method. However, the electrical method has limitations in the treatment of local resistance, which is difficult to deal with the network shunt, the gradual shrinkage or expansion of microchannels, etc. that can be solved in the algorithm of fluid network easily. Furthermore, the accuracy of its calculation can be improved by merely modifying the resistance term in the algorithm. The computational method of the microfluidic network that we proposed in this paper can be used as an alternative to the electrical method. The follow-up work is to study the resistance term of the model and how to deal with the various devices presented in microfluidic is an important research direction.

Author Contributions: Conceptualization, Y.Z.; methodology, Y.Z. and K.Z.; validation, Y.Z., K.Z., M.Y. formal analysis, K.Z.; writing—original draft preparation, Y.Z. and F.G.; writing—review and editing, K.Z. and M.Y. All authors have read and agreed to the published version of the manuscript.

Funding: This research was funded by the National Natural Science Foundation of China (Grant No. 11472260).

References

1. Gou, Y.; Jia, Y.; Wang, P.; Sun, C. Progress of Inertial Microfluidics in Principle and Application. *Sensors* **2018**, *18*, 1762. [CrossRef] [PubMed]
2. Zhang, K.; Jiang, L.; Gao, Z.; Zhai, C.; Yan, W.; Wu, S. Design and Numerical Study of Micropump Based on Induced Electroosmotic Flow. *J. Nanotechnol.* **2018**, *2018*, 4018503. [CrossRef]
3. Aracil, C.; Perdigones, F.; Moreno, J.M.; Luque, A.; Quero, J.M. Portable Lab-on-PCB platform for autonomous micromixing. *Microelectron. Eng.* **2014**, *131*, 13–18. [CrossRef]
4. Okkels, F.; Bruus, H. Design of Micro-Fluidic Bio-Reactors Using Topology Optimization. In Proceedings of the European Conference on Computational Fluid Dynamics, Mekelweg, The Netherlands, 5–8 September 2006.
5. Mathew, B.; John, T.J.; Hegab, T.J. Effect of Manifold Design on Flow Distribution in Multichanneled Microfluidic Devices. In Proceedings of the Asme Fluids Engineering Division Summer Meeting, Vail, CO, USA, 2–6 August 2009.
6. Delsman, E.R.; Pierik, A.; De Croon, M.H.J.M.; Kramer, G.J.; Schouten, J.C. Microchannel Plate Geometry Optimization for Even Flow Distribution at High Flow Rates. *Chem. Eng. Res. Des.* **2004**, *82*, 267–273. [CrossRef]
7. Alharbi, A.Y.; Pence, D.V.; Cullion, R.N. Fluid Flow Through Microscale Fractal-Like Branching Channel Networks. *J. Fluids Eng.* **2003**, *125*, 1051–1057. [CrossRef]
8. Jing, D.; Yi, S. Electroosmotic Flow in Tree-like Branching Microchannel Network. *Fractals* **2019**, *9*, 1950095. [CrossRef]

9. Chen, Y.; Cheng, P. Heat transfer and pressure drop in fractal tree-like microchannel nets. *Int. J. Heat Mass Transf.* **2002**, *45*, 2643–2648. [CrossRef]

10. Bassiouny, M.K.; Martin, H. Flow Distribution and Pressure Drop in Plate Heat Exchangers I U-Type Arrangement. *Chem. Eng. Sci.* **1984**, *39*, 693–700. [CrossRef]

11. Bassiouny, M.K.; Martin, H. Flow distribution and pressure drop in plate heat exchangers II Z-type arrangement. *Chem. Eng. Sci.* **1984**, *39*, 701–704. [CrossRef]

12. Kim, S.; Choi, E.; Cho, Y.I. The effect of header shapes on the flow distribution in a manifold for electronic packaging applications. *Int. Commun. Heat Mass Transf.* **1995**, *22*, 329–341. [CrossRef]

13. Wang, J. Pressure drop and flow distribution in parallel-channel configurations of fuel cells: Z-type arrangement. *Int. J. Hydrogen Energy* **2010**, *35*, 5498–5509. [CrossRef]

14. Idelchik, I.E. *Handbook of Hydraulic Resistance*, 4th Edition Revised and Augmented; Begell House Inc.: Moscow, Russia, 2008.

15. Bruus, H. *Theoretical Microfluidics*; Oxford University Press: Oxford, UK, 2008.

Article

Numerical Study on the Fluid Flow and Heat Transfer Characteristics of Al$_2$O$_3$-Water Nanofluids in Microchannels of Different Aspect Ratio

Huajie Wu [1,*] and Shanwen Zhang [2]

[1] College of Traffic Engineering, Yangzhou Polytechnic Institute, Yangzhou 225127, China
[2] College of Mechanical Engineering, Yangzhou University, Yangzhou 225127, China; swzhang@yzu.edu.cn
* Correspondence: whj_china@163.com

Abstract: The study of the influence of the nanoparticle volume fraction and aspect ratio of microchannels on the fluid flow and heat transfer characteristics of nanofluids in microchannels is important in the optimal design of heat dissipation systems with high heat flux. In this work, the computational fluid dynamics method was adopted to simulate the flow and heat transfer characteristics of two types of water-Al$_2$O$_3$ nanofluids with two different volume fractions and five types of microchannel heat sinks with different aspect ratios. Results showed that increasing the nanoparticle volume fraction reduced the average temperature of the heat transfer interface and thereby improved the heat transfer capacity of the nanofluids. Meanwhile, the increase of the nanoparticle volume fraction led to a considerable increase in the pumping power of the system. Increasing the aspect ratio of the microchannel effectively improved the heat transfer capacity of the heat sink. Moreover, increasing the aspect ratio effectively reduced the average temperature of the heating surface of the heat sink without significantly increasing the flow resistance loss. When the aspect ratio exceeded 30, the heat transfer coefficient did not increase with the increase of the aspect ratio. The results of this work may offer guiding significance for the optimal design of high heat flux microchannel heat sinks.

Keywords: microchannel; nanofluid; heat transfer enhancement; numerical simulation

Citation: Wu, H.; Zhang, S. Numerical Study on the Fluid Flow and Heat Transfer Characteristics of Al$_2$O$_3$-Water Nanofluids in Microchannels of Different Aspect Ratio. *Micromachines* **2021**, *12*, 868. https://doi.org/10.3390/mi12080868

Academic Editors: Junfeng Zhang and Ruijin Wang

Received: 17 June 2021
Accepted: 22 July 2021
Published: 24 July 2021

Publisher's Note: MDPI stays neutral with regard to jurisdictional claims in published maps and institutional affiliations.

1. Introduction

With the continuous miniaturization and integration of electronic devices, the heat flux density continues to increase, and conventional cooling methods are no longer effective. Therefore, the heat dissipation problem of high heat flux density has been a research hotspot in the field of heat transfer [1,2]. In 1981, Tuckerman and Pease [3] proposed a silicon microchannel heat sink, which uses deionized water as a liquid cooling medium and whose heat dissipation capacity can be as high as 790 W/cm^2. In 1995, Choi and Eastman [4] proposed the concept of nanofluids; a proportion of solid particles with diameters less than 100 nm were added into a base fluid with low thermal conductivity, and the resulting suspension with high thermal conductivity was found to be relatively stable. This special liquid can significantly improve the convective heat transfer capacity of cooling media. The addition of nanoparticles to high-Prandtl number liquids significantly increases the heat transfer performance of micro heat-sinks [5]. An increase in nanoparticle concentration can lead to an increase in thermal conductivity and viscosity and an increase in nanoparticle size [6]. Nanofluids contribute to the improvement of heat transfer processes and reduce and optimize thermal systems. Different properties, such as wettability and thermal conductivity, can be adjusted by altering the nanoparticles' concentration, thereby making nanofluids suitable for a wide range of applications [7]. Nanofluids contain metal or nonmetal particles with nanometer sizes and exhibit much greater thermal conductivity. M. Goodarzi's [8] expression for calculating the enhanced thermal conductivity of nanofluids has been derived from the general solution of the heat conduction equation in spherical coordinates and the equivalent hard-sphere fluid model representing the microstructure of particle/liquid mixtures. The cooling method of microchannel heat sinks combined with nanofluids has become one of the effective ways to solve the heat dissipation problem of high heat flux. By comparing the heat transfer characteristics of trapezoidal, semicircular and rectangular cross-section microchannels, Vinoth et al. [9] found that, compared with rectangular and semicircular cross-section microchannels, trapezoidal cross-section microchannels have the best heat transfer effect because of their larger wall area and effective inlet length, but a larger pressure drop. For some special-shaped cross-sections, Alfaryjat et al. [10] used numerical simulation methods to study hexagonal, circular and rhombic cross-section microchannels. The results show that the heat transfer coefficients of hexagonal cross-section microchannels are the highest, followed by circular and rhombic cross-section microchannels. Ahmed et al. [11] used the three-dimensional numerical simulation method to optimize the microchannel with triangular, trapezoidal and rectangular grooves. The results showed that the optimized microchannel with trapezoidal groove had the best heat transfer effect, the Nusselt number increased by 51.59% and the friction coefficient increased by 2.35% compared with the optimization. Kumar [12] used the finite volume method to simulate and optimize the trapezoidal microchannel. The results show that the heat transfer performance of trapezoidal microchannels with semicircular grooves is 16% higher than that of trapezoidal microchannels with a rectangular groove, but the friction coefficient is 18%. At present, the structure optimization of microchannels has some limitations; namely, the processing is very difficult, and the cost is high. Researchers gradually try to improve the heat transfer by changing the flow medium in the microchannel. Farsad et al. [13] numerically simulated the heat transfer performance of Al_2O_3, CuO and Cu-H_2O nanofluids in copper rectangular microchannels. The results show that the thermal conductivity of metal nanofluids is higher than that of metal oxide nanofluids. Researchers have conducted numerous studies to obtain the optimized aspect and states in order to improve the heat transfer of equipment and use various nanofluids. Shi Xiaojun et al. [14] carried out a multi-objective optimization design on a single-layer nanofluid rectangular microchannel. The results show that the pump power and thermal resistance are more sensitive to the channel width and spacing ratio than the aspect ratio. Naphon and Khonseur [15] used air as a cooling medium to conduct an experimental study on the flow and heat transfer characteristics of microchannels with different heights and widths in the Reynolds number range of 200–1000. The results showed that the height and width of rectangular microchannels exert a significant impact on their heat exchange effect and resistance loss. Studies of this research indicate that the fluid in the indented sections has a higher heat transfer with the heated wall. Karimipour et al. [16] numerically studied a two-dimensional indented rectangular microchannel. They concluded that by increasing the volume fraction of nanoparticles, the thermal efficiency of the nanofluid is enhanced. Yari Ghale et al. [17] numerically studied the laminar and forced flow of a Water/Al_2O_3 nanofluid in an indented microchannel by using two-phase or single-phase methods. Their results showed that the Nusselt numbers and friction factors in an indented microchannel are higher compared to the smooth microchannel, and therefore, these parameters can improve fluid flow efficiency by increasing the width of the rib. A segmental analysis pertaining to the heat exchanger takes place to evaluate the influence of nanofluid usage on the heat transfer coefficient, the exchanger's length and its pressure drop. When the volume fraction of Al_2O_3 nanofluids is 5%, the heat transfer coefficient is increased by 10% compared with pure water, and the pressure drop is significantly reduced [18]. T Raghuraman [19] used pure water as a cooling medium to study the effect of the microchannel aspect ratio on heat transfer performance; the study showed that the microchannel aspect ratio influences the heat transfer coefficient, pumping power, pressure drop and heat transfer performance at different Reynolds numbers. A large aspect ratio can enhance heat transfer, but it can also increase power consumption. Based on the computational fluid dynamics (CFD) method, Mohamadpour et al. [20] carried out a numerical simulation study on the heat transfer efficiency of the cooperative jet in the microchannel. It was found that increasing the jet frequency and pulse amplitude can significantly improve the heat transfer ability of the microchannel.

In the current work, a rectangular microchannel heat sink is used as the research object, and a three-dimensional flow and heat transfer numerical simulation study is conducted on the basis of the computational fluid dynamics method. Water/Al_2O_3 nanofluids of different concentrations are utilized as the cooling medium, and their influence on the heat transfer performance of microchannels is analyzed. This work also focuses on the evaluation of the flow resistance characteristics and heat transfer laws of microchannels with different aspect ratios. The purpose of the study is to provide theoretical guidance for the optimal design of high heat flux density microchannel heat sinks.

2. Numerical Method and Model Description

2.1. Mathematical Model

The single-phase fluid model is commonly used to study the flow and heat transfer of nanofluids in microchannels. The nanoparticles in nanofluids are considered to be uniformly distributed in base fluids, and they are in thermal equilibrium. No relative slip velocity exists between nanoparticles and base liquids, and the flow is regarded as an incompressible steady laminar flow.

On the basis of these assumptions, the governing equations of nanofluid flow and heat transfer can be expressed as:

$$\frac{\partial}{\partial x_j}\left(\rho_{nf}u_j\right) = 0 \tag{1}$$

$$\frac{\partial\left(\rho_{nf}u_iu_j\right)}{\partial x_j} = -\frac{\partial p}{\partial x_i} + \frac{\partial}{\partial x_j}\left[\mu_{nf}\left(\frac{\partial u_i}{\partial x_j} + \frac{\partial u_j}{\partial x_i}\right)\right] \tag{2}$$

$$\frac{\partial}{\partial x_j}\left(\rho_f C_{pf}Tu_j\right) = \frac{\partial}{\partial x_j}\left[\lambda_{nf}\frac{\partial T}{\partial x_j}\right] \tag{3}$$

where u_i and u_j are velocity components, x_i and x_j are Cartesian coordinate components, p is the pressure in the flow field, T is the temperature, ρ_{nf} is the density of nanofluids, and μ_{nf} is the dynamic viscosity. C_{pf} is the specific heat capacity, and λ_{nf} is the thermal conductivity. The calculation formula is as follows [21,22]:

$$\rho_{nf} = (1-\alpha)\rho_w + \alpha\rho_p \tag{4}$$

$$\mu_{nf} = \left(1 + 0.025\alpha + 0.015\alpha^2\right)\mu_w \tag{5}$$

$$C_{pf} = \left[(1-\alpha)(\rho C_p)_w + \alpha(\rho C_p)_p\right]/\rho_{nf} \tag{6}$$

$$\lambda_{nf} = \frac{\lambda_p + (n-1)\lambda_w - (n-1)\alpha(\lambda_w - \lambda_p)}{\lambda_p + (n-1)\lambda_w + \alpha(\lambda_w - \lambda_p)}\lambda_p \tag{7}$$

The subscripts w and p denote the corresponding thermophysical properties of the base fluid and nanoparticles, respectively. α represents the volume fraction of nanoparticles, and n is the shape factor of nanoparticles. In this work, nanoparticles are regarded as regular spheres with a value of $n = 3$.

The heat distribution in the solid area of the heat sink can be calculated by the following formula, T_s is the temperature of the solid region; λ_s is the thermal conductivity of the solid region:

$$\frac{\partial}{\partial x_i}\left(\lambda_s\frac{\partial T_s}{\partial x_i}\right) = 0 \tag{8}$$

2.2. D Model and Boundary Conditions

A microchannel heat sink is usually composed of more than 10, or even dozens, of microchannels, and complete modeling and simulation require a large amount of computing resources. As a result of the symmetry of the model, a typical microchannel heat sink unit can be extracted for simulation. The structure and size of the microchannel heat sink used in this study are shown in Figure 1. The width W of the heat sink unit is 1 mm, the length L is 50 mm, and the microchannel width W_c is 0.5 mm. The microchannel height W_h values are 5, 10, 15, 20 and 25 mm; they correspond to five different aspect ratios, that is, $HW = W_h/W_c$ with values of 10, 20, 30, 40 and 50, respectively. The bottom height of the heat sink is set at 6 mm. Symmetrical boundary conditions are set on both sides of the heat sink, and a uniform heat flux $q = 0.8$ MW/m^2 is set at the bottom. The top wall is set as the adiabatic boundary. The inlet of the microchannel fluid adopts the velocity inlet boundary, the inlet fluid temperature is set at 300 K, the outlet of the microchannel fluid is set as pressure outlet, the upper wall of the fluid domain adopts an adiabatic solid wall boundary condition and the interface between fluid and solid adopts a coupled heat flow boundary condition. In this paper, ICEM CFD software is used for 3D modeling and mesh generation, and the general CFD software FLUENT 15.0 is used for numerical simulation. According to the geometric models with different aspect ratios, three sets of grids are divided to analyze the grid independence under the maximum Reynolds number.

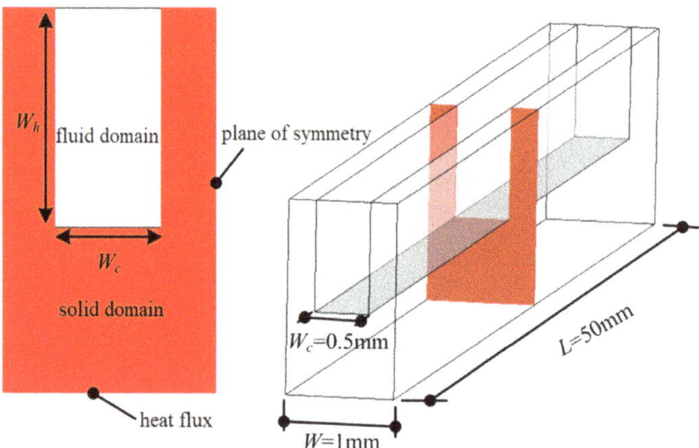

Figure 1. Schematic of heat sink with square cross-section.

2.3. Model Validation

The experimental data of Lei [23] are used for comparison to verify the prediction performance of the mathematical model. The microchannel width W_c is 0.1 mm, the height W_h is 0.5 mm, the length L is 10 mm, and the heat flux q at the bottom of the heat sink is 0.6 MW/m^2. The cooling medium is pure water. In the calculation, the volume fraction of the nanoparticles α is 0.

The characteristic scale of microchannels can be defined as:

$$D_h = \frac{2W_h W_c}{W_h + W_c} \tag{9}$$

The microchannel Reynolds number is:

$$Re = \frac{\rho_f U_{in} D_h}{\mu_f} \tag{10}$$

where U_{in} is the fluid inlet velocity.

The average temperature T_c of the heat transfer surface and the average temperature T_f of the whole microchannel fluid can be respectively defined as:

$$T_c = \frac{\int T dA}{\int dA} \tag{11}$$

$$T_f = \frac{\int \rho_f T dV}{\int \rho_f dV} \tag{12}$$

Therefore, the average heat transfer coefficient can be expressed as:

$$h = \frac{q A_b}{A_c(T_c - T_f)} \tag{13}$$

where A_b is the heating area of the bottom of the thermal sink and A_c is the heat exchange area between the fluid domain and the solid domain in the microchannel.

The average Nusselt number is defined as follows:

$$Nu = \frac{h D_h}{\lambda_f} \tag{14}$$

The comparison between the average Nusselt number predicted by the mathematical model and the experimental results is shown in Figure 2. The results show that the proposed model achieves certain accuracy and reliability in the prediction of fluid laminar flow and heat transfer in microchannels. The article only carried out simulation research, and there will be errors between the simulation results and the experiment. There are many sources of error, such as the flow may not be completely laminar, the number of grids, the finite element method and so on.

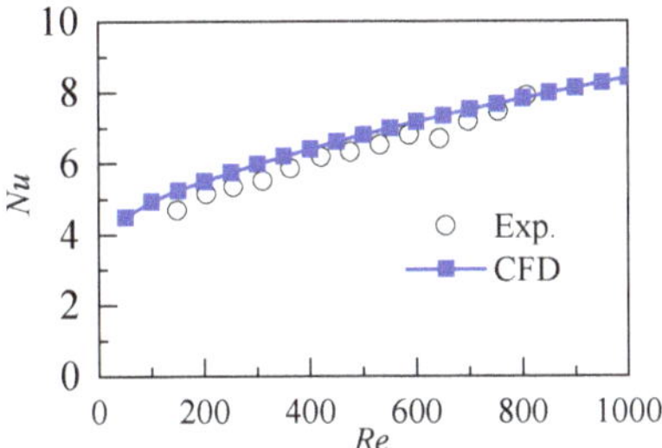

Figure 2. Model validation by comparing the present results with Lei et al. experiments [23].

3. Result Analysis and Discussion

3.1. Influence of Nanoparticle Volume Fraction

Take the microchannel heat sink with aspect ratio $HW = 10$ ($HW = W_h/W_c$) as an example. The three-dimensional flow and heat transfer of the nanofluids are simulated in the Reynolds number range of 100–500, and the nanoparticle volume fractions α are 0.5% and 5%, respectively. According to Formula (10), the Reynolds number is directly related to density, inlet velocity, characteristic scale and viscosity, while density and viscosity are related to the content of solid particles in nanofluids. In this paper, the physical properties of the fluid and the characteristic scale of the channel are determined by giving the solid particle content and aspect ratio in the nanofluid. Finally, the Reynolds number can be changed by adjusting the inlet velocity. The temperature distributions in the middle section of the heat sink are extracted for comparison, and the results are shown in Figure 3. The laws of the temperature distributions in the middle sections of two types of nanofluids at different Reynolds numbers are similar. The temperature at the bottom of the heat sink is the highest, and the temperature in the solid region decreases gradually along the height of the microchannel. The temperature of the fluid at the center of the fluid domain of the microchannel is relatively low. At the fluid structure coupling heat transfer surface, the fluid temperature is relatively high because of the heating of the solid surface. At the same time, with the increase of the Reynolds number, the heat sink temperature and fluid temperature decrease significantly. When the Reynolds number remains the same, the heat sink temperature and fluid temperature of the nanofluid with a nanoparticle volume fraction of 5% are significantly lower than those of the nanofluid with a nanoparticle volume fraction of 0.5%.

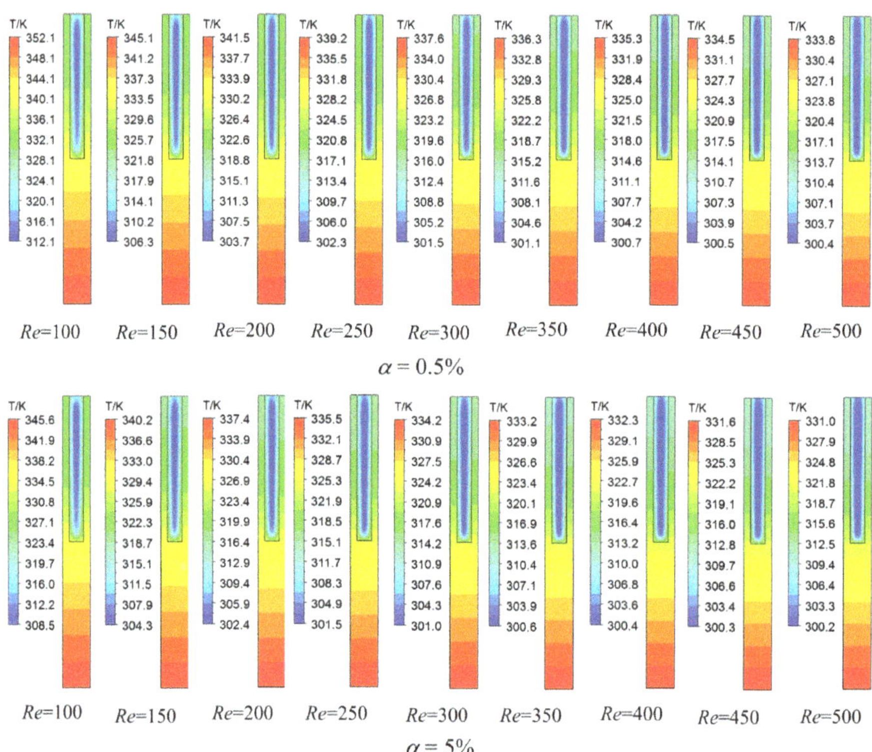

Figure 3. Temperature distributions of fluid and solid domains at the middle plane.

The average temperature on the heat exchange surface of the fluid and solid domains under each working condition is obtained. The results are shown in Figure 4. With the increase of the Reynolds number, the average temperature on the heat transfer surface of the two types of nanofluids decreases. At the same Reynolds number, increasing the volume fraction of nanoparticles in the nanofluid can reduce the temperature on the heat transfer surface. In the range of Reynolds numbers studied in this work, the average temperature difference on the heat transfer surface caused by the differences in nanoparticle volume fraction decreases with the increase of the Reynolds number. When the Re is 100, the average temperature difference between the nanofluid with a nanoparticle volume fraction of 5% and that with a nanoparticle volume fraction of 0.5% is 6.3 K. When the Re is 500, the average temperature difference of the heat exchange surface decreases to 2.6 K. This result shows that the average temperature of the heat exchange surface can be reduced by increasing the nanoparticle volume fraction under the condition with a low Reynolds number.

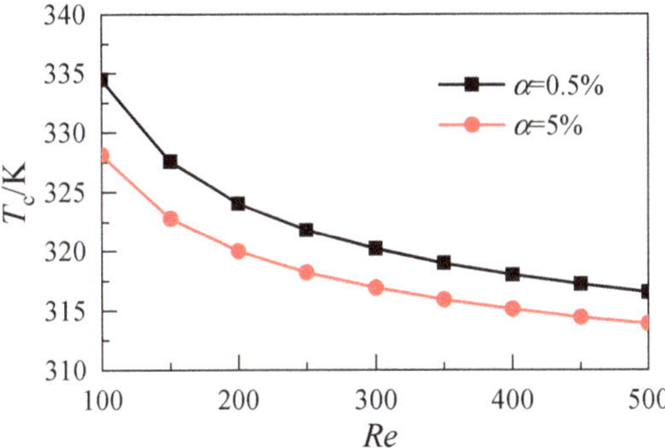

Figure 4. Average temperature distributions of the fluid-solid surface.

Figure 5 shows the effects of the nanoparticle volume fraction on heat transfer coefficients with different Reynolds numbers. The results show that the average heat transfer coefficient increases with the increase of the Reynolds number and that the increase of the nanoparticle volume fraction can improve the heat transfer ability of the nanofluids. As shown in Figure 6, in order to comprehensively consider the difference of fluid thermal conductivity caused by different volume fractions of nanoparticles, the Nusselt number is used as the evaluation index for comparative analysis, and the same conclusion can be obtained.

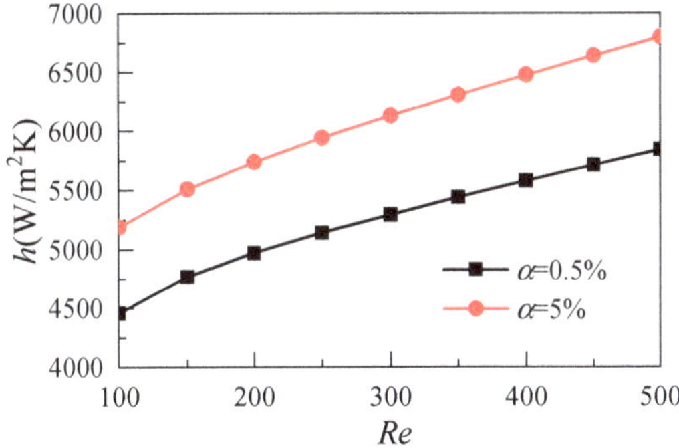

Figure 5. Effects of nanoparticle volume fraction on heat transfer coefficients.

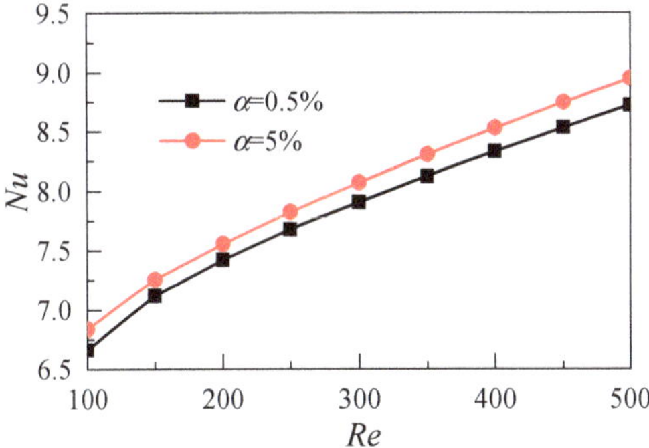

Figure 6. Influence of nanoparticle volume fraction on Nusselt number.

Figure 7 shows the distribution of the pressure difference between the inlet and outlet of the microchannel with different volume fractions. The drag loss of nanofluids through microchannels increases with the increase of the Reynolds number. However, the drag loss of nanofluids with a nanoparticle volume fraction of 5% is greater than that of nanofluids with a nanoparticle volume fraction of 0.5%. At the same time, because the volume fraction of nanoparticles affects the density and dynamic viscosity of nanofluids, the velocity of the microchannel inlet must be adjusted to keep the same Reynolds number. As shown in Figure 8, with the increase of the Reynolds number, the inlet velocity of the 5% nanofluid is greater than that of the 0.5% nanofluid. Pumping power is introduced as the evaluation index to reasonably evaluate the synergistic effect of inlet velocity and resistance loss. Its physical meaning is the external work required for nanofluids to pass through microchannels. Pumping power p is expressed as:

$$P = N \cdot U_{\text{in}} \cdot W_{\text{h}} \cdot W_{\text{c}} \cdot_\Delta p \tag{15}$$

N is the number of microchannels in the whole heat sink, and the value is $N = 1$.

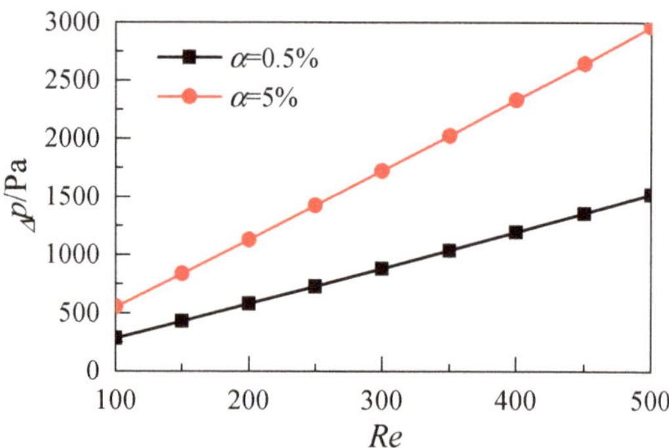

Figure 7. Influence of nanoparticle volume fraction on pressure difference.

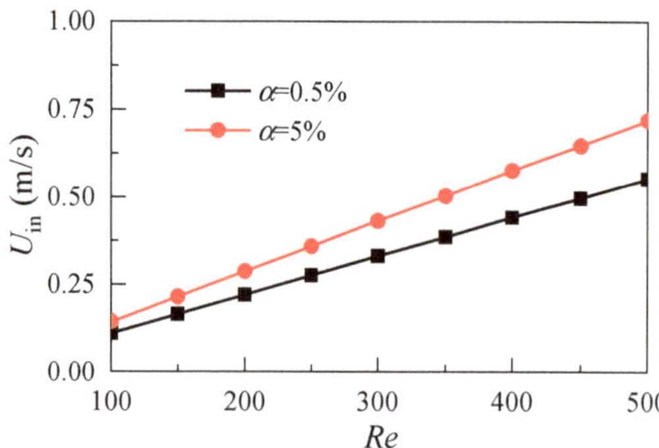

Figure 8. Effect of nanoparticle volume fraction on inlet velocity.

The variation of pumping power with the Reynolds number is shown in Figure 9. With the increase of the Reynolds number, the pumping power of the nanofluid with a nanoparticle volume fraction of 5% is significantly higher than that of the nanofluid with a nanoparticle volume fraction of 0.5%. This result shows that the heat transfer performance cannot be improved by increasing the nanoparticle volume fraction because doing so greatly increases the power consumption of the whole system. At the same time, a high volume fraction renders the nanoparticles in nanofluids unable to maintain a stable and uniform suspension state [23]. Therefore, in engineering applications, the nanoparticle volume fraction in nanofluids needs to be maintained at a low level.

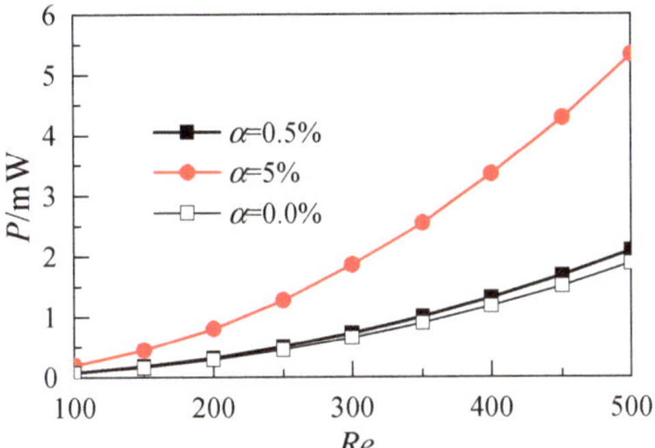

Figure 9. Effects of nanoparticle volume fraction on pumping power.

3.2. Influence of the Aspect Ratio of Microchannels

This work studies nanofluids with a nanoparticle volume fraction of 5%. The different aspect ratios (HW) of microchannels can be obtained by changing their height (W_h). The aspect ratios of the five microchannel heat sink models are 10, 20, 30, 40 and 50. A three-dimensional simulation of flow and heat transfer under different Reynolds numbers is conducted, and the flow and heat transfer characteristics of nanofluid microchannels with different aspect ratios are analyzed and compared. As shown in Figure 10, the pressure difference between the inlet and the outlet increases with the increase of the Reynolds number under different aspect ratios of the microchannel heat sink. The pressure difference decreases with the increase of the aspect ratio at the same Reynolds number. When the aspect ratio is 20–50, the pressure difference is not obvious. This result shows that the increase of the aspect ratio of the microchannel does not greatly enhance the flow resistance loss of nanofluids under the parameters studied in this work. Figure 11 shows the variation of the resistance coefficient f of the microchannel with the Reynolds number. The simulation results of the five different aspect ratios show that the drag coefficient decreases with the increase of the Reynolds number. Moreover, the differences in the drag coefficients caused by different aspect ratios decrease with the increase of the Reynolds number.

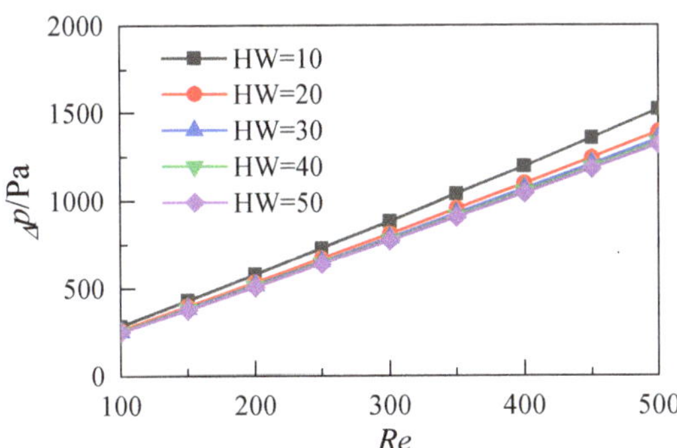

Figure 10. Distributions of pressure difference between the inlet and the outlet of the microchannel heat sink with different aspect ratios.

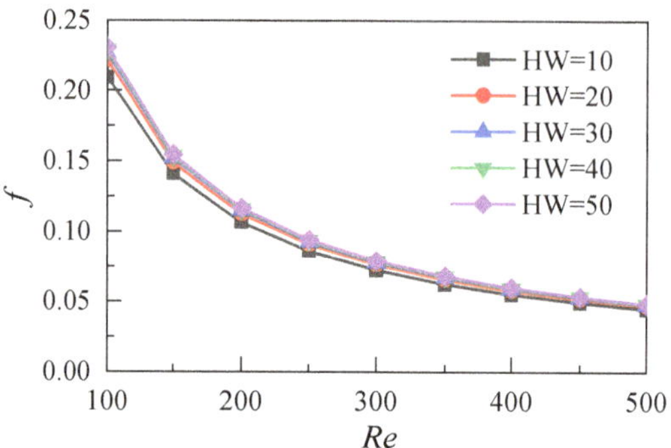

Figure 11. Distributions of friction coefficients.

Figure 12 shows the variation of the average temperature at the bottom of the microchannel heat sink with the Reynolds numbers at different aspect ratios. The results show that the mean bottom temperature, corresponding to the five microchannel heat sinks, decreases with the increase of the Reynolds number and that increasing the aspect ratio of the microchannel can reduce the bottom temperature of the heat sink. However, when the aspect ratio exceeds 20, the decrease of the heat sink's bottom temperature drops, caused by an increase in the aspect ratio of the microchannel. Furthermore, the average temperature values at the bottom of the three microchannels with aspect ratios 30, 40, and 50 almost coincide.

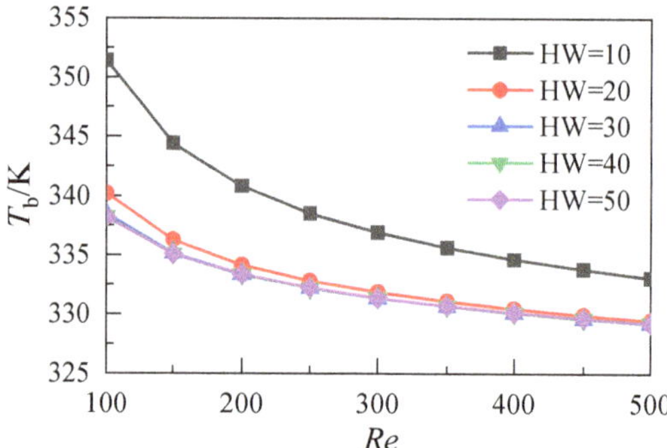

Figure 12. Temperature distributions of the heat sink bottom with different aspect ratios.

The average temperature distribution of the fluid-solid surface relative to the heat transfer in the microchannels with different aspect ratios is shown in Figure 13. With the increase of the Reynolds number, the average temperature of the heat transfer surface decreases. When the aspect ratio HW increases from 10 to 20, the corresponding temperature drop is 16.8 K at an Re of 100. When the aspect ratio HW increases from 20 to 30, the corresponding temperature drop is 5.7 K. When the aspect ratio HW increases from 30 to 40, the corresponding temperature drop decreases to 2.9 K. When the aspect ratio HW increases from 40 to 50, the corresponding temperature drop further decreases to 1.8 K. As shown in Figure 14, when the aspect ratio HW increases from 10 to 30, the Nusselt number increases. When the aspect ratio further increases from 30 to 50, the Nusselt number does not increase significantly. This result indicates that the increase of the aspect ratio does not significantly improve the heat transfer performance of the microchannel heat sink in this range. The above analysis shows that the change of the aspect ratio of the microchannel affects its heat transfer performance and resistance characteristics. In this work, the comprehensive heat transfer performance index is used to quantitatively evaluate the synergistic effect. It is defined as:

$$\eta = \frac{(Nu/Nu_0)}{(f/f_0)^{1/3}} \tag{16}$$

where f_0 is the resistance coefficient of the microchannel with aspect ratio $HW = 10$ and Nu_0 is the average Nusselt number of the microchannel with aspect ratio $HW = 10$.

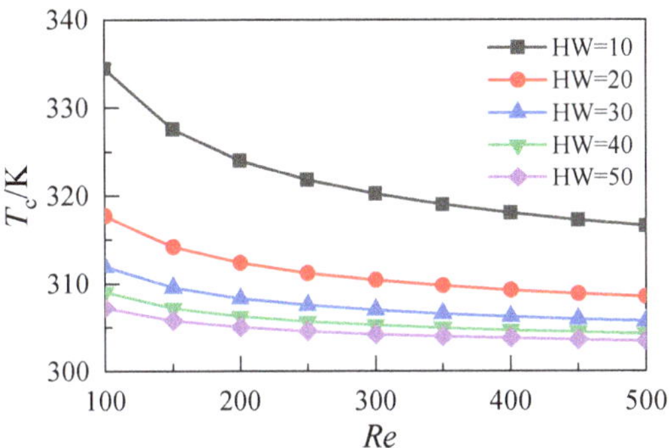

Figure 13. Temperature distributions of fluid-solid surface in microchannels with different aspect ratios.

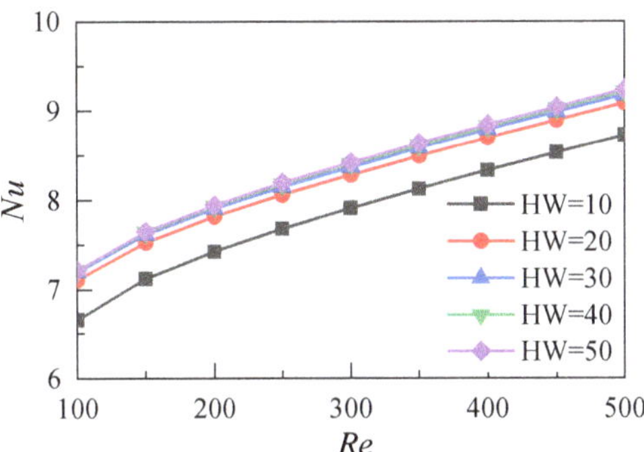

Figure 14. Nusselt number distributions of microchannels with different aspect ratios.

Figure 15 shows the variations of the comprehensive heat transfer performance parameters of the microchannel heat sinks with different aspect ratios and given different Reynolds numbers. In the range of the Reynolds numbers studied in this work, the comprehensive heat transfer performance parameters are greater than 1. The results indicate that for the microchannel heat sink with $HW = 10$, increasing the aspect ratio can improve its comprehensive heat transfer performance. When the aspect ratio is increased to 30, the comprehensive heat transfer performance of the microchannel heat sink does not continue to improve.

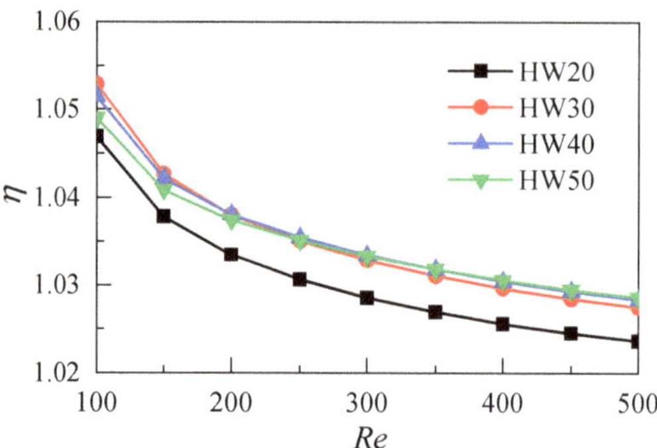

Figure 15. Comparisons of comprehensive heat transfer performances of microchannels with different aspect ratios.

4. Conclusions

In this work, a microchannel heat sink is studied on the basis of computational fluid dynamics. The flow and heat transfer of two nanofluids with different volume fractions and five microchannel heat sinks with different aspect ratios in the Reynolds number range of 100–500 are simulated. The flow and heat transfer characteristics of the microchannel heat sinks are compared, and the optimal parameters of the aspect ratio are analyzed. The conclusions are as follows:

(1) Increasing the volume fraction of nanoparticles can effectively reduce the average temperature of the heat transfer surface and improve the heat transfer capability of nanofluids. However, because of the dual increase of the inlet velocity and flow resistance, the power consumption of the whole system increases greatly;

(2) Increasing the aspect ratio of the microchannel does not cause significant flow resistance loss, and the resistance coefficient of the microchannel tends to be consistent with the increase of the Reynolds number at different aspect ratios;

(3) Increasing the aspect ratio of the microchannel can reduce the temperature of the heat sink. When the aspect ratio exceeds 30, the average temperature at the bottom of the microchannel does not decrease, and the heat transfer coefficient does not increase;

(4) In the range of the parameters studied in this paper, the aspect ratio of the microchannel heat sink with a thickness of 6 mm has an optimal value. Based on the comprehensive heat transfer performance parameters, the optimal value of the aspect ratio of the microchannel heat sink is 30.

This study shows that the aspect ratio of the heat sink has a significant impact on the heat transfer performance of a microchannel, and there is an optimal value in the range of Reynolds numbers under the condition of a given thickness. Further research will be carried out for different thicknesses of heat sinks under different Reynolds number conditions to obtain a universal empirical formula for guiding engineering practice.

Author Contributions: Conceptualization, H.W. and S.Z.; methodology, H.W.; software, H.W.; validation, S.Z.; formal analysis, S.Z.; investigation, S.Z.; resources, S.Z.; writing—original draft preparation, H.W.; writing—review and editing, H.W.; visualization, H.W.; supervision, S.Z.; funding acquisition, S.Z. All authors have read and agreed to the published version of the manuscript.

Funding: The authors acknowledge the support of the 14th batch High-level Talents Project for "Six Talents Peak" (Grant No. XCL-092), the Province Postdoctoral Foundation of Jiangsu (1501164B), the Technical Innovation Nurturing Foundation of Yangzhou University (2017CXJ024), China Postdoctoral Science Foundation (2016M600447), Yangzhou Innovative Capacity Building Plan Project (YZ2017275) and Yangzhou University Science Foundation Project (x20180290).

Conflicts of Interest: The authors declare no conflict of interest.

Nomenclature

u_i	velocity components in direction i
u_j	velocity components in direction j
x_i	cartesian coordinate components
x_j	cartesian coordinate components
p	pressure in the flow field
T	temperature
ρ_{nf}	density of nanofluids
μ_{nf}	dynamic viscosity
C_{pf}	specific heat capacity
λ_{nf}	thermal conductivity
w	corresponding thermophysical properties of the base fluid
p	corresponding thermophysical properties of the nanoparticles
α	volume fraction of nanoparticles
n	shape factor of nanoparticles
U_{in}	fluid inlet velocity
q	heat flux
Nu	Nusselt number
N	the number of microchannels
T_s	the temperature of solid region
λ_s	the thermal conductivity of the solid region.

References

1. Wu, R.; Hong, T.; Cheng, Q.; Zou, H.; Fan, Y.; Luo, X. Thermal modeling and comparative analysis of jet impingement liquid cooling for high power electronics. *Int. J. Heat Mass Transf.* **2019**, *137*, 42–51. [CrossRef]
2. Choi, T.J.; Kim, S.H.; Jang, S.P.; Yang, D.J.; Byeon, Y.M. Heat transfer enhancement of a radiator with mass-producing nanofluids (EG/water-based Al$_2$O$_3$ nanofluids) for cooling a 100 kW high power system. *Appl. Therm. Eng.* **2020**, *180*, 115780. [CrossRef]
3. Tuckerman, D.B.; Pease, R. High-performance heat sinking for VLSI. *IEEE Electron. Device Lett.* **1981**, *2*, 126–129. [CrossRef]
4. Choi, S.; Eastman, J.A. Enhancing thermal conductivity of fluids with nanoparticles. In Proceedings of the 1995 International Mechanical Engineering Congress and Exhibition, San Francisco, CA, USA, 12–17 November 1995.
5. Koo, J.; Kleinstreuer, C. Laminar nanofluid flow in microheat-sinks. *Int. J. Heat Mass Transf.* **2005**, *48*, 2652–2661. [CrossRef]
6. Apmann, K.; Fulmer, R.; Soto, A.; Vafaei, S. Thermal conductivity and viscosity: Review and optimization of effects of nanoparticles. *Materials* **2021**, *14*, 1291–1362. [CrossRef] [PubMed]
7. Malvandi, A.; Safaei, M.R.; Kaffash, M.H.; Ganji, D.D. MHD mixed convection in a vertical annulus filled with Al$_2$O$_3$-water nanofluid considering nanoparticle migration. *J. Magn. Magn. Mater.* **2015**, *3*, 296–306. [CrossRef]
8. Goodarzi, M.; Amiri, A.; Goodarzi, M.S.; Safaei, M.R.; Karimipour, A.; Languri, E.M.; Dahari, M. Investigation of heat transfer and pressure drop of a contour flow corrugated plate heat exchanger using MWCNT based nanofluids. *Int. Commun. Heat Mass Transf.* **2015**, *66*, 172–179. [CrossRef]
9. Vinoth, R.; Kumar, D.S. Experimental investigation on heat transfer characteristics of an oblique finned microchannel heat sink with different channel cross sections. *Heat Mass Transfer* **2018**, *54*, 3809–3817. [CrossRef]
10. Alfaryjat, A.A.; Mohammed, H.A.; Adam, N.M.; Ariffin, M.K.A.; Najafabadi, M.I. Influence of geometrical parameters of hexagonal, circular, and rhombus microchannel heat sinks on the thermohydraulic characteristics. *Int. Commun. Heat Mass Transf.* **2014**, *52*, 121–131. [CrossRef]
11. Ahmed, H.E.; Ahmed, M.I. Optimum thermal design of triangular, trapezoidal and rectangular grooved microchannel heat sinks. *Int. Commun. Heat Mass Transf.* **2015**, *66*, 47–57. [CrossRef]
12. KumaR, P. Numerical investigation of fluid flow and heat transfer in trapezoidal microchannel with groove structure. *Int. J. Therm. Sci.* **2019**, *136*, 33–43. [CrossRef]
13. Farsad, E.; Abbasi, S.P.; Zabihi, M.S.; Sabbaghzadeh, J. Numerical simulation of heat transfer in a micro channel heat sinks using nanofluids. *Heat Mass Transf.* **2010**, *47*, 479–490. [CrossRef]
14. Shi, X.; Li, S.; Wei, Y.; Gao, Y. Multi objective optimization of structural parameters of nanofluid rectangular microchannel heat sink. *J. Xi'an Jiaotong Univ.* **2018**, *52*, 61–66.
15. Ghale, Z.Y.; Haghshenasfard, M.; Esfahany, M.N. Investigation of nanofluids heat transfer in a ribbed microchannel heat sink using singlephase and multiphase CFD models. *Int. Commun. Heat Mass Transf.* **2015**, *68*, 122–129. [CrossRef]
16. Manca, O.; Nardini, S.; Ricci, D. A numerical study of nanofluid forced convection in ribbed channels. *Appl. Therm. Eng.* **2012**, *37*, 280–292. [CrossRef]
17. Xie, H.; Fujii, M.; Zhang, X. Effect of interfacial nanolayer on the effective thermal conductivity of nanoparticle-fluid mixture. *Int. J. Heat Mass Transf.* **2004**, *48*, 2926–2932. [CrossRef]

18. Gkountas, A.A.; Benos, L.T.; Nikas, K.S.; Sarris, I.E. Heat transfer improvement by an Al2O3-water nanofluid coolant in printed-circuit heat exchangers of supercritical CO2 Brayton cycle. *Therm. Sci. Eng. Prog.* **2020**, *20*, 1025–1051. [CrossRef]
19. Sohel, M.R.; Khaleduzzaman, S.S.; Saidur, R.; Hepbasli, A.; Sabri, M.F.M.; Mahbubul, I.M. An experimental investigation of heat transfer enhancement of a minichannel heat sink using Al_2O_3–H_2O nanofluid. *Int. J. Heat Mass Transf.* **2014**, *74*, 164–172. [CrossRef]
20. Mohammadpour, J.; Salehi, F.; Sheikholeslami, M.; Masoudi, M.; Lee, A. Optimization of nanofluid heat transfer in a microchannel heat sink with multiple synthetic jets based on CFD-DPM and MLA. *Int. J. Therm. Sci.* **2021**, *167*, 107008. [CrossRef]
21. Khanafer, K.; Vafai, K. A critical syn study of thermophysical characteristics of nanofluids. *Int. J. Heat Mass Transf.* **2011**, *54*, 4410–4428. [CrossRef]
22. Chai, L.; Xia, G.; Wang, L.; Zhou, M.; Cui, Z. Heat transfer enhancement in microchannel heat sinks with periodic expansion-constriction cross-sections. *Int. J. Heat Mass Transf.* **2013**, *62*, 741–751. [CrossRef]
23. Lee, J.; Mudawar, I. Assessment of the effectiveness of nanofluids for single-phase and two-phase heat transfer in micro-channels. *Int. J. Heat Mass Transf.* **2007**, *50*, 452–463. [CrossRef]

Article

Formation and Elimination of Satellite Droplets during Monodisperse Droplet Generation by Using Piezoelectric Method

Zejian Hu [1], Shengji Li [2,*], Fan Yang [1], Xunjie Lin [1], Sunqiang Pan [3], Xuefeng Huang [1,*] and Jiangrong Xu [1]

[1] Institute of Energy, Department of Physics, Hangzhou Dianzi University, Hangzhou 310018, China; zejianhu@hdu.edu.cn (Z.H.); yf@hdu.edu.cn (F.Y.); 17072215@hdu.edu.cn (X.L.); jrxu@hdu.edu.cn (J.X.)

[2] College of Materials and Environmental Engineering, Hangzhou Dianzi University, Hangzhou 310018, China

[3] Division of Biological and Chemical Metrology, Zhejiang Institute of Metrology, Hangzhou 310018, China; pansunqiang@hotmail.com

* Correspondence: shengjili@hdu.edu.cn (S.L.); xuefenghuang@hdu.edu.cn (X.H.); Tel.: +86-571-8687-8677 (X.H.)

Citation: Hu, Z.; Li, S.; Yang, F.; Lin, X.; Pan, S.; Huang, X.; Xu, J. Formation and Elimination of Satellite Droplets during Monodisperse Droplet Generation by Using Piezoelectric Method. *Micromachines* 2021, *12*, 921. https://doi.org/10.3390/mi12080921

Academic Editors: Junfeng Zhang and Ruijin Wang

Received: 10 July 2021
Accepted: 28 July 2021
Published: 31 July 2021

Publisher's Note: MDPI stays neutral with regard to jurisdictional claims in published maps and institutional affiliations.

Abstract: One of the key questions in the generation of monodisperse droplets is how to eliminate satellite droplets. This paper investigates the formation and elimination of satellite droplets during the generation of monodisperse deionized water droplets based on a piezoelectric method. We estimated the effects of two crucial parameters—the pulse frequency for driving the piezoelectric transducer (PZT) tube and the volume flow rate of the pumping liquid—on the generation of monodisperse droplets of the expected size. It was found that by adjusting the pulse frequency to harmonize with the volume flow rate, the satellite droplets can be eliminated through their coalescence with the subsequent mother droplets. An increase in the tuning pulse frequency led to a decrease in the size of the monodisperse droplets generated. Among three optimum conditions (OCs) (OC1: 20 mL/h, 20 kHz; OC2: 30 mL/h, 30 kHz; and OC3: 40 mL/h, 40 kHz), the sizes of the generated monodisperse deionized water droplets followed a bimodal distribution in OC1 and OC2, whereas they followed a Gaussian distribution in OC3. The average diameters were 87.8 μm (OC1), 85.9 μm (OC2), and 84.8 μm (OC3), which were 8.46%, 6.14%, and 4.69% greater than the theoretical one (81.0 μm), respectively. This monodisperse droplet generation technology is a promising step in the production of monodisperse aerosols for engineering applications.

Keywords: monodisperse droplet generation; satellite droplets; piezoelectric method; droplet coalescence

1. Introduction

In recent years, monodisperse droplet generation technology has been widely used in the fields of additive manufacturing [1,2], inkjet printing [3,4], electronic packaging [5,6], bioengineering [7–9], instrument calibration [10,11], etc. It has prompted many researchers to become engaged in developing droplet generation techniques to meet new requirements such as droplets that are highly uniform and monodisperse in terms of their size, shape, density, and surface characteristics, with a variety of solutes and solvents.

Many attempts have been made to generate monodisperse droplets, such as hot bubble [12], mechanical [13], pneumatic [14,15], piezoelectric [16], electromagnetic [17], and droplet-based microfluidic [18–20] technologies. Among these, the piezoelectric droplet generation method is one of the best choices to obtain monodisperse droplets. On the basis of the piezoelectric method, the generation of droplets depends on the control of the pulse waveform for the driving of the PZT tube. Li et al. [21] reported that monodisperse droplets can be obtained through adjusting the frequency and amplitude of a rectangular pulse waveform at a high operating pressure of 3.5 MPa. Fan et al. [22] reported that monodisperse droplets were generated by controlling the upper and lower limits of the pulse amplitude. However, during droplet generation, many satellite droplets were produced, leading to a wide size distribution of the droplets. Shin et al. [23] also reported that, for low viscosity liquid, satellites were generated when a single-pulse waveform was applied, whereas when a double-pulse waveform was utilized, the satellites were eliminated. A simple change in the time separation can precisely control the droplet size. If the time separation is shortened, the droplets' size becomes smaller, since the second pulse reduces the mass and momentum of the ejected liquid. However, the double-pulse waveform can possibly result in the coalescence of two adjacent mother droplets. Lin et al. [24] further studied the influence of pulse frequency, positive voltage time, and voltage magnitude on droplet ejection velocity using a double-pulse voltage pattern, but they estimated it to have a scarce effect on droplet size.

Regardless of the control associated with the single-pulse waveform or the double-pulse waveform, the breakup of fluid filaments ejected from the nozzle leads to an array of uniformly spaced large droplets (called mother droplets) with smaller droplets (known as satellite droplets) in between them. Therefore, to generate uniform mother droplets based on piezoelectric method, the problem of eliminating a large number of satellite droplets has to be solved. Otherwise, this situation would result in a nonuniform size distribution of the breakup droplets. There is no doubt that one of the key steps involved in generating uniform and monodisperse droplets is eliminating the satellite droplets and avoiding the coalescence of two adjacent mother droplets.

In this work, we have aimed to investigate the formation and elimination of satellite droplets during the generation of monodisperse deionized water droplets based on the piezoelectric method. The emphasis of this study was on harmonizing the relationship between the frequency of the square-pulse waveform and the volume flow rate in order to obtain monodisperse droplets of the expected size. Therefore, this work involves: (1) observing the formation and elimination processes of satellite droplets during the generaton of monodisperse deionized water droplets through a high-resolution imaging system; (2) estimating the effect of pulse frequency and volume flow rate parameters on the droplet size and its distribution; (3) revealing the mechanisms involved in the generation and elimination of satellite droplets related to these two crucial parameters.

2. Experimental Methods

A schematic of the experimental setup is shown in Figure 1a. It was mainly composed of a micropump and syringe, filter, nozzle, controller, high-speed camera, and an LED lamp. The micropump was used to feed the deionized water stored in the syringe and to control the volume flow rate with an accuracy of $\pm$ 2%. The deionized water (the properties of which are listed in Table 1) was transported to the nozzle through the connecting pipe and filter. The droplets were then extruded by means of the PZT tube in the nozzle, controlled by the controller. The high-speed camera (Phantom M310, Vision Research Inc., Wayne, NJ, USA) with a lens (AT-X M100 AF PRO, Tokina, Japan) was utilized to record the pinch-off process of the liquid filaments, the formation of mother droplets, and the generation of satellite droplets. The recording frame was set to 5000 fps. The LED lamp with an adjustable luminance was used to illuminate the field of view for clear imaging using the high-speed camera.

The nozzle (MDG100) was purchased from TSI Co. Ltd., USA, and was based on a squeeze mode design. A PZT tube was wrapped outside a glass tubular reservoir, shown in Figure 1b. The orifice diameter at the bottom of the nozzle was 50 μm. The PZT tube was driven by the controller using a transistor–transistor logic (TTL) signal to modulate the width and frequency of periodic rectangle-wave pulse as shown in Figure 2. The period of the pulse waveform is T, and its voltage amplitude is V. The duty ratio is T_p/T, where T_p is the applied time of the pulse. The single-pulse waveform is a rectangle wave, which is defined by a pulse width, a rising edge, and a falling edge. The applying time of the pulse can be divided into three parts—the rising time (t_r), dwelling time (t_d), and falling time (t_f). When the rising (falling) edge of the waveform is applied to the PZT tube, a negative (positive) pressure wave is produced, expanding (contracting) the glass tubular reservoir, so that the fluid filaments will be ejected and broken up. These parameters of the driving pulse for the PZT tube are listed in Table 2.

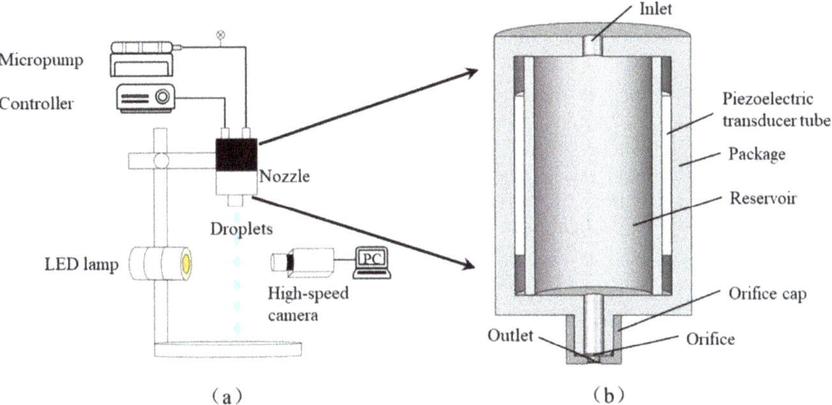

Figure 1. (**a**) Schematic of experimental setup; (**b**) schematic of nozzle.

Table 1. Physical properties of deionized water.

Items	Density (kg/m^3)	Surface Tension (N/m) @25 °C	Viscosity (Pa·s) @15 °C	Acoustic Velocity (m/s)
Value	1.0×10^3	0.072	1.14×10^{-3}	1435 [24]

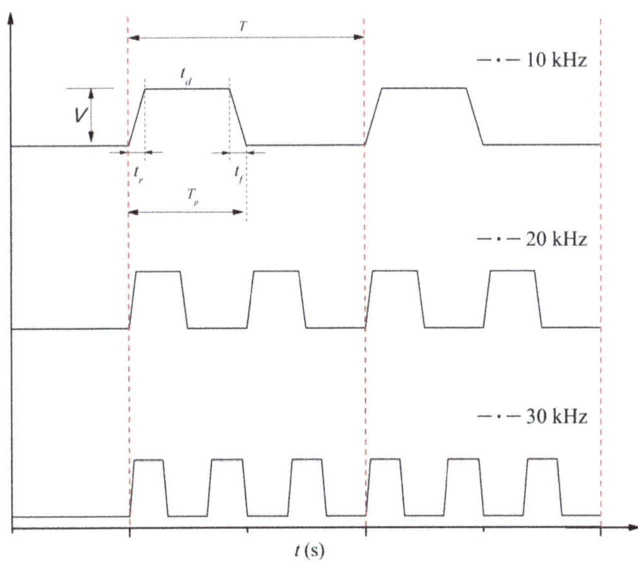

Figure 2. Schematic of periodic rectangle-wave pulse waveform.

Table 2. Parameters of pulse waveform for driving the PZT tube.

Items	Amplitude	Frequency	Duty Ratio
Value	4.2 V	5~40 kHz	0.5

The measurement of droplet size is crucial to estimate performance in the generation of monodisperse droplets. In this work, a digital image processing method was used. The elaborate operation procedures have been presented in our previous work [25]. Briefly, (1) a magnified image of the calibrating ruler was acquired; (2) pictures of the breakup droplets were taken under the same imaging conditions; (3) a self-programming digital imaging treatment program was then used to compare the images of the breakup droplets with calibrated pixel pitches. If the droplets appeared ellipsoidal or non-spherical due to deformation, a characteristic dimension d_{21} (called the equivalent diameter) was calculated by $d_{21} = 4S_d/L_d$, where S_d and L_d represent the area and the perimeter of each droplet, respectively. Finally, the statistical average diameter of multiple droplets (d) was determined by the equation $d = \frac{\sum n_i d_i}{\sum n_i}$.

3. Results and Discussion

3.1. Pinch-Off of Liquid Filament and Generation of Satellite Droplets

Figure 3 demonstrates a snapshot taken during the pinch-off process of the liquid filament at a pulse frequency of 10 kHz and a volume flow rate of 30 mL/h. It can be observed that surface waves cover the liquid filament, and a mother droplet has been produced; another droplet is being generating at the liquid neck (the narrowest position along the liquid filament). At the bottom of the picture, when a mother droplet is detached from the liquid filament, a satellite droplet is simultaneously generated.

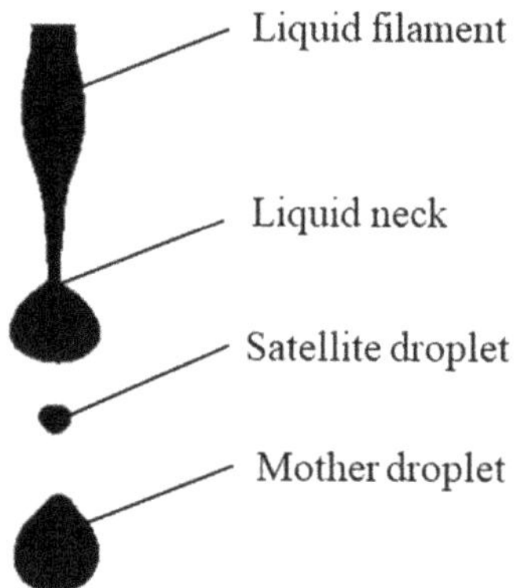

Figure 3. Snapshot of the pinch-off of a deionized water filament and the generation of satellite droplets.

The Reynolds and Weber numbers can be expressed as $Re = \rho v d/\mu$ and $We = \rho v^2 d/\sigma$, in which ρ, v, d, σ and μ represent the density, velocity, orifice diameter, surface tension, and viscosity of the deionized water, respectively. In this case, the Reynolds number was ~186.2 and the Weber number was ~1.4. This suggests that the pinch-off of the deionized water filament generally exhibited a laminar regime, and the surface tension played a key role in the detachment.

The evolution of the liquid filament was considered as a function of the viscosity ratio (p) of the fluids and the initial wavenumber of the interface perturbation, and the satellite droplets were generated around the neck region of a highly deformed filament. Tjahjadt et al. [26] numerically and experimentally explained that, in low-viscosity ratio systems, $p < O$ (0.1), the breakup mechanism depends on self-repetition in multiple breakup sequences in which every pinch-off is always associated with the formation of a neck; the neck undergoes pinch-off, and the process repeats. In this case, unlike multiple breakup sequences, at each period, we observed one breakup sequence and a generated satellite droplet. This means that, at each corresponding period of the controlled rectangle-wave pulse of PZT tube, the duty ratio, the frequency, and the sequence of time applied will be crucial in order to affect the formation of mother and satellite droplets.

3.2. Elimination of Satellite Droplets

Theoretically, if the liquid filament breaks up into monodisperse mother droplets without the generation of satellite droplets, the input volume per unit time should be equal to the total volume of the monodisperse droplets (mother droplets) generated per unit time:

$$q \times \frac{10^{-6}}{3600} = \frac{\pi}{6} d_{th}^3 \times 10^{-18} \times f \times 10^3 \tag{1}$$

In which d_{th}, q, and f represent the theoretical average diameter of the generated monodisperse droplets (μm), the volume flow rate of the feeding liquid (mL/h), and the pulse frequency for the driving of the PZT tube (kHz), respectively.

Thus, the theoretical average droplet diameter of the generated monodisperse droplets in an ideal state can be calculated as follows:

$$d_{th} = \sqrt[3]{\frac{q}{600\pi f}} \times 10^3 \tag{2}$$

The volume flow rate of deionized water was still set to 30 mL/h. As the pulse frequency was adjusted from 10 kHz to 30 kHz, we observed the coalescence of satellite droplets and mother droplets, as shown in Figure 4. During the pinch-off process of the liquid filament, a satellite droplet was generated (Figure 4a), but it immediately merged with the subsequent mother droplet (Figure 4b,c). Finally, monodisperse droplets were generated with a highly uniform size distribution (Figure 4d). The average diameter of the generated monodisperse droplets was 85.9 μm in this case, which is 6.14% larger than the theoretical one. This suggests that the adjustment of the pulse frequency could eliminate the satellite droplets by means of coalescence, validating the feasibility of this methodology.

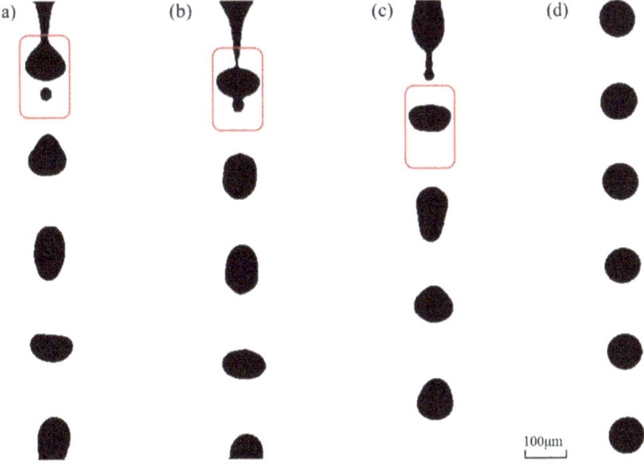

Figure 4. Snapshots of the pinch-offs of liquid filaments and the elimination of satellite droplets. (**a**) Formation of satellite droplets. (**b,c**) Satellite droplets merged by a subsequent mother droplet and eliminated; (**d**) a highly uniform size distribution of monodisperse droplets were obtained (volume flow rate of 30 mL/h; pulse frequencies of 30 kHz).

It is worthy of notice that the number of monodisperse droplets is theoretically equal to the pulse frequency, i.e., only a single droplet is generated as a pulse is applied to the PZT tube. Thus, the pulse width and the duty ratio of each pulse need to be negotiated for the detachment of each droplet from the liquid filament, as shown in Figure 5. The generation time of monodisperse droplets (t_{gen}) is equal to the theoretical pulse width, or the applied time of the practical pulse, i.e., $t_{gen}=T_p=t_r + t_d + t_f$, in which t_r, t_d, and t_f are the rising time, dwelling time, and falling time, respectively. The flight time of the monodisperse droplets (t_{fly}) is the same as the vacant time of the pulse, i.e., $t_{fly}=T - T_p$, in which T and T_p are the period of the pulse waveform and the applied time of the pulse, respectively. Lin et al. [24] suggested that a sufficient time of t_r and t_f is necessary to reach the desired voltage amplitude and to make the PZT tube expand and contract enough. The empirical expression of the voltage amplitude of the pulse waveform must meet the following requirements [24]:

$$V/t_r \leq 15, V/t_f \leq 15 \tag{3}$$

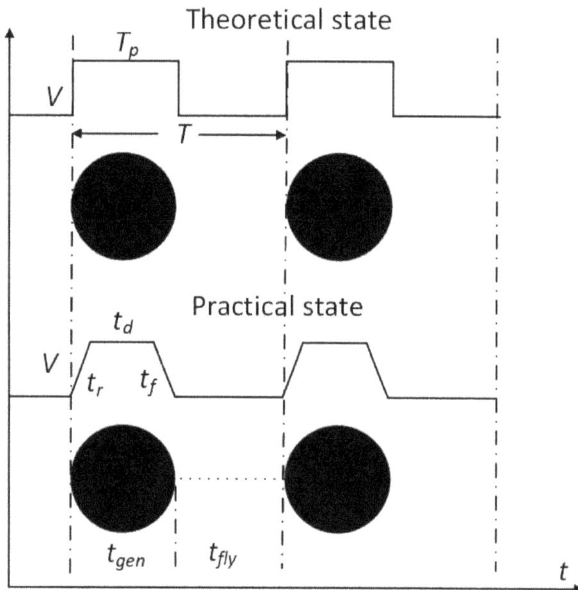

Figure 5. Schematic of the theoretical and practical generation of monodisperse droplets.

This indicates that the slope of the voltage variation should be less than 15 V/μs. Additionally, the voltage amplitude should be high enough to ensure that the droplet is ejected, but not so high that the ejection becomes chaotic and it becomes difficult to obtain monodisperse droplets.

To generate monodisperse droplets, the rising time (t_r) and the falling time (t_f) should also harmonize with the detaching time of each droplet. Under the same operating conditions, the detaching time depends mainly on the viscosity of the liquid. The larger the viscosity, the longer the detaching time becomes. The dwelling time (t_d) has been given by Bogy et al. [27], i.e., the optimum pulse width (t_{opt}), which can be calculated as follows:

$$t_d = t_{opt} = L/v_{aco,liq} \tag{4}$$

in which L and $v_{aco,liq}$ are the length of nozzle and the velocity of acoustic wave propagation in the liquid, respectively. In this work, the theoretical value of the optimum dwelling time (t_d) was ~12 μs. In Figure 4d, the central spacing between two adjacent droplets was twice as large as the droplet diameter, since the duty ratio of the rectangular pulse was 0.5. To avoid the coalescence of two spherical mother droplets without satellite droplets, the central spacing must be greater than the droplet diameter. However, since the droplets exhibited severe deformation after being detached from the liquid filament (Figure 4a–c), the maximum ratio of droplet length to width reached 1.8, and the central spacing was close to the droplet diameter. This suggests that the duty ratio should be below 0.8.

3.3. Effect of Pulse Frequency on the Elimination of Satellite Droplets

Figure 6 demonstrates the snapshots of droplet generation as the pulse frequency ranged from 5 to 45 kHz at the volume flow rate of 30 mL/h. At the pulse frequencies of 5 kHz, 10 kHz, and 15 kHz, it was observed that the satellite droplets were difficult to eliminate (Figure 6b3,c2), and the two adjacent mother droplets coalesced (Figure 6a5). When the pulse frequency was enhanced to 20 kHz or 25 kHz, satellite droplets with greater size were generated (Figure 6d4,e3). At 30 kHz and 35 kHz, the satellite droplets were almost eliminated (Figure 6f3,g3). However, as the pulse frequency exceeded 40 kHz, the generated droplets were chaotic, and they had a wide size distribution.

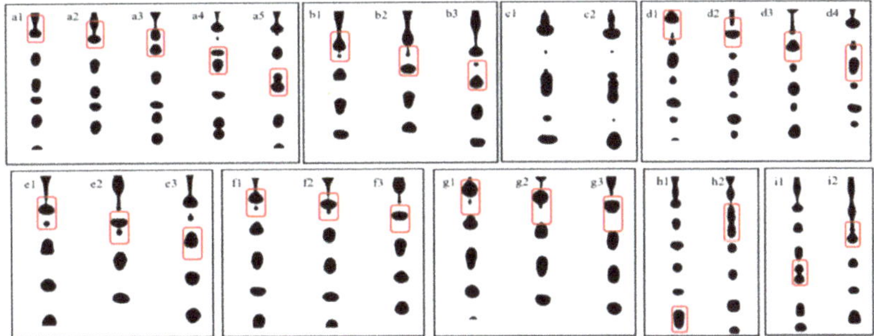

Figure 6. Snapshots of droplet generation at the volume flow rate of 30 mL/h at different pulse frequencies ((**a1–a5**): 5 kHz; (**b1–b3**): 10 kHz; (**c1–c2**): 15 kHz; (**d1–d4**): 20 kHz; (**e1–e3**): 25 kHz; (**f1–f3**): 30 kHz; (**g1–g3**): 35 kHz; (**h1–h2**): 40 kHz; (**i1–i2**): 45 kHz).

Experiments on the influence of pulse frequency on the formation and elimination of satellite droplets were also carried out at the volume flow rates of 20 mL/h and 40 mL/h, respectively. According to the results, it can be concluded that, at a certain volume flow rate, the pulse frequency must be adjusted to an optimum range to generate monodisperse droplets. At the volume flow rate of 20 mL/h, the optimum range of the pulse frequency was 18–22 kHz. For the volume flow rates of 30 mL/h and 40 mL/h, the optimum ranges of the pulse frequency were 25–35 kHz and 35–50 kHz, respectively.

3.4. Effect of Pulse Frequency on the Average Diameter of Droplets

According to Equation (2), the diameter of the generated droplets depends on the volume flow rate and the pulse frequency. As the volume flow is fixed, the desired diameter of the droplets can be obtained through adjusting the pulse frequency. Figure 7 shows the average diameters of the generated droplets at the volume flow rate of 30 mL/h, at different pulse frequencies of 25–40 kHz, with an interval of 5 kHz. The theoretical diameter of the droplets was calculated using Equation (2), and these are also shown in Figure 7. In general, the experimental result displayed a similar trend to the theoretical one. As the pulse frequency for driving the PZT tube was enhanced, the average diameter of the generated droplets decreased. The average droplet diameters based on the experiments were 5–11% greater than those based on the theoretical calculation. This difference can be attributed to the random error associated with dealing with the images and the systematic calibration error.

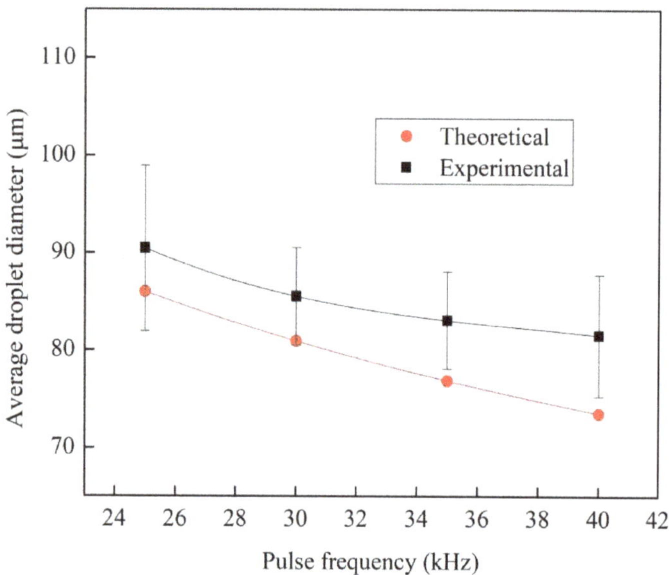

Figure 7. Average droplet diameters at the volume flow rate of 30 mL/h at different pulse frequencies of 25 kHz–40 kHz.

3.5. Size Distribution of Monodisperse Droplets under Optimum Operating Conditions

If the ratio of the volume flow rate and the optimum pulse frequency are kept constant, based on Equation (2), the diameter of the generated droplet would be theoretically the same. This was also supported by the experimental results. Figure 8 illustrates the snapshots of the generated monodisperse droplets under three optimum conditions (OCs) (OC1: 20 mL/h, 20 kHz; OC2: 30 mL/h, 30 kHz; and OC3: 40 mL/h, 40 kHz). In the optimum pulse frequency range, no satellite droplets were observed.

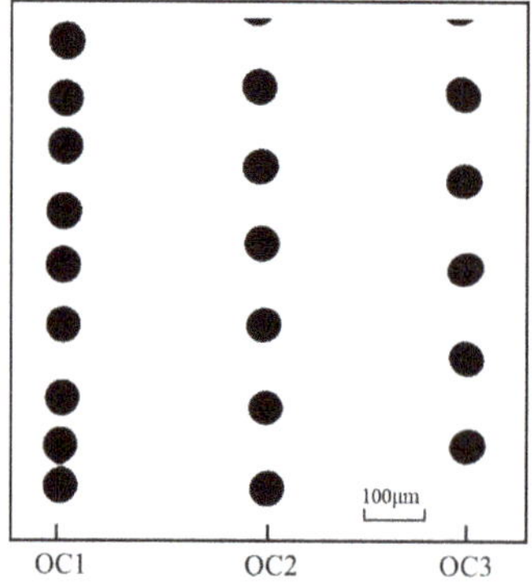

Figure 8. Snapshots of monodisperse droplets generated under three optimum conditions (OC1: 20 mL/h, 20 kHz; OC2: 30 mL/h, 30 kHz; and OC3: 40 mL/h, 40 kHz).

The size distributions of the monodisperse droplets generated under the three optimum conditions were further extracted and analyzed in detail, as shown in Figures 9–11. Under OC1, the size of generated monodisperse droplets were mainly distributed in the range of 80–95 μm, with a small number of droplets of 95–110 μm (Figure 9). As for OC2, the size distribution of generated monodisperse droplets ranged from 80 to 93 μm

(Figure 10). In OC3, the droplet size was mainly distributed in the range of 75–95 μm (Figure 11). We observed two-peak profiles for both OC1 and OC2. Compared with the bimodal size distribution in OC1 and OC2, under OC3, the generated monodisperse droplets had a better dispersion uniformity, and the droplet size followed a Gaussian distribution.

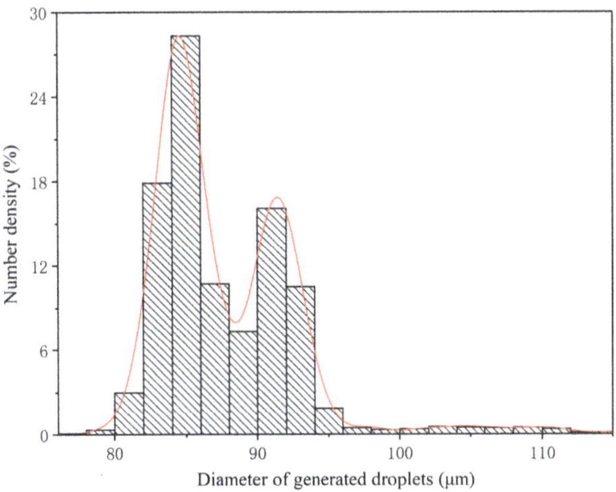

Figure 9. Size distribution of generated monodisperse droplets under OC1 (q = 20 mL/h, f = 20 kHz).

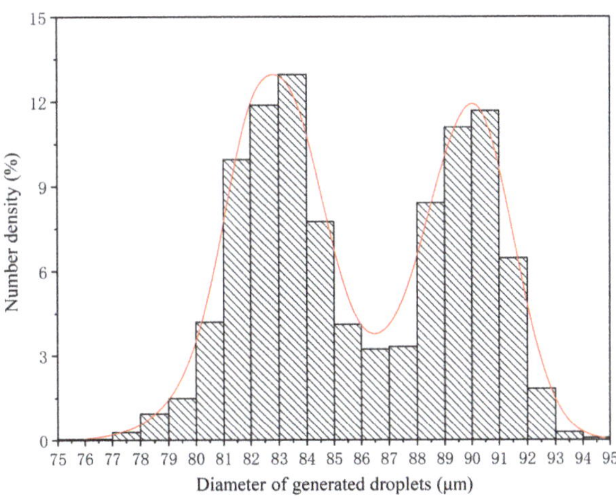

Figure 10. Size distribution of generated monodisperse droplets under OC2 (q = 30 mL/h, f = 30 kHz).

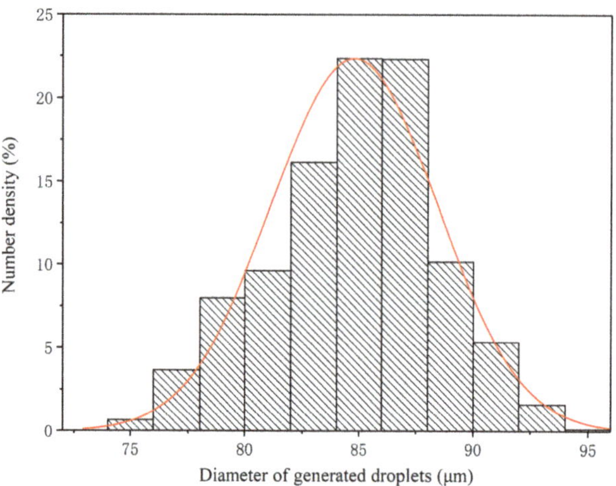

Figure 11. Size distribution of generated monodisperse droplets under OC3 (q = 40 mL/h, f = 40 kHz).

The droplets generated under the three optimum conditions were generally well dispersed and had a good uniformity of dispersion. The average diameters of the generated monodisperse droplets were 87.8 μm (OC1), 85.9 μm (OC2), and 84.8 μm (OC3), respectively, as shown in Figure 12. Compared to the theoretical diameter (81.0 μm), the experimental results were 8.46%, 6.14%, and 4.69% greater than the theoretical one, respectively.

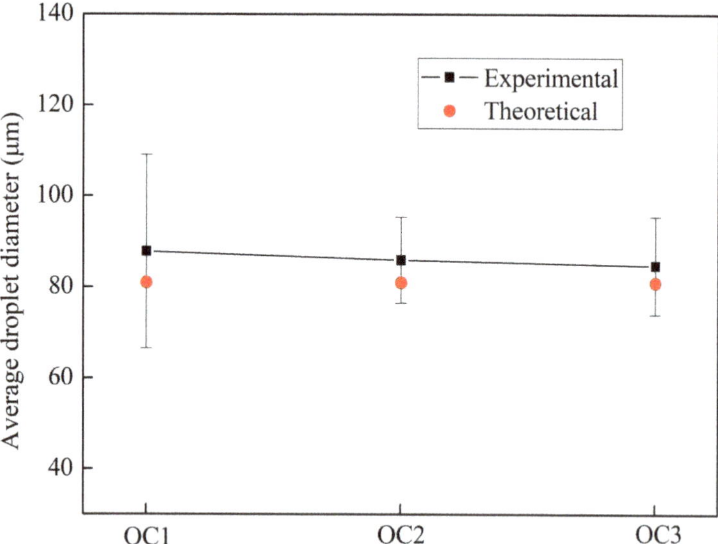

Figure 12. Average diameters of theoretically and experimentally generated monodisperse droplets.

4. Conclusions

Experiments on the generation of monodisperse deionized water droplets based on the piezoelectric method were conducted. Our observations demonstrated that satellite droplets were generated because the mother droplets detached from the liquid filament with surface waveform perturbations. As the frequency of the square-wave pulse was tuned to match the volume flow rate of the pumping liquid, the satellite droplets merged with the subsequent mother droplets, and we were thus able to generate monodisperse droplets with a uniform size distribution. If the pulse frequency is relatively low, the satellite droplets cannot merge with the subsequent mother droplets. It is also possible that the two adjacent mother droplets can coalesce if the pulse frequency is so high that the generated droplets are chaotic. This results in a wide size distribution. Under different volume flow rates, the optimal ranges of pulse frequency required to generate monodisperse droplets are different. The size of the monodisperse droplets decreases with the increase in the pulse frequency, as the volume flow rate is given. The sizes of the monodisperse droplets generated follow bimodal and Gaussian distributions under optimum operating conditions.

Author Contributions: Conceptualization, X.H. and S.L.; methodology, X.H. and Z.H.; experiments, Z.H., F.Y., and X.L.; writing, X.H. and S.L.; review and editing, S.P. and S.L.; supervision, J.X. All authors have read and agreed to the published version of the manuscript.

Funding: This work was financially supported by the National Natural Science Foundation of China (51976050 and 41805108), the Fundamental Research Funds for the Provincial Universities of Zhejiang (GK209907299001-004), and the Scientific Research Projects of Zhejiang Administration for Market Regulation (20200103).

References

1. Yamaguchi, K.; Sakai, K.; Yamanaka, T. Generation of Three-Dimensional Micro Structure Using Metal Jet. *Precis. Eng.* **2000**, *24*, 2–8. [CrossRef]
2. Behera, D.; Cullinan, M. Current Challenges and Potential Directions Towards Precision Microscale Additive Manufacturing—Part I: Direct Ink Writing/Jetting Processes. *Precis. Eng.* **2021**, *68*, 326–337. [CrossRef]
3. Zeng, H.; Yang, J.; Katagiri, D.; Ying, R.; Xue, S.; Nakajim, H.; Uchiyama, K. Investigation of Monodisperse Droplet Generation in Liquids by Inkjet. *Sensor. Actuat. B Chem.* **2015**, *220*, 958–961. [CrossRef]
4. Kim, C.S.; Park, S.J.; Sim, W. Modeling and Characterization of an Industrial Inkjet Head for Micro-Patterning on Printed Circuit Boards. *Comput. Fluids* **2009**, *38*, 602–612. [CrossRef]
5. Carter, J.C.; Alvis, R.M.; Brown, S.B. Fabricating optical Fiber Imaging Sensors Using Inkjet Printing Technology: A pH Sensor Proof-Of-Concept. *Biosens. Bioelectron.* **2006**, *21*, 1359–1364. [CrossRef] [PubMed]
6. Liu, Q.; Orme, M. High Precision Solder Droplet Printing Technology and the State-Of-The-Art. *J. Mater. Process. Technol.* **2001**, *115*, 271–283. [CrossRef]
7. Hossain, S.; Luckham, R.E.; Smith, A.M. Development of a Bioactive Paper Sensor for Detection of Neurotoxins Using Piezoelectric Inkjet Printing of Sol? Gel-Derived Bioinks. *Anal. Chem.* **2009**, *81*, 5474–5483. [CrossRef] [PubMed]
8. Sackmann, E.K.; Fulton, A.L.; Beebe, D.J. The Present and Future Role of Microfluidics in Biomedical Research. *Nat. Cell Biol.* **2014**, *507*, 181–189. [CrossRef]
9. Hochstetter, A. Lab-on-a-Chip Technologies for the Single Cell Level: Separation, Analysis and Diagnostics. *Micromachines* **2020**, *11*, 468. [CrossRef] [PubMed]
10. Delene, D.J.; Deshler, T. Calibration of a Photometric Cloud Condensation Nucleus Counter Designed for Deployment on a Balloon Package. *J. Atmos. Ocean. Technol.* **2000**, *17*, 459–467. [CrossRef]
11. Kalantarifard, A.; Alizadeh-Haghighi, E.; Saateh, A.; Elbuken, C. Theoretical and Experimental Limits of Monodisperse Droplet Generation. *Chem. Eng. Sci.* **2021**, *229*, 116093. [CrossRef]
12. Chen, J.; Wise, K.D. A High-Resolution Silicon Monolithic Nozzle Array for Inkjet Printing. *IEEE Trans. Electron Devices* **1997**, *44*, 1401–1409. [CrossRef]
13. Jadamson, S.; Manager, S.P.; Wang, A.D. A Change in Dispensing Technology-Jetting Takes Off. *Semicond. Tech.* **2004**, *3*, 67–70.
14. Cheng, S.; Chandra, S. A Pneumatic Droplet-On-Demand Generator. *Exp. Fluids* **2003**, *34*, 755–762. [CrossRef]
15. Cheng, S.X.; Li, T.; Chandra, S. Producing Molten Metal Droplets with a Pneumatic Droplet-On-Demand Generator. *J. Mater. Process. Technol* **2005**, *159*, 295–302. [CrossRef]
16. Sun, J.M.; Wei, X.F.; Huang, B.Q. Influence of the Viscosity of Edible Ink to Piezoelectric Ink-Jet Printing Drop State. *Appl. Mech. Mater.* **2012**, *200*, 676–680. [CrossRef]
17. Tseng, A.A.; Lee, M.H.; Zhao, B. Design and Operation of a Droplet Deposition System for Freeform Fabrication of Metal Parts. *J. Eng. Mater. Tech.* **2001**, *123*, 74–84. [CrossRef]
18. Teo, A.J.T.; Li, K.-H.H.; Nguyen, N.-T.; Guo, W.; Heere, N.; Xi, H.-D.; Tsao, C.-W.; Li, W.; Tan, S.H. Negative Pressure Induced Droplet Generation in a Microfluidic Flow-Focusing Device. *Anal. Chem.* **2017**, *89*, 4387–4391. [CrossRef]
19. Filatov, N.A.; Evstrapov, A.A.; Bukatin, A.S. Negative Pressure Provides Simple and Stable Droplet Generation in a Flow-Focusing Microfluidic Device. *Micromachines* **2021**, *12*, 662. [CrossRef]

20. Chung, C.H.Y.; Cui, B.; Song, R.; Liu, X.; Xu, X.; Yao, S. Scalable Production of Monodisperse Functional Microspheres by Multilayer Parallelization of High Aspect Ratio Microfluidic Channels. *Micromachines* **2019**, *10*, 592. [CrossRef]
21. Li, H.; Liu, F.; Li, Y. The Characteristics of 20 μm-Diameter Single Droplets. *Acta Phys. Sin.* **2007**, *56*, 5926–5930.
22. Fan, K.C.; Chen, J.Y.; Wang, C.H. Development of Drop-On-Demand Droplet Generator for One-Drop-Filling Technology. *Sens. Actuator A Phys.* **2008**, *147*, 649–655. [CrossRef]
23. Shin, P.; Sung, J. The Effect of Driving Waveforms on Droplet Formation in a Piezoelectric Inkjet Nozzle. In Proceedings of the 2009 11th Electronics Packaging Technology Conference, Singapore, 9–11 December 2009.
24. Lin, H.; Wu, H.; Shan, T. The Effects of Operating Parameters on Micro-Droplet Formation in a Piezoelectric Inkjet Printhead Using a Double Pulse Voltage Pattern. *Mater. Trans.* **2006**, *47*, 375–382. [CrossRef]
25. Li, S.; Zhuo, Z.; He, L.; Huang, X. Atomization Characteristics of Nano-Al/Ethanol Nanofluid Fuel in Electrostatic Field. *Fuel* **2019**, *236*, 811–819. [CrossRef]
26. Tjahjadi, M.; Stone, H.A.; Ottino, J.M. Satellite and Subsatellite Formation in Capillary Breakup. *J. Fluid Mech.* **1992**, *243*, 297–317. [CrossRef]
27. Bogy, D.B.; Talke, F.E. Experimental and Theoretical Study of Wave Propagation Phenomena in Drop-On-Demand Ink Jet Devices. *IBM J. Res. Dev.* **1984**, *28*, 322–337. [CrossRef]

Article

Neutrally Buoyant Particle Migration in Poiseuille Flow Driven by Pulsatile Velocity

Lizhong Huang , Jiayou Du and Zefei Zhu *

School of Mechanical Engineering, Hangzhou Dianzi University, Hangzhou 310018, China;
ricky@hdu.edu.cn (L.H.); abc@hdu.edu.cn (J.D.)
* Correspondence: zzf_3691@163.com; Tel.: +86-571-8691-9007

Abstract: A neutrally buoyant circular particle migration in two-dimensional (2D) Poiseuille channel flow driven by pulsatile velocity is numerical studied by using immersed boundary-lattice Boltzmann method (IB-LBM). The effects of Reynolds number ($25 \leq Re \leq 200$) and blockage ratio ($0.15 \leq k \leq 0.40$) on particle migration driven by pulsatile and non-pulsatile velocity are all numerically investigated for comparison. The results show that, different from non-pulsatile cases, the particle will migrate back to channel centerline with underdamped oscillation during the time period with zero-velocity in pulsatile cases. The maximum lateral travel distance of the particle in one cycle of periodic motion will increase with increasing Re, while k has little impact. The quasi frequency of such oscillation has almost no business with Re and k. Moreover, Re plays an essential role in the damping ratio. Pulsatile flow field is ubiquitous in aorta and other arteries. This article is conducive to understanding nanoparticle migration in those arteries.

Keywords: lattice Boltzmann method; inertial migration; Poiseuille flow; pulsatile velocity

Citation: Huang, L.; Du, J.; Zhu, Z. Neutrally Buoyant Particle Migration in Poiseuille Flow Driven by Pulsatile Velocity. *Micromachines* **2021**, *12*, 1075. https://doi.org/10.3390/mi12091075

Academic Editor: Myung-Suk Chun

Received: 4 August 2021
Accepted: 3 September 2021
Published: 6 September 2021

Publisher's Note: MDPI stays neutral with regard to jurisdictional claims in published maps and institutional affiliations.

1. Introduction

Particle two-phase flow is a very complex problem, which ubiquitously exists in nature, industry, hemodynamics, such as the formation and movement of sand dunes, haze (PM2.5), ventilation dusting system, spread of virus (COVID-19), inertial microfluidics, drug delivery in blood, etc. Numerous researches have revealed the behavior of the particles in fluid flow depends on Reynolds number ($Re = U_m H/\nu$, where U_m is the maximum inlet velocity, H is the channel width, ν is the fluid kinematic viscosity) and blockage ratio ($k = D_p/H$, donates the ratio of particle diameter D_p and the channel width), whether or not the particles are neutrally buoyant [1–6]. The density of the neutrally buoyant particles is the same as the suspension fluid, which means the particles will suspend in the fluid.

Segré and Silberberg [7] first discovered experimentally neutrally buoyant spherical particles would migrate to a radial equilibrium position in a pipe flow and form the Segré and Silberberg (SS) annulus, which is known as SS effect. This phenomenon prompted a lot of correlation research to reveal the underlying mechanism. Ho and Leal [8] theoretically studied particle migration in two-dimensional (2D) Poiseuille flow. Asmolov [9] interpreted that the particles migration was due to the effect of inertial lift by using the matched asymptotic expansions method. The results indicated the wall induced inertial lift became significant in the thin layers near the channel wall, and such lift could be neglected when the particles are far away from the wall. Matas et al. [10] found that the particles would move closer to the circular tube wall as Re increased and revealed additional inner annulus when Re was greater than 600. Moreover, if Re exceeded 700, the particles in the inner annulus accounted for the majority. Matas et al. [11] also utilized the matched asymptotic expansions method to calculate the lateral force in the pipe geometry (used to be in the plane geometry), but they did not find the second zero lateral force intersection point which indicates the inner annulus. Thus, they concluded the inner annulus was most likely due to finite-size effect. Hood et al. [12] calculated the lateral forces by a perturbation analysis. Morita et al. [13] predicted that all particles in the inner annulus would return to the SS annulus according to their experimental results when Re is less than 1000 and the tube is long enough. Due to their equipment limitations, the tube length was only 500 times the particle diameter. Then, Nakayama et al. [14] increased the length of the tube to 1000 times of the particle diameter, and the results showed that three regimes were eventually formed: only the SS annulus, only the inner annulus or those two annuli exist at the same time. The transition between these three regimes was determined by the critical Re which decreases with increasing of k.

Besides the aforementioned theoretical and experimental methods, computational fluid dynamics (CFD) simulation has become a powerful tool in analyzing particle-fluid interaction. Feng et al. [15] adopted finite-element method to investigate a circular particle migration in Poiseuille flow. Their simulations agreed qualitatively with the results of perturbation theories and pertinent experiments. By using LBM [16–19], the same problem was studied, and the SS effect was reconstructed. Shao et al. [20] found the inner annulus for elevated Re by using the fictitious domain method. Abbas et al. [21] mentioned that the equilibrium position (TEP) depends exclusively on Re and k. Recently, inertial microfluidics can precisely separate particles with or without extra external force field by realizing SS effect [22–28].

All the research works mentioned above are based on a non-pulsatile flow field, but in the artery, the blood pumped by the heart behaves as a pulsatile flow field. To the best of our knowledge, there are no articles related to particle migration in the pulsatile flow field.

Cancer is one of the leading causes of death in the world; nanoparticles are widely used for cancer therapy. Ideally, the therapeutic nanoparticles system should be able to deliver drugs just to the tumor and have no severe side effects on the body [29]. However, nanoparticle movement in non-pulsatile blood flow field is quite different from that in the pulsatile blood flow field.

In this study, we perform a series of CFD simulations to investigate the migration of one neutrally buoyant circular particle in 2D Poiseuille flow driven by pulsatile velocity. The influence of Re and k on the particle migration is analyzed in detail. The difference between particle migration driven in pulsatile and non-pulsatile velocity is illustrated. Engineering precision therapeutic nanoparticles system in artery [30] can be attainable by understanding the particle migration in pulsatile blood flow field. Furthermore, it can help to optimize therapeutic nanoparticles system for achieving precise medical against cancer near arteries.

The organization of this article is as follows. The numerical method, boundary conditions, and problem description are introduced in Section 2. In Section 3, we simulated TEPs of one particle at several specific parameters, and our computational code is validated by comparing with the published paper. Afterwards, the simulation results are presented and discussed in Section 4. Finally, conclusions are provided in Section 5.

2. Method and Problem

LBM is widely used to simulate particle migration, turbulence, flow in porous media, multiphase flow, non-Newtonian rheology, and so on [31–43]. Because of its advantages in efficiency, easy to code, and parallel run, LBM has become a very popular CFD numerical tool.

2.1. Lattice Boltzmann Method

In this work, the single-relaxation time (SRT) lattice Bhatnagar–Gross–Krook (LBGK) Boltzmann method is adopted to solve particles migration in incompressible viscous flow [44]:

$$f_i(\boldsymbol{x} + \mathbf{e}_i \Delta t, t + \Delta t) = f_i(\boldsymbol{x}, t) - \frac{1}{\tau}\left[f_i(\boldsymbol{x}, t) - f_i^{(eq)}(\boldsymbol{x}, t)\right], \tag{1}$$

where $f_i(x, t)$ is the distribution function at space coordinate $x = (X, Y)$ and time t in the ith direction; $f_i^{(eq)}(x, t)$ is the correspond equilibrium distribution function; τ is the SRT; Δt is the time step; the discrete velocities e_i of 2D nine-velocity (D2Q9) model are shown in Figure 1; $c = \Delta x / \Delta t$ is the lattice velocity; Δx is the lattice spacing. For coding conveniently, both the time step and the lattice spacing are set to be equal to 1, which result in $\Delta t = \Delta x = c = 1$.

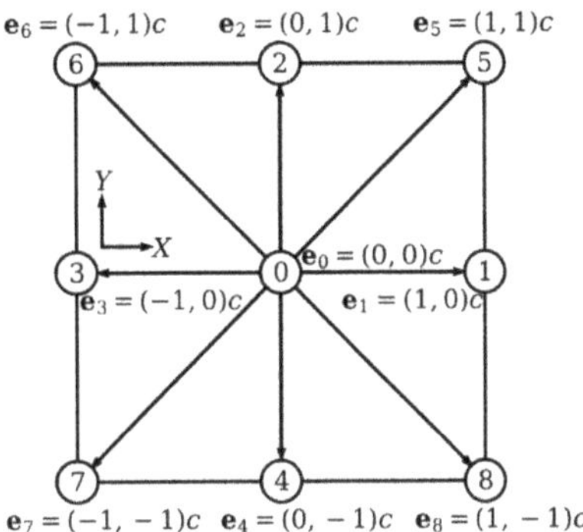

Figure 1. D2Q9 Cartesian lattice and discrete velocities.

The equilibrium distribution function can be calculated by [44]:

$$f_i^{(eq)}(x, t) = w_i \rho_f \left[1 + \frac{3 \mathbf{e}_i \cdot u}{c^2} + \frac{4.5 (\mathbf{e}_i \cdot u)^2}{c^4} - \frac{1.5 u^2}{c^2} \right], \tag{2}$$

where w_i is the weight factor with $w_0 = 4/9$, $w_{1\sim4} = 1/9$ and $w_{5\sim8} = 1/36$, ρ_f is the fluid density and u is fluid velocity which can be determined by:

$$\rho_f = \sum f_i(x, t), \quad u = \frac{1}{\rho_f} \sum \mathbf{e}_i f_i(x, t). \tag{3}$$

For low Mach number, the Navier–Stokes equations can be derived from the lattice Boltzmann equation by utilizing Chapman–Enskog expansion [45].

2.2. Improved Bounce-Back Scheme

In the LBM simulations, improved bounce-back scheme is of particular importance, and it allows to implement no-slip boundary condition on the surface of moving particle [46], which will be explained briefly below.

As shown in Figure 2, the sky-blue circles represent the fluid nodes, and the red thin diamonds donate the solid nodes. The orange triangles indicate uncovered fluid nodes means that those nodes are located inside the particle at time t and will locate outside the particle when the particle travels after one lattice time step. Similarly, the purple squares donate covered fluid nodes represent that those nodes will change from fluid nodes to solid nodes after the particle moves.

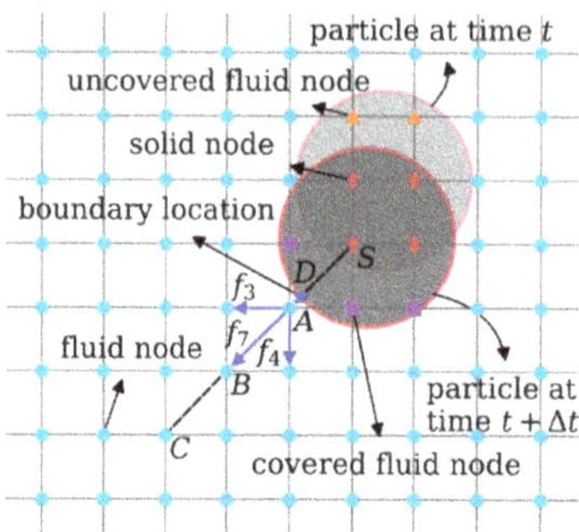

Figure 2. Improved bounce-back scheme boundary conditions. Lighter gray circle represents the location of particle at time t and darker gray circle at time t + Δt, which result in two orange triangles uncover to fluid node, three purple squares cover to solid node and four red thin diamonds remain solid node. The other sky-blue circles denote the fluid nodes. The distribution function f_7 can be determined by the information of node S, D, A, B, C.

Take fluid node A as an example, after the streaming step, three unknown distribution functions (f_3, f_7 and f_4 denoted by three blue arrows shown in Figure 2) need to be determined by applying improved bounce-back scheme. Therefore, the distribution function f_7 can be calculated by:

$$f_7(A) = \begin{cases} q(1+2q)f_5(S) + (1-4q^2)f_5(A) - q(1-2q)f_5(B) - 2\omega_5\rho_f \frac{\mathbf{e_5}\cdot\mathbf{u}_D}{c_s^2}, & q < \frac{1}{2}, \\ \frac{1}{q(2q+1)}f_5(S) + \frac{2q-1}{q}f_7(B) - \frac{2q-1}{2q+1}f_7(C) - \frac{2\omega_5\rho_f}{q(2q+1)}\frac{\mathbf{e_5}\cdot\mathbf{u}_D}{c_s^2}, & q \geqslant \frac{1}{2}, \end{cases} \tag{4}$$

where S is the nearest solid node along $\mathbf{e}_5$ direction; B and C are the two nearest fluid nodes along $\mathbf{e}_7$ direction; the blue star D denotes the boundary location which is the intersection point of particle boundary and line AS, meanwhile, q can be determined by $q = |AD|/|AS|$; u_D is the velocity of the boundary location D; $c_s = c/\sqrt{3}$ is the speed of sound. The other unknown distribution functions of fluid nodes around particle boundary can be solved in the similar method.

2.3. Force, Torque, and Particle Motion

Let us continue to take the boundary location D as an example. As shown in Figure 2, the hydrodynamic force and torque acting on the solid particle migrating in fluid can be integrated by adopting momentum exchange algorithm [46,47] as follows:

$$\begin{aligned} F^{(h)}(x + q\mathbf{e}_5, t) &= \mathbf{e}_5[f_5(x + \mathbf{e}_5, t) + f_7(x, t)], \\ T^{(h)}(x + q\mathbf{e}_5, t) &= (x + q\mathbf{e}_5 - x_p) \times F^{(h)}(x, t), \end{aligned} \tag{5}$$

where x_p is the position of the particle.

When the particle is moving in the lattice grid from time t to $t + \Delta t$, as shown in Figure 2, the additional force and torque due to uncovered fluid node and covered fluid node exerted on the particle can be computed by [48]:

$$\begin{aligned} F^{(c)}(x, t) &= \rho_f(x, t)u(x, t), \\ T^{(c)}(x, t) &= (x - x_p) \times F^{(c)}(x, t), \\ F^{(u)}(x, t) &= -\rho_f(x, t)u(x, t), \\ T^{(u)}(x, t) &= (x - x_p) \times F^{(u)}(x, t). \end{aligned} \tag{6}$$

Moreover, in order to avoid unphysical overlapping between the particle and the channel wall, the extra lubrication force model is needed [49]:

$$F^{(l)} = \begin{cases} 0, & h \geq h_c, \\ -1.5\pi\rho_f\nu\left[D_p\left(\frac{1}{h} - \frac{1}{h_c}\right)\right]^{1.5}U_p, & h < h_c, \end{cases} \tag{7}$$

where $\nu = c_s^2(\tau - 0.5)\Delta t$ is the kinematic viscosity of the fluid; D_p is the diameter of the particle; U_p is the particle velocity towards the wall; h is the minimum gap between the particle and the wall; $h_c = 1.5\Delta x$ is the cutoff distance whether to consider the lubrication force or not.

By summation of the forces and torques acting on the particle, the movement of the particle can be solved explicitly using Newton's second law:

$$\begin{aligned}
a_p^{t+\Delta t} &= \left(\sum F^{(h)} + \sum F^{(c)} + \sum F^{(u)} + F^{(l)}\right)/m_p, \\
u_p^{t+\Delta t} &= u_p^t + 0.5\left(a_p^{t+\Delta t} + a_p^t\right)\Delta t, \\
x_p^{t+\Delta t} &= x_p^t + 0.5\left(u_p^{t+\Delta t} + u_p^t\right)\Delta t, \\
\alpha_p^{t+\Delta t} &= \left(\sum T^{(h)} + \sum T^{(c)} + \sum T^{(u)}\right)/I_p, \\
w_p^{t+\Delta t} &= w_p^t + 0.5\left(\alpha_p^{t+\Delta t} + \alpha_p^t\right)\Delta t, \\
\theta_p^{t+\Delta t} &= \theta_p^t + 0.5\left(w_p^{t+\Delta t} + w_p^t\right)\Delta t,
\end{aligned} \tag{8}$$

where a_p, u_p, α_p, w_p, θ_p, m_p, and I_p are the translational acceleration, velocity, rotational acceleration, rotational velocity, angle, mass and moment inertia of the particle.

2.4. Problem

The configuration of one circular particle migrating in Poiseuille flow is shown in Figure 3. A parabolic velocity profile with the maximum velocity U_m is set at the left inlet boundary in the positive X direction, and the velocity in Y direction is zero. For U_m, there are two cases: non-pulsatile or pulsatile. If pulsatile, U_m will change over time, otherwise, it will be a constant. For considering reproducibility of this work, patient-specific velocity profile [50] will not be adopted to impose at the inlet boundary. The half-period of the sinusoidal function at time interval $[t_0, t_1]$ (systolic period) and zero at $[t_1, t_2]$ (diastolic period) is utilized which can be seen from Figure 3. In the systolic period, $t_1 - t_0 = 0.3$ s, and in the diastolic period, $t_2 - t_1 = 0.5$ s, which means one cardiac cycle lasts 0.8 s. The unit conversion factor from lattice time to physical time is 10^{-5}. At the right boundary, the normal derivative of the velocity is zero and the pressure is set to be $p_{out} = \rho_f c_s^2$ [19]. No-slip boundary conditions are imposed at the top and bottom channel walls.

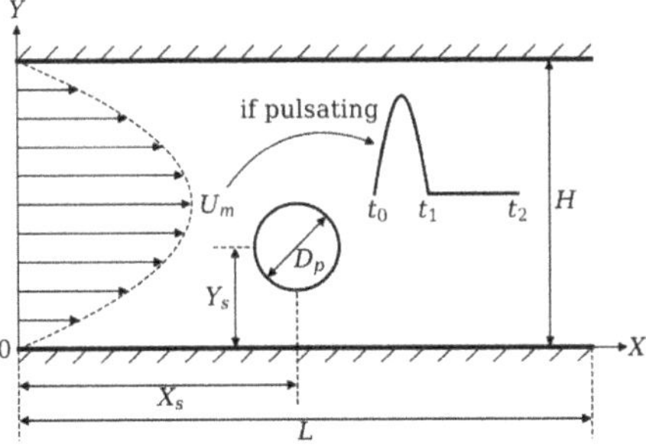

Figure 3. Configuration of a particle migrating in Poiseuille channel flow. The origin coordinate locates at left down corner, X in horizontal direction and Y in vertical direction. The simulation domain size is fixed as $L \times H$, while the particle is located at $[X_s, Y_s]$ initially. The diameter of the particle is D_p, U_m represents the maximum velocity. For pulsatile case, U_m will change by time.

If there are no additional statements, the channel length L is $500\Delta x$ and the height H is $100\Delta x$. TEP is basically unchanged for different channel length ($300\Delta x$, $400\Delta x$, $500\Delta x$, $600\Delta x$, $700\Delta x$ and $2000\Delta x$). So $L = 500\Delta x$ is adopted same as Wen et al. [17]. As shown in Figure 3, the circular particle center is located at $[X_s, Y_s]$ initially. The diameter of the circular particle in this work is $15\Delta x$, $20\Delta x$, $25\Delta x$, $30\Delta x$, $35\Delta x$, and $40\Delta x$, which means $k = D_p/H = 0.15$, 0.20, 0.25, 0.30, 0.35, 0.40, respectively. Meanwhile, $Re = U_m H/v = 25$, 50, 75, 100, 150, 200 is studied by changing the kinematic viscosity of the fluid.

Driven by the fluid velocity, the particle will always travel along X axis in positive direction, so an infinite channel is needed to avoid the particle moving out of the simulation domain which can be achieved by moving domain technique [6,51,52]. When the X-coordinate of the particle exceeds $X_s + \Delta x$, the fluid field and the particle need to shift on lattice spacing left, which ensures that the particle will never travel too far from its original position. Meanwhile, it should be noted that the actual moving distance along X axis in positive direction should add up one lattice spacing when performing this shift once.

3. Validation

To validate the accuracy of our LBM code, two benchmark cases are implemented below.

In the first case, Re is set to be 50, and the terminal particle Reynolds number $Re_p = U_{x,p} D_p/v$ is about 9.63, where $U_{x,p}$ is the terminal particle velocity in X direction. It is very close to Re_p while the particle is driven by pressure difference [17]. Our simulating results with $k = 0.25$ and $k = 0.35$ are plotted in Figure 4a,b, respectively. The SS effect is quite obviously found and TEP of the particle has nothing to do with the initial horizontal position ($Y_s/H = 0.20$, 0.25, 0.35, 0.40, 0.45) which only changes the trajectory from initial position to TEP. All the results, even the curve shape shown in Figure 4a,b are consistent with those of Wen et al. [17].

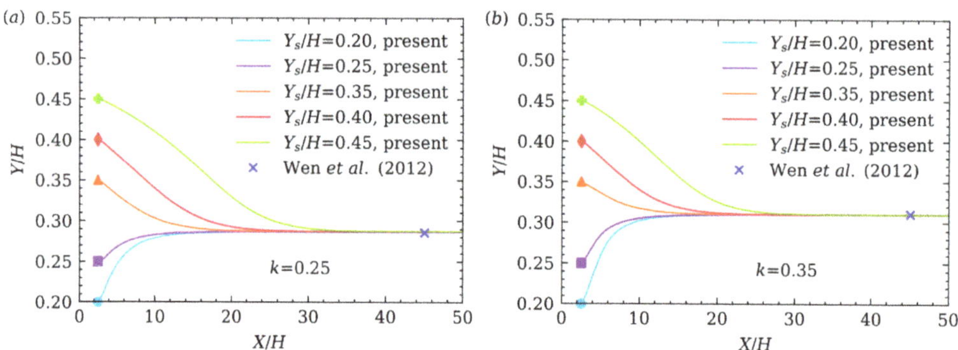

Figure 4. Lateral migration of the particle released from different initial positions in Poiseuille flow with (**a**) $k = 0.25$, (**b**) $k = 0.35$.

The foregoing results are validated only when Re is 50. The second case is adopted to validate over the entire range of $Re = 20$, 40, 100, 200. The particle diameter is $D_p = 22$, the channel width is $H = 200$, and the channel length is $L = 1000$, which leads to the blockage ratio being $k = 0.11$. Figure 5 shows the comparison of the present results with previous ones simulated by Di Chen et al. [53] The comparison shows TEPs are in good agreement and the particle will be closer to channel centerline with increasing Re in 2D Poiseuille flow.

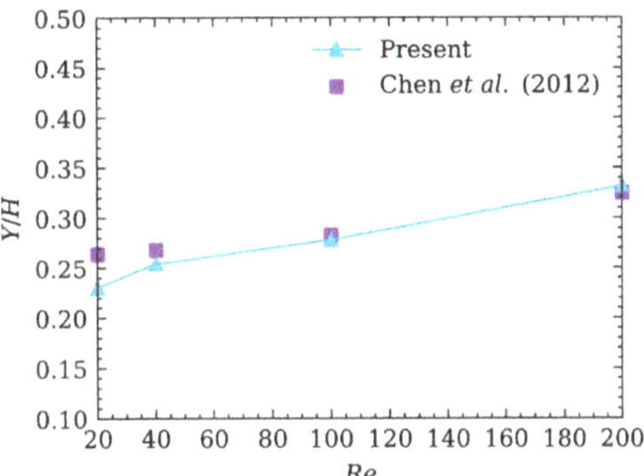

Figure 5. Comparison of TEPs of the particle migrating in Poiseuille flow at different *Re*.

4. Results and Discussions

4.1. Fluid and Particle Interaction

After validation, we first simulate the particle migrating in non-pulsatile flow at $k = 0.25$ and $Re = 50$. Figure 6a–c, shows the contour view of the dimensionless pressure $p' = (p - p_{out})/p_{out}$, the dimensionless fluid velocity in X direction $U'_x = U_x/U_m$ and in Y direction $U'_y = U_y/U_m$, respectively. Due to the moving domain method adopted here, the particle will always locate around the middle of the simulation domain which is significantly disturbed by the present of the particle. Obviously, the pressure strip can be observed upstream and downstream of the particle. The pressure at the upper left and down right side of the particle is higher than the upper right and down left corner, and hence generates a particle rotation in clockwise direction illustrated by the black arrow in Figure 6a. Due to non-slip boundary condition at the boundary location of particle, clockwise rotation of the particle will induce fluid flowing upward at left and downward at right as shown in Figure 6c. Moreover, it can be seen from Figure 6b that the particle follows with fluid movement quite well [54].

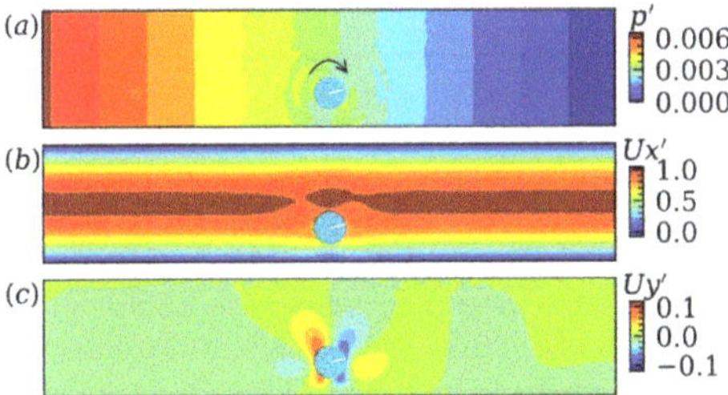

Figure 6. The contour view of (**a**) the dimensionless pressure p', (**b**) the dimensionless fluid velocity in X direction U'_x and (**c**) in Y direction U'_y in stable state of particle migrating in non-pulsatile flow at $k = 0.25$ and $Re = 50$.

4.2. Trajectory

Several dominant forces acting on the particle drive it to TEP in Y direction. The first force is shear gradient lift force due to the parabolic velocity profile, which points to the wall. The second force is wall induced lift force due to the interaction between the particle and the channel wall calculated by using added lubrication force model introduced before which directs towards the channel centerline. The third force is called Magnus force [55] due to the rotation of the particle migrating in fluid. As demonstrated above, the particle in non-pulsatile flow travels in X direction and at the same time it rolls in clockwise direction. Thus, the Magnus force is directed to the channel centerline. Certainly, when the particle moves in fluid, it will be affected by the drag force. In summary, there are at least four forces governing particle migration in fluid suspension.

Figure 7 shows the particle center trajectory along the channel for $k = 0.25$ and $k = 0.35$ at $Re = 50$. Obviously, several cardiac cycles later, the particle driven by pulsatile velocity migrates around TEP in non-pulsatile flow periodically. In the systolic period, the particle laterally migrates towards wall mainly affected by shear gradient lift force. While in the diastolic period, the shear gradient lift force disappears, and the particle still rotates because of inertia, so it will migrate back to the channel centerline under the Magnus force. The moving direction is illustrated by black arrows which are shown in the subplots of Figure 7. The particle oscillates in spiral-shaped structures and will travel towards the wall again in another systolic period.

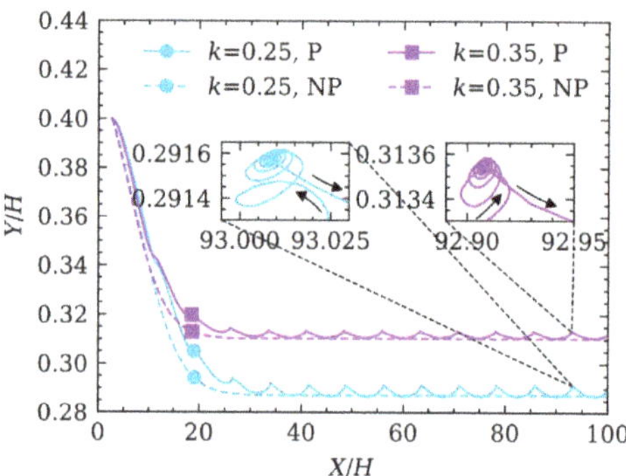

Figure 7. The particle center trajectory for $k = 0.25$ (two sky-blue curves with circle marker), $k = 0.35$ (two purple curves with square marker) at $Re = 50$. Solid curve represents the case driven by pulsatile velocity, which is abbreviated as P, while NP is the acronym for non-pulsatile flow donated by dash curve.

Figure 8a,c gives the particle center trajectory versus time $\left(t' = t/10^5\right)$ variation with Re at $k = 0.25$, and (b, d) donates different k at $Re = 50$. Moreover, (c, d) is the enlarged view of (a, b) in one stable cardiac cycle $(t' = [8.8, 9.6])$ illustrated by black dash line box, respectively. Overall, TEPs are all closer to the channel centerline when Re or k increases, whether or not in pulsatile flow. Consequently, the particle will take longer (or more cardiac cycles) to reach TEP. As shown in Figure 8c, we define Δ as the dimensionless distance between the highest and lowest position in this stable cardiac cycle, δ is the signed dimensionless distance between TEP in non-pulsatile flow and the lowest position (negative value means the particle did not exceed TEP in non-pulsatile flow), and t^* is time in this cardiac cycle when the particle locates the lowest position. In general, the influence of Re on δ, Δ, t^* is greater than k. As Re increases, the viscosity of the fluid decreases, thus the drag force acts on the particle decreases. The particle, therefore, can laterally migrate farther in systolic period, even exceeds TEP in non-pulsatile flow. Meanwhile, the particle will take longer to migrate from the highest position to the lowest position for its longer travel distance. As a result, Figure 8e shows that δ, Δ and t^* all increase monotonously with increasing Re. The effect of k on δ, Δ, t^* is a little more complicated. As k increases, the particle inertia increases, and the drag force also increases. Shear gradient force increases with increasing k, but decreases when the particle is closer to channel center. Consequently, it can be seen from Figure 8f, δ, Δ increases at first and then decreases when k increases, the maximum of δ, Δ occurs at $k = 0.2$. Moreover, t^* was mostly unchanged with increasing k.

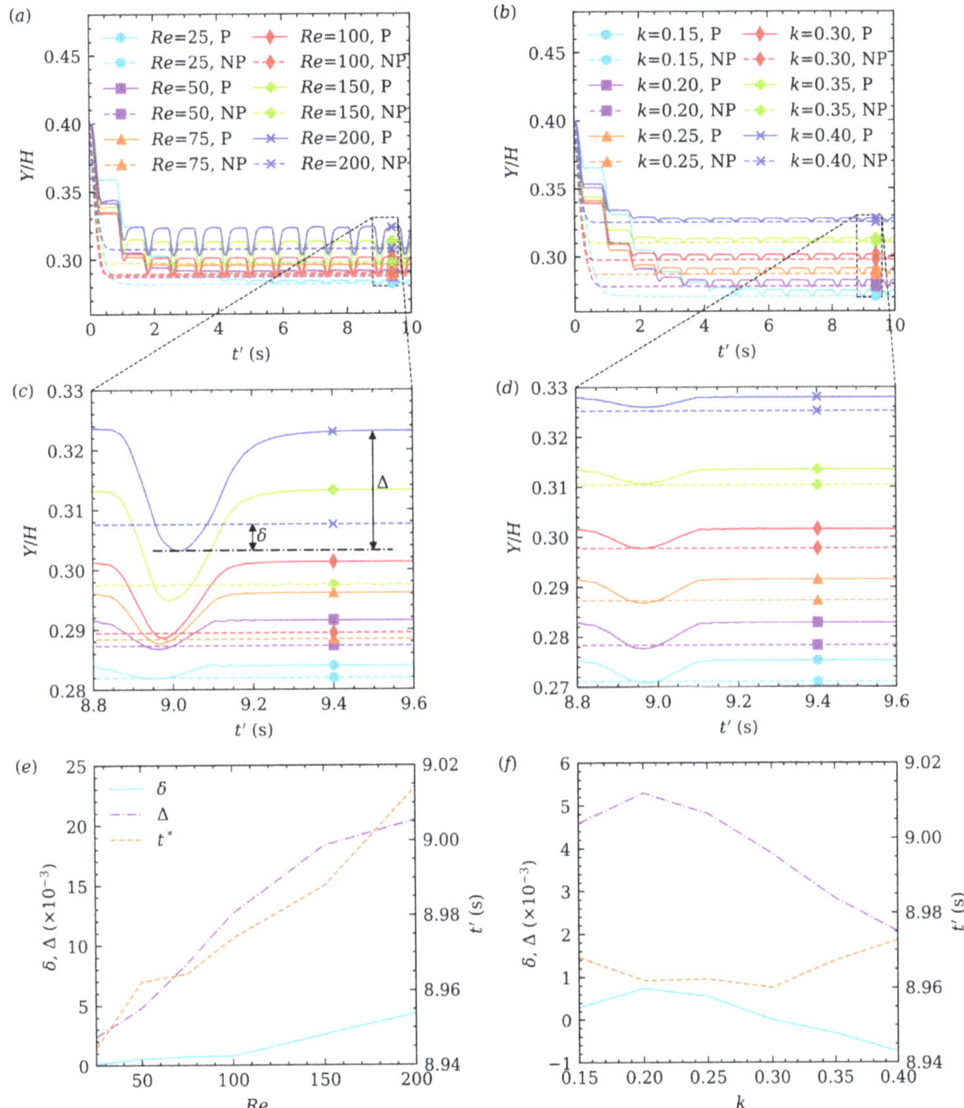

Figure 8. Time history of the particle center trajectory at different *Re* (**a,c**) at $k = 0.25$, and different k (**b,d**) at $Re = 50$. The variation of δ, Δ, t^* with *Re* (**e**) and k (**f**).

4.3. Orientation

Observed in Figure 9, the particle always rotates clockwise while migrating in the channel, irrespective of *Re*, *k*, and whether or not in pulsatile flow. In the diastolic period, the particle rotates much slower but still in the clockwise direction. If the particle is closer to the channel wall, it will experience larger gradient of fluid velocity. As a result, the smaller Re or k is, the faster the particle rotates. Figure 8 shows that, when $Re < 100$, TEP of the particle is very similar, so they rotate almost at the same speed (w) which is shown in the subplot of Figure 9a. In addition, it can be seen in the subplot of Figure 9b, the particle rotate speed is almost linearly coherent with k. Furthermore, the moment inertia of the particle is proportional to the square of the particle diameter, i.e., k. Consequently, by comparing Figure 9a,b, the influence of k on w is much larger than Re.

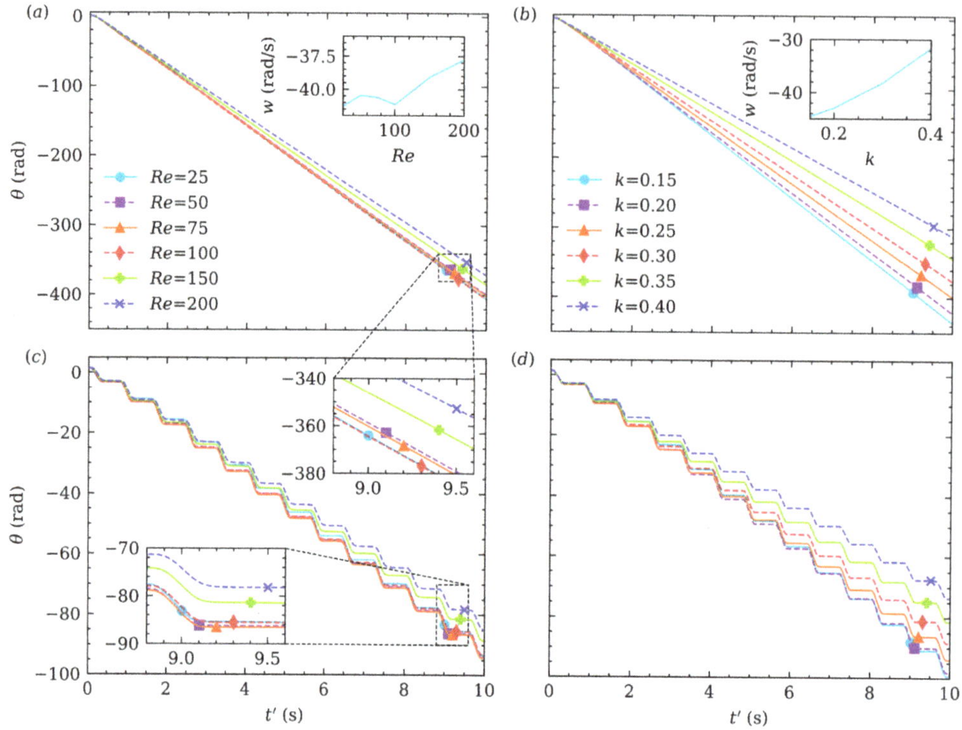

Figure 9. The particle orientation at different Re (**a,c**) and k (**b,d**). Meanwhile, (**a,b**) is in non-pulsatile and (**c,d**) in pulsatile flow.

4.4. Damping

As mentioned before, the particle oscillates in the diastolic period which is shown in Figure 7. We choose the particle velocity which indicates the interaction between the particle and fluid [56] for analysis. Moreover, the particle velocity in X direction (U_x) is preferred, because the particle velocity curve is not center symmetry in Y direction.

For better illustration, the particle velocity in X direction is normalized by: $U_x' = U_x / U_m$. The curves of U_x' in Figure 10 are much like a spring-mass system which is underdamped. Figure 10a shows, when Re is small, U_x' damps out rapidly after several quasi periods. If Re is small enough, like $Re \approx 1$, over damped is expected. Obviously, in Figure 10a–f, the damping effect becomes weaker with increasing Re. We fit the upside and downside envelope curve by using exponential function for purpose. For example, $A_{up} = e^{-15.99t' + 142.39}$ in Figure 10a, and the decay rate $\lambda = 15.99$. The constants of fitting exponential function of the upside and downside are almost the same for each Re. And the absolute values of those constants decrease with increasing Re. The influence of k is also analyzed, but it can be seen from Figure 11a–f that k has almost no effect on damping.

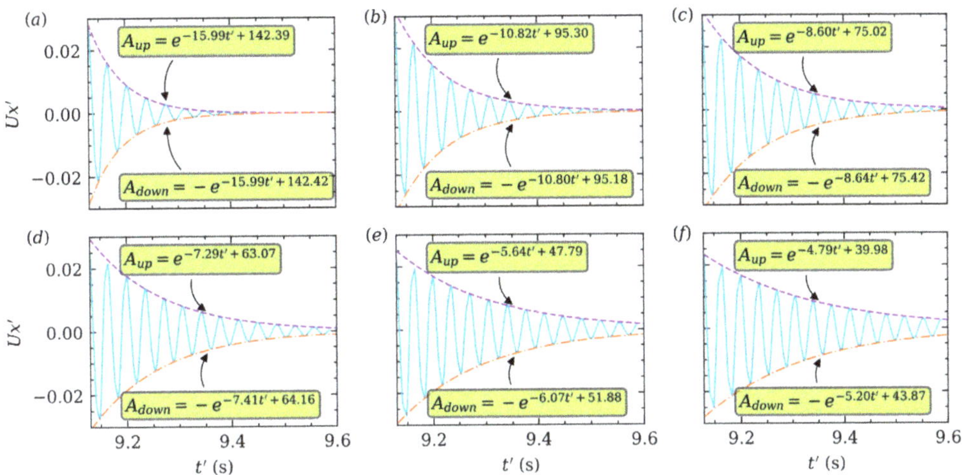

Figure 10. Damping of U'_x (solid line in sky-blue color) of the particle at $k = 0.25$ and different $Re =$ (**a**) 25, (**b**) 50, (**c**) 75, (**d**) 100, (**e**) 150, (**f**) 200. The upside (dashed line in purple color) and downside (dash dotted line in orange color) envelope curves are fitted by exponential function.

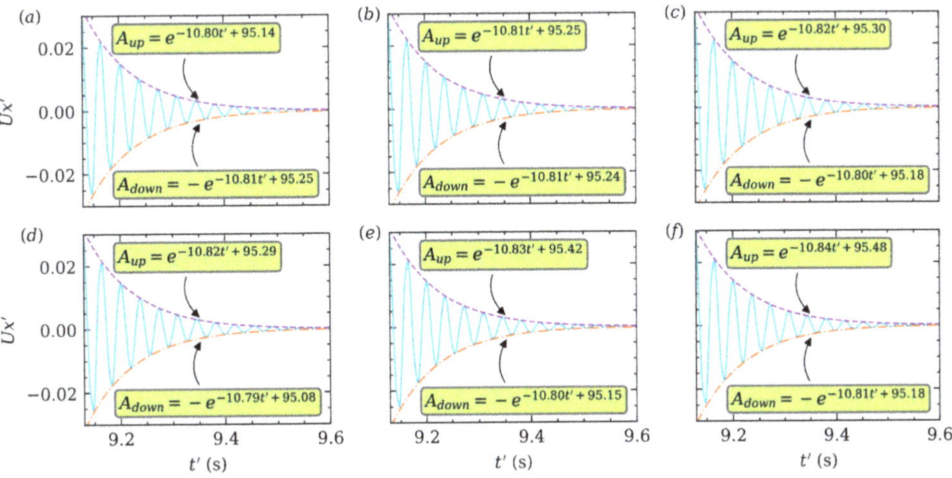

Figure 11. Damping of U'_x at $Re = 50$ and different $=$ (**a**) 0.15, (**b**) 0.20, (**c**) 0.25, (**d**) 0.30, (**e**) 0.35, (**f**) 0.40. Envelope curves are fitted in the same way.

Figure 12a gives the quasi frequency (f_q) of U'_x oscillation at different Re and k. It can be seen from the subplot that the quasi frequency increases as increasing k, but the impact of k on the quasi frequency is relatively small. However, the quasi frequency will not change when Re increases except for $Re = 25$. As mentioned earlier, in Figure 10a, when $Re = 25$, U'_x damps out rapidly after several quasi periods. The first few quasi periods are longer than the last ones, which will result in smaller quasi frequency when $Re = 25$.

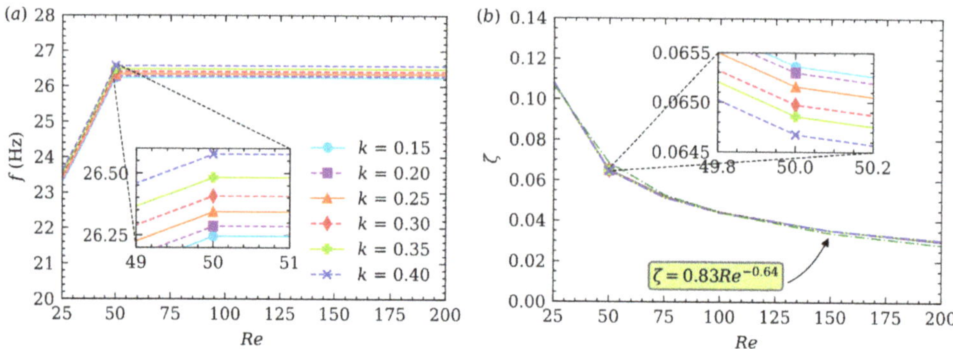

Figure 12. The quasi frequency (**a**) and damping ratio (**b**) at different *Re* and *k*.

As shown in Figure 12b, the damping ratio $\zeta \approx \lambda/(2\pi f_q)$, in our cases, is always smaller than 1, which determines this is an underdamped system, and this system will die out slower when ζ decreases. Additionally, ζ of U'_x oscillation basically does not change with variation of k, but it decreases with increasing Re, i.e., Re is the decisive parameter on ζ [57]. A similar conclusion is made by C.A. Coulomb in 1784 by using several material plates hanging by a metal wire with an initial torsion angle and then released to start timing until the plate oscillation until dying out. ζ is only related to the viscosity of fluid and not the material of plate. The reason for damping is the friction inside the fluid which is called Newton's law of friction, not the interaction friction between the plate and the fluid. Finally, in Figure 12b, the relation between ζ and Re is fitted as: $\zeta = 0.83Re^{-0.64}$.

5. Conclusions

IB-LBM was used to simulate one neutrally buoyant circular particle migration in 2D Poiseuille channel flow driven by pulsatile and non-pulsatile velocity. The moving domain technique is adopted to achieve that the particle can migrate in infinite channel. The results show that the particle moving in pulsatile flow is slightly different from that in non-pulsatile one. It will laterally migrate back to the channel centerline with small oscillations in the spiral-shaped structure during the diastole and move back toward TEP during the systole. The effect of Re on ζ is decisive. This research may shed some light on understanding the particle behavior in Poiseuille flow driven by pulsatile velocity for optimizing therapeutic nanoparticles system in arteries.

The limitation of this work is that Re varies only from 25 to 200. Smaller Re leads to bigger SRT, thus the simulation will not be accurate. In contrast, bigger Re results in smaller SRT which the simulation will diverge. Furthermore, k range only from 0.15 to 0.40. Because the channel height is set to be 100 lattice units, if k is smaller than 0.15, the simulation resolution will not be adequate. On the other hand, if k is bigger than 0.4, half channel width will be blocked by the particle. Parameters need to choose carefully to get a more varied range of Re and k. This will be a future direction of this work. Another future direction will be considering more particles or simulating particle migration in three-dimensional pipe pulsatile flow.

Author Contributions: Conceptualization, L.H. and Z.Z.; methodology, L.H.; software, L.H.; validation, L.H. and J.D.; writing—original draft preparation, L.H.; writing—review and editing, J.D. and Z.Z.; funding acquisition, J.D. and Z.Z. All authors have read and agreed to the published version of the manuscript.

Funding: This research was funded by the Fundamental Research Funds for the Provincial Universities of Zhejiang (GK199900299012-026) and National Natural Science Foundation of China (Grant Nr. 11572107).

Data Availability Statement: The data that support the findings of this study are available from the first author upon reasonable request.

Conflicts of Interest: The authors report that they have no conflict of interest.

References

1. Hu, X.; Lin, J.Z.; Chen, D.M.; Ku, X.K. Influence of non-Newtonian power law rheology on inertial migration of particles in channel flow. *Biomicrofluidics* **2020**, *14*, 14105. [CrossRef]
2. Hu, X.; Lin, J.Z.; Chen, D.M.; Ku, X.K. Stability condition of self-organizing staggered particle trains in channel flow. *Microfluid Nanofluidics* **2020**, *24*, 25. [CrossRef]
3. Jiang, R.J.; Lin, J.Z.; Tu, C.X. Poiseuille flow-induced vibrations of a cylinder in subcritical conditions. *J. Fluids Struct.* **2018**, *82*, 272–286. [CrossRef]
4. Li, G.J.; McKinley, G.H.; Ardekani, A.M. Dynamics of particle migration in channel flow of viscoelastic fluids. *J. Fluid Mech.* **2015**, *785*, 486–505. [CrossRef]
5. Yu, Z.S.; Wang, P.; Lin, J.Z.; Hu, H.H. Equilibrium positions of the elasto-inertial particle migration in rectangular channel flow of Oldroyd-B viscoelastic fluids. *J. Fluid Mech.* **2019**, *868*, 316–340. [CrossRef]
6. Zhang, J.; Yuan, D.; Zhao, Q.B.; Teo, A.J.T.; Yan, S.; Ooi, C.H.; Li, W.H.; Nguyen, N.T. Fundamentals of Differential Particle Inertial Focusing in Symmetric Sinusoidal Microchannels. *Anal. Chem.* **2019**, *91*, 4077–4084. [CrossRef] [PubMed]
7. Segré, G.; Silberberg, A. Behaviour of macroscopic rigid spheres in Poiseuille flow Part 2. Experimental results and interpretation. *J. Fluid Mech.* **1962**, *14*, 136–157. [CrossRef]
8. Ho, B.P.; Leal, L.G. Migration of rigid spheres in a two-dimensional unidirectional shear flow of a second-order fluid. *J. Fluid Mech.* **1976**, *76*, 783–799. [CrossRef]
9. Asmolov, E.S. The inertial lift on a spherical particle in a plane Poiseuille flow at large channel Reynolds number. *J. Fluid Mech.* **1999**, *381*, 63–87. [CrossRef]
10. Matas, J.P.; Morris, J.F.; Guazzelli, É. Inertial migration of rigid spherical particles in Poiseuille flow. *J. Fluid Mech.* **2004**, *515*, 171–195. [CrossRef]
11. Matas, J.P.; Morris, J.F.; Guazzelli, É. Lateral force on a rigid sphere in large-inertia laminar pipe flow. *J. Fluid Mech.* **2009**, *621*, 59–67. [CrossRef]
12. Hood, K.; Lee, S.; Roper, M. Inertial migration of a rigid sphere in three-dimensional Poiseuille flow. *J. Fluid Mech.* **2015**, *765*, 452–479. [CrossRef]
13. Morita, Y.; Itano, T.; Sugihara Seki, M. Equilibrium radial positions of neutrally buoyant spherical particles over the circular cross-section in Poiseuille flow. *J. Fluid Mech.* **2017**, *813*, 750–767. [CrossRef]
14. Nakayama, S.; Yamashita, H.; Yabu, T.; Itano, T.; Sugihara Seki, M. Three regimes of inertial focusing for spherical particles suspended in circular tube flows. *J. Fluid Mech.* **2019**, *871*, 952–969. [CrossRef]
15. Feng, J.; Hu, H.H.; Joseph, D.D. Direct simulation of initial value problems for the motion of solid bodies in a Newtonian fluid. Part 2. Couette and Poiseuille flows. *J. Fluid Mech.* **1994**, *277*, 271–301. [CrossRef]
16. Li, H.B.; Lu, X.Y.; Fang, H.P.; Qian, Y.H. Force evaluations in lattice Boltzmann simulations with moving boundaries in two dimensions. *Phys. Rev. E* **2004**, *70*, 26701. [CrossRef]
17. Wen, B.H.; Li, H.B.; Zhang, C.Y.; Fang, H.P. Lattice-type-dependent momentum-exchange method for moving boundaries. *Phys. Rev. E* **2012**, *85*, 16704. [CrossRef]
18. Tao, S.; Hu, J.J.; Guo, Z.L. An investigation on momentum exchange methods and refilling algorithms for lattice Boltzmann simulation of particulate flows. *Comput. Fluids* **2016**, *133*, 1–14. [CrossRef]
19. Inamuro, T.; Maeba, K.; Ogino, F. Flow between parallel walls containing the lines of neutrally buoyant circular cylinders. *Int. J. Multiph. Flow* **2000**, *26*, 1981–2004. [CrossRef]
20. Shao, X.M.; Yu, Z.S.; Sun, B. Inertial migration of spherical particles in circular Poiseuille flow at moderately high Reynolds numbers. *Phys. Fluids* **2008**, *20*, 103307. [CrossRef]
21. Abbas, M.; Magaud, P.; Gao, Y.; Geoffroy, S. Migration of finite sized particles in a laminar square channel flow from low to high Reynolds numbers. *Phys. Fluids* **2014**, *26*, 123301. [CrossRef]
22. Wang, R.J.; Sun, S.S.; Wang, W.; Zhu, Z.F. Investigation on the thermophoretic sorting for submicroparticles in a sorter with expansion-contraction microchannel. *Int. J. Heat Mass Transf.* **2019**, *133*, 912–919. [CrossRef]
23. Wang, R.J.; Du, J.Y.; Guo, W.C.; Zhu, Z.F. Investigation on the Thermophoresis-Coupled Inertial Sorting of Submicrometer Particles in a Microchannel. *Nanoscale Microscale Thermophys. Eng.* **2016**, *20*, 51–65. [CrossRef]
24. Du, J.Y.; Li, L.; Zhuo, Q.Y.; Wang, R.J.; Zhu, Z.F. Investigation on Inertial Sorter Coupled with Magnetophoretic Effect for Nonmagnetic Microparticles. *Micromachines* **2020**, *11*, 566. [CrossRef] [PubMed]
25. Zhao, Q.B.; Yuan, D.; Zhang, J.; Li, W.H. A Review of Secondary Flow in Inertial Microfluidics. *Micromachines* **2020**, *11*, 461. [CrossRef] [PubMed]
26. Razavi Bazaz, S.; Mashhadian, A.; Ehsani, A.; Saha, S.C.; Krüger, T.; Ebrahimi Warkiani, M. Computational inertial microfluidics: A review. *Lab Chip* **2020**, *20*, 1023–1048. [CrossRef] [PubMed]
27. Mutlu, B.R.; Edd, J.F.; Toner, M. Oscillatory inertial focusing in infinite microchannels. *Proc. Natl. Acad. Sci. USA* **2018**, *115*, 7682–7687. [CrossRef]

28. Di Carlo, D. Inertial microfluidics. *Lab Chip* **2009**, *9*, 3038–3046. [CrossRef]
29. Dai, Y.L.; Xu, C.; Sun, X.L.; Chen, X.Y. Nanoparticle design strategies for enhanced anticancer therapy by exploiting the tumour microenvironment. *Chem. Soc. Rev.* **2017**, *46*, 3830–3852. [CrossRef]
30. Mitchell, M.J.; Billingsley, M.M.; Haley, R.M.; Wechsler, M.E.; Peppas, N.A.; Langer, R. Engineering precision nanoparticles for drug delivery. *Nat. Rev. Drug Discov.* **2021**, *20*, 101–124. [CrossRef] [PubMed]
31. Aidun, C.K.; Clausen, J.R. Lattice-Boltzmann Method for Complex Flows. *Annu. Rev. Fluid Mech.* **2010**, *42*, 439–472. [CrossRef]
32. Elghobashi, S. Direct Numerical Simulation of Turbulent Flows Laden with Droplets or Bubbles. *Annu. Rev. Fluid Mech.* **2019**, *51*, 217–244. [CrossRef]
33. Ladd, A.J.C.; Verberg, R. Lattice-Boltzmann Simulations of Particle-Fluid Suspensions. *J. Stat. Phys.* **2001**, *104*, 1191–1251. [CrossRef]
34. Rettinger, C.; Rüde, U. A comparative study of fluid-particle coupling methods for fully resolved lattice Boltzmann simulations. *Comput. Fluids* **2017**, *154*, 74–89. [CrossRef]
35. Eshghinejadfard, A.; Hosseini, S.A.; Thévenin, D. Fully-resolved prolate spheroids in turbulent channel flows: A lattice Boltzmann study. *AIP Adv.* **2017**, *7*, 95007. [CrossRef]
36. Rettinger, C.; Rüde, U. A coupled lattice Boltzmann method and discrete element method for discrete particle simulations of particulate flows. *Comput. Fluids* **2018**, *172*, 706–719. [CrossRef]
37. Eshghinejadfard, A.; Abdelsamie, A.; Hosseini, S.A.; Thévenin, D. Immersed boundary lattice Boltzmann simulation of turbulent channel flows in the presence of spherical particles. *Int. J. Multiph. Flow* **2017**, *96*, 161–172. [CrossRef]
38. Karimnejad, S.; Amiri Delouei, A.; Nazari, M.; Shahmardan, M.M.; Mohamad, A.A. Sedimentation of elliptical particles using Immersed Boundary – Lattice Boltzmann Method: A complementary repulsive force model. *J. Mol. Liq.* **2018**, *262*, 180–193. [CrossRef]
39. Thorimbert, Y.; Marson, F.; Parmigiani, A.; Chopard, B.; Lätt, J. Lattice Boltzmann simulation of dense rigid spherical particle suspensions using immersed boundary method. *Comput. Fluids* **2018**, *166*, 286–294. [CrossRef]
40. Jebakumar, A.S.; Magi, V.; Abraham, J. Lattice-Boltzmann simulations of particle transport in a turbulent channel flow. *Int. J. Heat Mass Transf.* **2018**, *127*, 339–348. [CrossRef]
41. Peng, C.; Ayala, O.M.; Wang, L.P. A comparative study of immersed boundary method and interpolated bounce-back scheme for no-slip boundary treatment in the lattice Boltzmann method: Part I, laminar flows. *Comput. Fluids* **2019**, *192*, 104233. [CrossRef]
42. Geneva, N.; Peng, C.; Li, X.M.; Wang, L.P. A scalable interface-resolved simulation of particle-laden flow using the lattice Boltzmann method. *Parallel Comput.* **2017**, *67*, 20–37. [CrossRef]
43. Jebakumar, A.S.; Premnath, K.N.; Magi, V.; Abraham, J. Fully-resolved direct numerical simulations of particle motion in a turbulent channel flow with the lattice-Boltzmann method. *Comput. Fluids* **2019**, *179*, 238–247. [CrossRef]
44. Qian, Y.H.; D'Humières, D.; Lallemand, P. Lattice BGK Models for Navier-Stokes Equation. *EPL* **1992**, *17*, 479–484. [CrossRef]
45. He, X.Y.; Luo, L.S. Lattice Boltzmann Model for the Incompressible Navier–Stokes Equation. *J. Stat. Phys.* **1997**, *88*, 927–944. [CrossRef]
46. Lallemand, P.; Luo, L.S. Lattice Boltzmann method for moving boundaries. *J. Comput. Phys.* **2003**, *184*, 406–421. [CrossRef]
47. Mei, R.W.; Da Yu, Z.; Shyy, W.; Luo, L.S. Force evaluation in the lattice Boltzmann method involving curved geometry. *Phys. Rev. E* **2002**, *65*, 41203. [CrossRef] [PubMed]
48. Aidun, C.K.; Lu, Y.N.; Ding, E.J. Direct analysis of particulate suspensions with inertia using the discrete Boltzmann equation. *J. Fluid Mech.* **1998**, *373*, 287–311. [CrossRef]
49. Yuan, X.F.; Ball, R.C. Rheology of hydrodynamically interacting concentrated hard disks. *J. Chem. Phys.* **1994**, *101*, 9016–9021. [CrossRef]
50. Pirola, S.; Cheng, Z.; Jarral, O.A.; O'Regan, D.P.; Pepper, J.R.; Athanasiou, T.; Xu, X.Y. On the choice of outlet boundary conditions for patient-specific analysis of aortic flow using computational fluid dynamics. *J. Biomech.* **2017**, *60*, 15–21. [CrossRef] [PubMed]
51. Hu, X.; Lin, J.Z.; Ku, X.K. Inertial migration of circular particles in Poiseuille flow of a power-law fluid. *Phys. Fluids* **2019**, *31*, 73306.
52. Nie, D.M.; Lin, J.Z. Simulation of sedimentation of two spheres with different densities in a square tube. *J. Fluid Mech.* **2020**, *896*, A12. [CrossRef]
53. Di Chen, S.; Pan, T.W.; Chang, C.C. The motion of a single and multiple neutrally buoyant elliptical cylinders in plane Poiseuille flow. *Phys. Fluids* **2012**, *24*, 103302. [CrossRef]
54. Qian, S.Z.; Jiang, M.Q.; Liu, Z.H. Inertial migration of aerosol particles in three-dimensional microfluidic channels. *Particuology* **2021**, *55*, 23–34. [CrossRef]
55. Barkla, H.M.; Auchterlonie, L.J. The Magnus or Robins effect on rotating spheres. *J. Fluid Mech.* **1971**, *47*, 437–447. [CrossRef]
56. Mahmoud, G.M.; Ahmed, M.E. Chaotic and Hyperchaotic Complex Jerk Equations. *Int. J. Mod. Nonlinear Theory Appl.* **2012**, *01*, 6–13. [CrossRef]
57. Schaaf, C.; Stark, H. Particle pairs and trains in inertial microfluidics. *Eur. Phys. J. E* **2020**, *43*, 50. [CrossRef]

 *micromachines*

Article

Visualization Experimental Study on Silicon-Based Ultra-Thin Loop Heat Pipe Using Deionized Water as Working Fluid

Wenzhe Song, Yanfeng Xu, Lihong Xue, Huajie Li and Chunsheng Guo [1,*]

School of Mechanical, Electrical & Information Engineering, Shandong University, Weihai 264209, China; 201800800305@mail.sdu.edu.cn (W.S.); 201916564@mail.sdu.edu.cn (Y.X.); 202017437@mail.sdu.edu.cn (L.X.); 201800800220@mail.sdu.edu.cn (H.L.)
* Correspondence: guo@sdu.edu.cn; Tel.: +86-631-56887762

Abstract: As a type of micro flat loop heat pipe, s-UTLHP (silicon-based ultra-thin loop heat pipe) is of great significance in the field of micro-scale heat dissipation. To prove the feasibility of s-UTLHP with high heat flux in a narrow space, it is necessary to study its heat transfer mechanism visually. In this paper, a structural design of s-UTLHP was proposed, and then, to realize the working fluid charging and visual experiment, an experimental system including a holding module, heating module, cooling module, data acquisition module, and vacuum chamber was proposed. Deionized water was selected as a working fluid in the experiment. The overall and micro phenomena of s-UTLHP during startup, as well as the evaporation and condensation phenomena of s-UTLHP during stable operation, were observed and analyzed. Finally, the failure phenomenon of s-UTLHP was analyzed, and several solutions were proposed. The observed phenomena and experimental conclusions can provide references for further related experimental research.

Keywords: loop heat pipe; deionized water; two-phase flow; visualization; heat transfer experiment

Citation: Song, W.; Xu, Y.; Xue, L.; Li, H.; Guo, C. Visualization Experimental Study on Silicon-Based Ultra-Thin Loop Heat Pipe Using Deionized Water as Working Fluid. *Micromachines* **2021**, *12*, 1080. https://doi.org/10.3390/mi12091080

Academic Editors: Junfeng Zhang and Ruijin Wang

Received: 19 July 2021
Accepted: 2 September 2021
Published: 7 September 2021

Publisher's Note: MDPI stays neutral with regard to jurisdictional claims in published maps and institutional affiliations.

1. Introduction

With the development of electronic technology, as the function of the chip becomes more and more powerful, its power consumption is also rises. At the same time, electronic products are developing in the direction of small, light, and thin, which has caused the heat flux of electronic components to rise sharply. Therefore, the evolution of mobile terminals has led to more stringent requirements on the size of electronic cooling components [1]. Taking smartphones as an example, in recent years, the power consumption of system chips of smartphones has increased to 3–5 W, while the thickness of smartphones has been reduced to about 6 mm [2]. When the equipment runs at high power, it produces a great deal of heat. This heat accumulation leads to uneven temperature distribution, and the resulting thermal stress causes thermal deformation of internal electronic devices [3]. There is evidence indicating that the micro flat loop heat pipe can maximally reduce the temperature of the chip and improve the overall temperature uniformity of the chip. Furthermore, its required heat dissipation space is very small. Thus, the micro flat loop heat pipe appears to be an ideal solution for high-intensity heat dissipation of micro-scale components [4].

The heat pipe is characterized by high thermal conductivity, long service life, convenient maintenance, and compact and flexible structure. Recently years, research and experiments on heat pipes have attracted more attention [5]. The focus of prior studies mainly differ from the following four aspects: microchannel design, working fluid selection, heat pipe performance, and visualization. In terms of microchannel research, Lim J. [6] proposed a channel layout for flat plate micro heat pipe under local heating conditions that can effectively overcome the limitations of local heating conditions. For working fluid research, Kim J. [7] used ethanol, FC-72, HFE-7000, R-245fa, and R-134a as experimental working fluids to study the selection criteria of working fluids in micro heat pipes. In addition, Narayanasamy M. [8] mixed acetone, deionized water, and tetrahydrofuran fluids with graphene oxide nanoparticles by ultrasonication and prepared nanofluids as working fluids for micro heat pipes for lithium-ion batteries. In terms of performance research, Zhao Y. N. [9] deduced the heat transfer performance of a micro heat pipe array, the influence of inclination angle on heat transfer characteristics, and the pressure limit of the heat pipe structure by using ammonia as a working fluid. Chen G. [10] researched the effects of heat load, cooling water temperature, inclination angle, and other factors on the thermal response performance, temperature distribution, thermal resistance, and other heat transfer characteristics of ultra-thin flat heat pipes. Wang G. [11] concentrated on the effects of working fluid type, charging rate, and inclination angle on the performance of flat plate micro heat pipes through experiments. In terms of visualization research, Kamijima C. [12] studied the relationship between the thermal characteristics and the internal flow characteristics of the micro heat pipe with FC-72 as the working fluid by measuring the effective thermal conductivity, visualizing the flow pattern, and simulating heat transfer according to the flow pattern. The study by Kim Y. B. [13,14] investigated the thermal flow and rapid thermal oscillatory flow in an asymmetric micro pulsation heat exchanger utilizing FC-72 and ethanol as working fluids. It also clarified the optimal charging rate of the two working fluids, evaluated their thermal resistances, identified their key mechanisms of circulating motion, and observed two different flow modes (oscillatory eruption mode and circulating mode). Our research group performed an in-depth study of the characteristics of microchannel flow using numerical simulation [15]. On the basis of our previous research and the four types of research listed above, we conducted in-depth research focusing on visualization and expanded the literature on the design of microchannels and the performance of heat pipes.

Microchannel systems can significantly improve energy transfer efficiency and reduce the overall size of chips. In various technical fields, microchannel technology is one of the most promising areas for the development of equipment [16]. As a kind of phase change heat transfer, gas–liquid two-phase flow heat transfer has very high efficiency [17]. The microchannel is etched on a silicon substrate, and the heat transfer is carried out by gas–liquid two-phase flow, which can realize a flat loop heat pipe with small size and high efficiency. Compared with other working fluids, deionized water has stronger polarity and stronger affinity with silicon chips, which are suitable for silicon-based heat pipes. The phase transition process, flow process, and vapor–liquid distribution of the working fluid can be observed by visualization experiment [18]. Therefore, in this study, a micro ultra-thin loop heat pipe was designed with silicon as the mainboard material, deionized water as the working fluid, and capillary force provided by the microchannel array. The phase transition behavior and vapor–liquid interface of the heat pipe during startup and stable operation were visualized through experiments. The results have an important reference value for the theoretical analysis of the heat transfer mechanism of ultra-thin loop heat pipes.

2. Structure and Experimental Setup of s-UTLHP

2.1. Structure of s-UTLHP

The main structure of the s-UTLPH designed in this study, as shown in Figure 1a,b, included a silicon-based motherboard and a Pyrex 7740 glass upper cover plate. The high boron glass upper cover plate was packaged with the silicon-based motherboard by bonding, which allowed us to observe the flow pattern's evolution process in the microchannel visually and intuitively and to understand its physical mechanism. The silicon-based motherboard comprised an evaporation chamber, a condensation chamber, a vapor channel, and a liquid channel connecting the evaporation chamber and the condensation chamber. The overall s-UTLPH size was 40 mm × 20 mm × 1.45 mm. The capillary force of s-UTLPH was provided by parallel microchannels. Both the depth of the microchannel and the whole etching depth were 180 μm. The width of the microchannel was 30 μm, and the aspect ratio was 6:1. In addition, a rectangular vapor overflow cavity was etched on the corresponding glass cover plate above the parallel microchannel. The height of the vapor overflow cavity was 240 μm, and its structure is displayed in Figure 2.

The evaporation chamber converted the working fluid into vapor by heating in the evaporation chamber, and the capillary core inside provided the capillary driving force to ensure the unidirectional circulation of the working fluid. The main function of the condensing chamber was to condense the vapor working fluid back to the liquid working fluid to release heat. The compensation chamber can store the liquid working fluid of cooling backflow, which was convenient to supplement the liquid working fluid in time and prevent the "dry burning" of the cutoff flow. Eight vapor channels and four liquid channels were located between the evaporation chamber and the condensation chamber. The vapor working fluid produced by phase change flowed to the condensation chamber through the vapor channel. When the condensation chamber became cold, the working fluid turned into liquid and then flowed back to the evaporation chamber through the liquid channel. A hollow insulating slot was also arranged on the s-UTLHP to reduce heat leakage. The size of the s-UTLHP devised in this study is presented in Table 1.

(**a**)

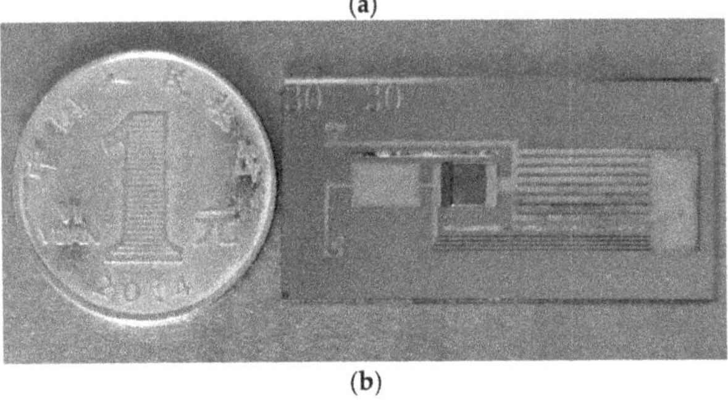

(**b**)

Figure 1. s-UTLHP prototype designed in this paper. (**a**) SolidWorks assembly sketch of s-UTLHP; (**b**) Physical drawing of s-UTLHP.

Figure 2. Structure diagram of the evaporator.

Table 1. s-UTLHP dimension parameters.

s-UTLHP Components	Area (mm^2)
Evaporation chamber	24
Liquid storage chamber	26
Condensing chamber	42.315
Vapor channel	57.6
Liquid channel	11.73

2.2. Experimental Device

To meet the requirements of the working fluid charging, the visualized heat transfer experiment, and the s-UTLHP measurement, it was necessary to arrange the experimental system around the s-UTLHP with limited space. As shown in Figure 3, the visualized heat transfer experimental system involved an s-UTLHP holding module, heating module, cooling module, data acquisition module, vacuum chamber, etc. The data acquisition module consisted of a temperature data acquisition module and a visualization module.

In this study, the silicon-based bottom plate of the s-UTLHP was set with two outlets for charging and pumping. First, a mechanical pump was used for primary air extraction, and an Edward molecular pump was utilized for secondary air extraction. When the pressure was below 1.5×10^{-3} Pa, the valve at the air extraction port was closed, and then the valve on the other side was opened to fill the working fluid. Deionized water was applied as the working fluid in this paper. The physical parameters of this working fluid are presented in Table 2. After vacuumizing the inner part of the s-UTLHP and its connecting pipelines, the charging operation of the s-UTLHP was carried out. The experimental results demonstrated that high vacuum degree and good charging effect could be obtained by the vacuum charging method, which provides a guarantee for the rapid startup and stable operation of s-UTLHP.

The s-UTLHP, after charging and sealing, was accurately heated by a ceramic heating plate combined with a heat transfer copper block. An ITECH programmable power supply was used of the model IT6121B. Furthermore, a copper cooling block was attached to the condensing end. Constant-temperature cooling water was continuously supplied by a constant-temperature circulating water cooler. The temperature of the condensing end was adjusted by controlling the temperature of the circulating water. In the experiment, the cooling water temperature was set to 5 °C. During the experiment, the Omega T thermocouple Was employed to detect the temperature of each key node, and Fluke2638A was used to record and store the temperature data. The arrangement of the thermocouple is displayed in Figure 4. The temperature measuring point was also set to detect the room temperature.

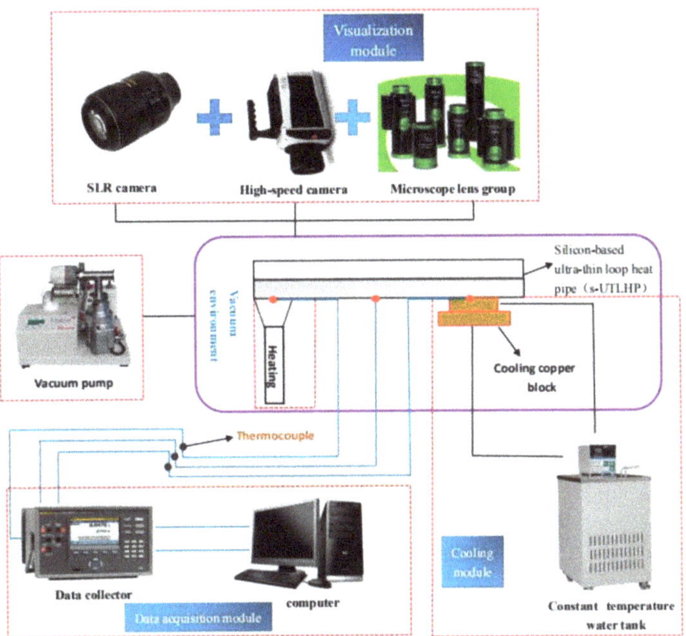

Figure 3. Visualization of the heat transfer experiment system.

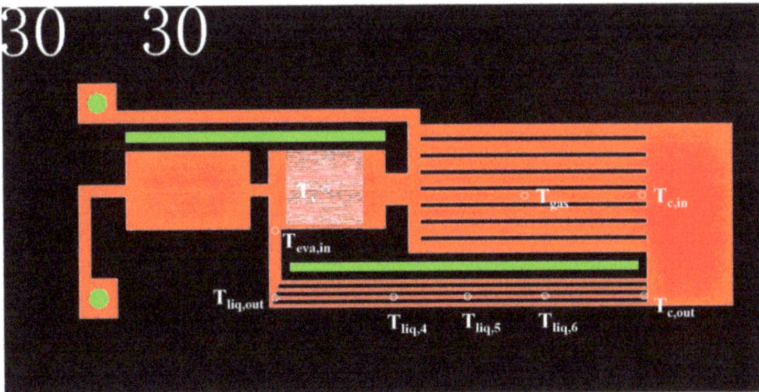

Figure 4. The layout of the thermocouple for temperature measurement. $T_{eva,\,in}$—evaporation chamber inlet, $T_{liq,\,out}$—liquid line outlet, $T_{liq,4,5,6}$—liquid pipeline, $T_{c,out}$—condensation chamber outlet, T_{gas}—gas channel, T_v—evaporation chamber, $T_{c,in}$—condensation chamber inlet.

Table 2. Parameters of each liquid working fluid at standard atmospheric pressure and room temperature.

Fluid Working Fluid	Melting Point (°C)	Boiling Point (°C)	Surface Tension Coefficient (n/M)	Viscosity Coefficient (mPa*s)	Belgium KJ/(kg*K)
H_2O	0	100	0.075	0.083	4.18

2.3. Error Analysis and Data Processing

2.3.1. Error Analysis

In this experiment, contact and non-contact temperature measurement methods were adopted through a combination of thermocouple and infrared temperature measurement. In contact measurement, thermocouples were pasted on the s-UTLHP silicon substrate, and thermally conductive silicone grease was coated between the heating block and the silicon substrate to reduce the contact thermal resistance. Meanwhile, the PTFE fixture was designed to reduce the heat loss of the insulation section. The parameters or data to be measured included charge statistics, surface temperature measurement, and input heat calculation. Hence, the main experimental error sources were the temperature measurement point, charge mode, test error, and DC power supply instrument.

(1) Charge uncertainty

A Hamilton 250 μL injector was used to achieve quantitative charging. The minimum calibration interval was 2.5 μL, and the error was ±1.25 μL.

(2) Temperature uncertainty

To ensure the repeatability and effectiveness of the experimental phenomenon, the method of repeated testing was utilized for the experiment under the same conditions. The directly measured parameter value (temperature) in the experiment was calculated according to the following formula:

$$U(T_i) = \sqrt{\frac{1}{I(J-1)} \sum_{j=1}^{J} \sum_{i=1}^{I} (T_{ij} - \overline{T_i})}$$

In the above formula, J is the serial number of the repeated experiment and i is the serial number of the temperature point tested. In the test results, the maximum deviation of temperature measured by the thermocouple was 1.56 °C, and the corresponding maximum uncertainty was 1.93%.

(3) Uncertainty of DC power supply

The uncertainty caused by the instrument is given by class B uncertainty $U(E) = \frac{A}{k}$, where A is the nominal error in the instruction manual of the instrument and $k = \sqrt{3}$. In the experiment of s-UTLHP heat transfer characteristics, the input heat load was changed by adjusting the voltage of the DC regulated power supply. The nominal error was ±1%, and the standard uncertainty of voltage was $U(V) = 1\%\sqrt{3} = 0.577\%$. According to the principle of error transfer, the uncertainty of input heat load Q_{in} was $U(Q_{in}) = \frac{\sqrt{(I \cdot u(V))^2 + (V \cdot u(I))^2}}{V \cdot I} = 2.341\%$.

2.3.2. Data Processing

The s-UTLHP charging rate is defined as the ratio of the volume of the working fluid to the total volume of the system. The specific value is computed by the following formula:

$$\Phi = \frac{V_{wf}}{V_{sys}}$$

where Φ is the charging rate, V_{wf} is the volume filled with liquid working fluid, and V_{sys} is the volume of the s-UTLHP system.

3. Analysis and Discussion of Experimental Results

3.1. Startup of the s-UTLHP

3.1.1. Overall Phenomena at Startup

For the operation of traditional heat pipes, a suitable working fluid should have high latent heat of vaporization and surface tension, low viscosity, and good wetting performance. Additionally, a working fluid with different critical points should be selected to conform to the temperature range [19]. In this study, deionized water was selected as a working fluid.

The startup process of the s-UTLHP filled with deionized water working fluid is presented in Figure 5. The voltage of the regulated power supply was set at 3.4 V, and the current change was recorded after stable operation. The heating power was 5.54 W, the effective heat transfer area was 4 mm $\times$ 3 mm, and the heat flux was about 46 W/cm^2.

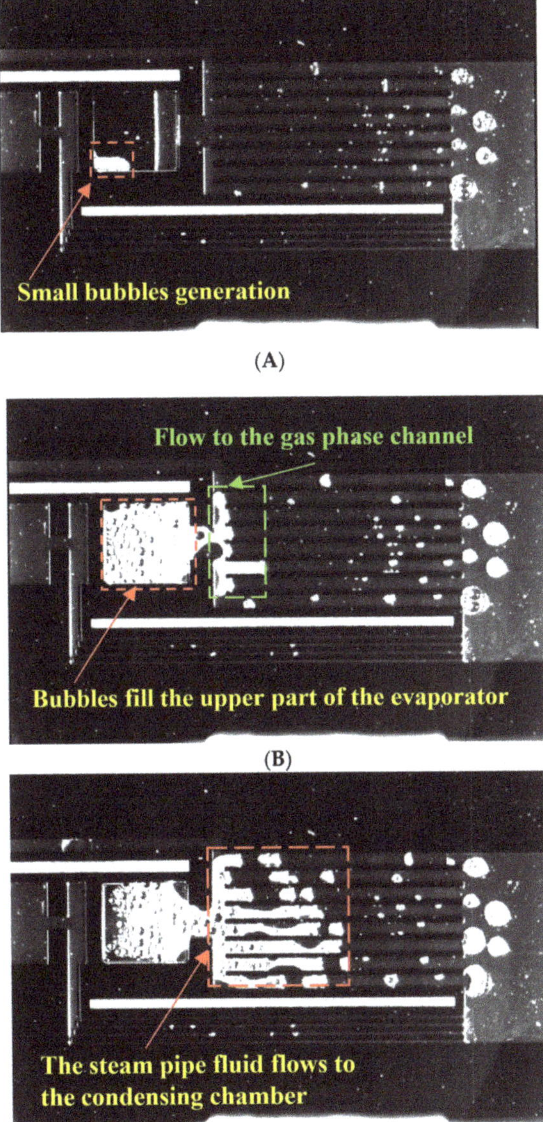

Figure 5. Startup process of the s-UTLHP with two different charging working fluids: (**A**) bubble generation stage at microchannel; (**B**) beginning stage of vapor outflow; (**C**) stable stage after vapor propulsion.

Small bubbles appeared when t = 5 s (Figure 5A). During the experiment, the liquid working fluid in the s-UTLHP evaporation chamber absorbed the external heat through the chip substrate and increased in temperature. After reaching the saturation temperature, nucleate boiling occurred in the parallel microchannel. The vapor bubbles in the vapor overflow cavity above the microchannel expanded after their appearance to occupy the whole vapor overflow cavity within 5 s (Figure 5B). With a continuous heating process or with an increase in heating power, the vapor bubbles above the vapor overflow cavity decreased and were replaced by pure working fluid vapor. The reason was that the vapor bubbles on the upper cover plate of the evaporation chamber were mixed with the surrounding droplets (liquid film), which was the final result of the dynamic process of evaporation and condensation in the evaporation chamber. Higher heat input made the evaporation effect far greater than the condensation effect, so that purer vapor appeared in the evaporation chamber. In the meantime, the working fluid vapor began to flow from the vapor overflow chamber to the vapor pipe, and the vapor continuously pushed the liquid in the vapor pipe to flow to the condensation chamber (Figure 5C). From Figure 5B,C, it can be seen that the continuous generation of vapor bubbles in the evaporation chamber was weakened, the nucleate boiling was replaced by thin-film evaporation, and the liquid pipe and compensation chamber were filled with liquid and continuously replenished the evaporation chamber. At this time, the s-UTLHP started up and ran stably. In the initial stage of heating and evaporation, it took only 14 s for the deionized water to advance to the vapor channel from the generation of bubbles to the evaporation drying of the vapor overflow chamber.

At the primary stage of startup, the area above the parallel microchannel array of the evaporation chamber evaporated first. During the phase transition of the deionized water working fluid, numerous bubbles were generated, and then a large number of fine droplets were formed, as displayed in Figure 5B,C. Moreover, nucleate boiling emerged in the parallel microchannel at the initial stage of operation.

3.1.2. Micro Phenomena during Startup

To have a more comprehensive understanding of the internal startup and evaporation boiling phenomena for the s-UTLHP, the microscopic phenomena of the s-UTLHP under variable power were observed, and the temperature data during the experiment were recorded. Deionized water has the advantages of being non-toxic and easy to obtain and of having a high latent heat of vaporization, which is commonly used as a working fluid for loop heat pipes [20]. The startup and variable power loading experiments for the s-UTLHP were carried out with deionized water as the working fluid, and the charging rate was 90%. In the experiment, we increased the input power at equal intervals. After each increase of the input power, different heat transfer phenomena occurred. According to these phenomena, we divided the stage into different regions. In addition, each time we heating at the regular rate is a variable power test after reaching steady state at a specific power. The sign of reaching steady state is that the temperature changes very little (gradually flattening from the temperature curve, that is, the temperature is stable or fluctuates within a certain range with the passage of time at a certain power). In this process, the relationship between the overall temperature and power within the experimental time is shown in Figure 6. Ten variable power loading experiments were carried out on the s-UTLHP, and the resulting curve was divided into 11 regions according to the variable power loading time. It can be seen from the figure that the s-UTLHP could respond quickly to different power and heat loads. For the evaporation chamber outlet temperature curve (the highest temperature curve in the figure), after changing the heating power, the temperature first rose rapidly, then gradually decreased and stabilized, and then fluctuated within a certain range. This was because the fluctuation of vapor flow resulted in the fluctuation of heat carried away, which gave rise to the fluctuation of temperature. The temperature of the liquid phase pipeline gradually increased along the fluid return path, because the position near the end of the return pipeline was close to the heating source, which could conduct part of the heat.

According to the above two startup judgment marks for the s-UTLHP, the startup power of the s-UTLHP with deionized water as the working fluid was 3.5517 W, which corresponds to phase E in the curve. The temperature difference between the inlet and outlet of the condensation chamber was 3.18 °C. The whole change process of the bubbles in the evaporation chamber during startup (shown in Figure 7) occurred in the area without the shallow cavity and microchannel in the evaporator. The change of bubble volume in stages A–D was not obvious and was accompanied by expansion–contraction oscillation with smaller volume change amplitude, but the magnification was different across the four stages. There were also both large and small droplets, which were constantly generated, merged, grown, and disappeared. The disappearance here was integration either with the liquid around the bubble or with the liquid below. At the E stage, the bubble volume increased rapidly and almost covered the whole area without the shallow cavity and microchannel in the evaporator.

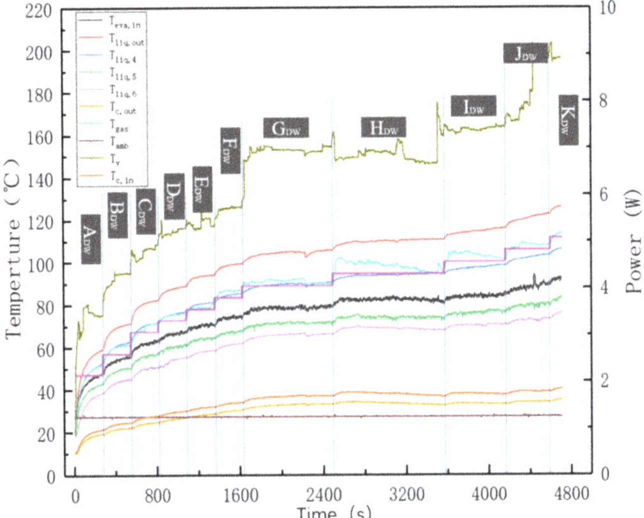

Figure 6. Relationship among the temperature and power of s-UTLHP and the time. $T_{eva,in}$ is the inlet temperature of the evaporation chamber; $T_{liq,out}$ is the outlet temperature of the liquid pipeline; $T_{liq,4,5,6}$ is the temperature of the liquid pipeline, evenly arranged along the liquid return direction; $T_{c,out}$ is the outlet temperature of condensation chamber; T_{gas} is the temperature of the vapor phase pipeline; T_{amb} is the ambient temperature; T_v is the temperature of the evaporation chamber; and $T_{c,in}$ is the inlet temperature of the condensation chamber.

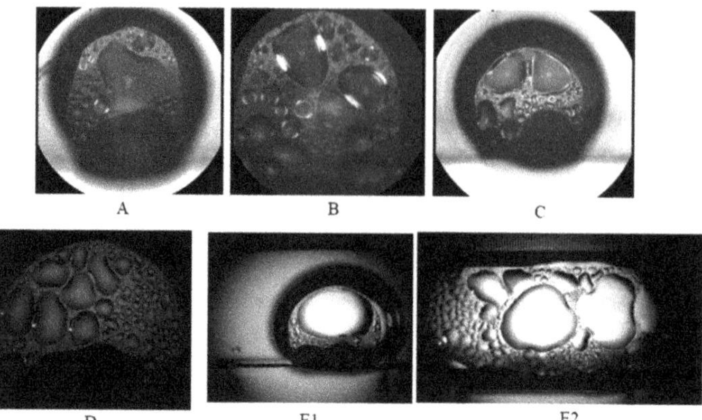

Figure 7. Startup process of the s-UTLHP.

3.2. Stable Operation of the s-UTLHP

3.2.1. Evaporation Phenomena in Stable Operation

The steady-state performance of the heat pipe was reflected in the variable power operation experiment to a certain extent [21], and the micro phenomena of vapor–liquid distribution, evaporation, and condensation were also captured. In the variable power operation of s-UTLHP, it was found that, as shown in Figure 6, when the power increased, the temperature responded in time, showing a ladder shape on the whole. The temperature curves of the single stages showed a trend of first increasing and then becoming basically stable.

As presented in Figure 8, when deionized water was applied as a working fluid, the s-UTLHP operated stably from stage F. In the subsequent stages, the inlet temperature curve of the vapor phase pipeline was generally stepped, and for each stage, it first increased, then decreased, and then generally stabilized. This was because, for the ceramic heating plate under a given voltage, over time, the current gradually decreased and then stabilized. The microscopic image of stage F at the junction of the evaporation chamber and vapor phase pipeline is presented in Figure 8a. Many droplets can be seen; these droplets constantly repeated the process of generation, growth, disappearance, and washing away. As the power increased, the process changed faster and faster. Figure 8b shows the phenomena in the evaporation chamber. Figure 8c,d show the bubbles at the inlet of the vapor phase pipeline during the stable operation time. When the deionized water working fluid shrank and broke at the inlet of the vapor phase pipeline, slug-like flow appeared, and adjacent bubbles did not easily fuse.

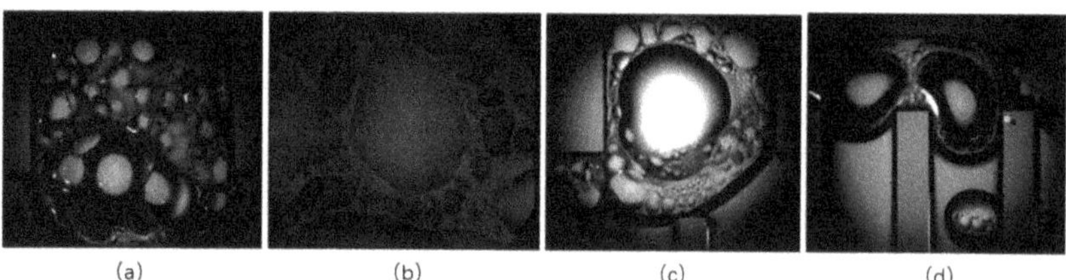

Figure 8. Microscopic images of the s-UTLHP at different stages of steady-state operation: (a) Image of the connection between the evaporator and the vapor phase line; (b) Image inside the evaporator; (c,d) The image of the inlet of the vapor phase pipeline during stable operation.

3.2.2. Condensation Phenomena in Stable Operation

Like an ordinary loop heat pipe, the s-UTLHP transfers heat through the evaporation–condensation phase transition of working fluid. Figure 9 displays the bubble condensation process in the condensation chamber during stable operation. The positions of Figure 9b–d are illustrated in Figure 9a. These images were obtained under the same magnification and field of view to observe and compare the change of bubble volume. Taking the NCG bubble in focus clearly within the field of view as the initial timing, the 0 s bubble phenomenon is shown in Figure 9b. The condensation process occurred when the heat flux was 37.5575 W/cm^2.

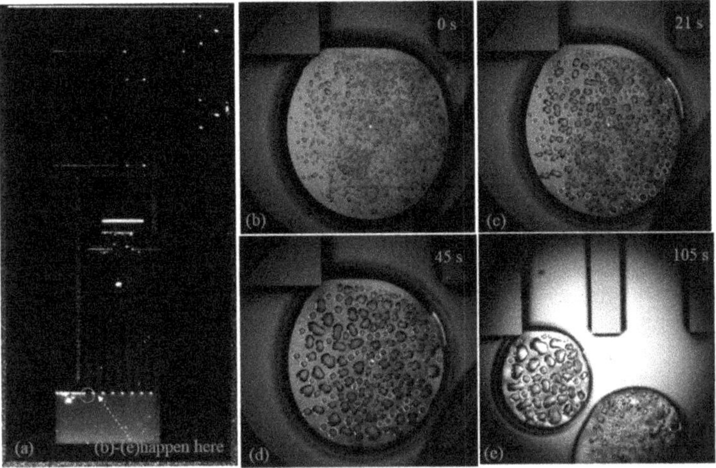

Figure 9. Bubble condensation process in the condensation chamber: (**a**) illustrates where (**b–e**) occur in the condenser; (**b–e**) are the images at 0 s, 21 s, 45 s and 105 s respectively.

There were many obvious droplets in the deionized water working fluid bubbles. At first, the droplets were small, round, separated from each other, and independent in shape, as shown in Figure 9b. When the edges of the two bubbles grew together as the condensation continued, the condensable components in the bubble gradually condensed. As a result, the droplets were produced or merged with the existing droplets, leading to an increase in the droplet density in the bubble. As shown in Figure 9c,d, the edges of the two bubbles tended to gradually separate. The reason is that the volume of the bubble decreased and the volume of the droplet inside the bubble became larger than depicted in Figure 9c. The two bubbles were completely separated at the time of Figure 9e. The bubble edge of the object was visible, the bubble volume was significantly smaller, and the internal droplet volume became larger.

It was observed that the condensation phenomenon of this s-UTLHP could occur in the gas-phase pipeline when the heat flux was moderate. Bubbles of deionized water working fluid in the gas-phase pipeline during stable operation are shown in Figure 10a–c. The deionized water was ejected into independent bubbles in the gas-phase pipeline, and it was difficult for those bubbles to fuse, even if they shared part of a bubble boundary, as shown in Figure 10a. This was related to the higher surface tension and lower viscosity of deionized water. The volumes of the four bubbles displayed in Figure 10a were different. The lower one had experienced a long time of condensation in the vapor line, while the upper one had just left the column. In the condensation process, slug flow occurred; that is, the vapor column broke at a certain distance, forming independently isolated bubbles. As described above for the condensation chamber, the bubbles were not always completely condensed, and the condensation rate was slow. During operation, the separating vapor column was positioned at the entrance of the rightmost vapor phase pipeline, as shown in Figure 10c, because of the stress at the corner.

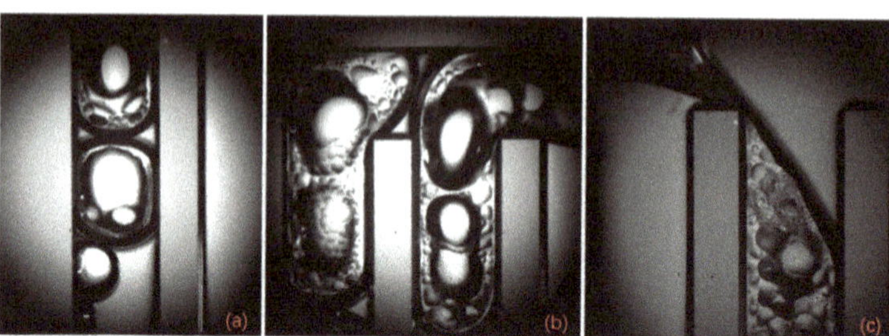

Figure 10. Condensation process in vapor phase pipeline.

3.3. Failure of s-UTLHP

When the heat flux continued to increase, the equilibrium between the refrigerant reflux and phase transition overflow was broken, and the capillary core was considered to be burnt out [22]. Compared with capillary cores made of porous media, the design of a large number of high-aspect-ratio microchannel arrays can make a system have higher stability, effectively reduce the flow resistance, improve the permeability of the evaporation chamber, ensure that the liquid can flow back in time under high heat flux and have higher mass flow rate, and delay the occurrence of flow interruption and drying phenomenon [23]. In this design, the compensation chamber compensated the backflow of liquid in time before the system started up and when the heat flow changed rapidly, so as to improve the stability of the continuous operation of the s-UTLHP. Furthermore, the high-aspect-ratio microchannel array was equipped with a wide pure microchannel area (as shown in area A in Figure 11). This area could generate a great capillary pressure difference through the microchannel under the condition of high heat flux and gas–liquid two-phase transition, thus forming a "hydraulic lock" and effectively preventing backflow. In area B, there was also a microchannel array with a high aspect ratio on the silicon substrate, but unlike in area A, there was a vapor overflow cavity above area B. Figure 11, area C shows the outlet of the vapor overflowing from the evaporation chamber to the gas-phase pipeline, and area D shows the inlet of the liquid working fluid reflux. By setting a reasonable proportion, the circulating pressure change of the working fluid in the s-UTLHP could be effectively balanced.

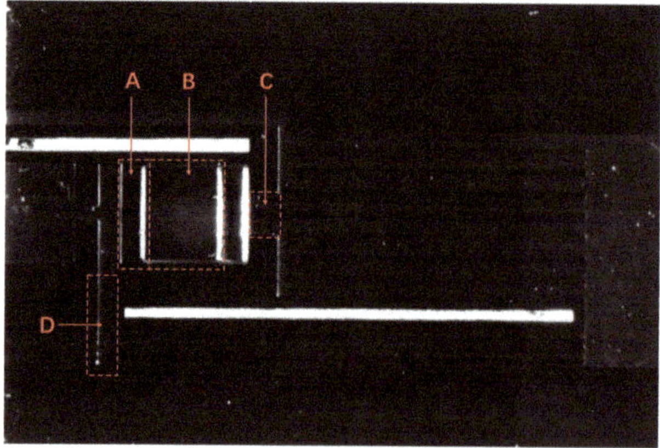

Figure 11. s-UTLHP structure diagram.

An image of the s-UTLHP filled with deionized water working fluid under high heat flux is shown in Figure 12. When the image was taken, the heat input was 41.3 W/cm^2. It can be observed that under relatively high heat flux, the gas phase pipeline appeared foggy because of two-phase flow. This phenomenon occurred because a great number of gaseous working fluids produced by phase transformation carried small droplets when flowing, and the pressure in the pipeline was high. Furthermore, there was liquid film attached to the wall of the gas phase pipeline. The vapor phase working fluid was produced by continuous phase change in the evaporation chamber. The vapor phase working fluid continuously advanced to occupy most of the gas phase pipeline and was in a relatively stable state.

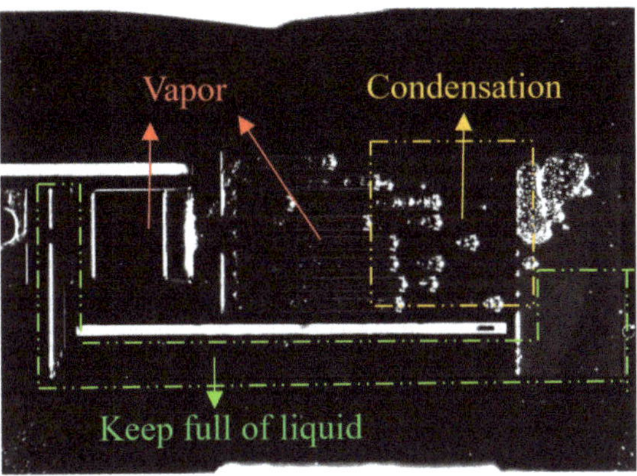

Figure 12. Imaging of the s-UTLHP under high heat flux.

The s-UTLHP in this study had a certain operating limit. When the heat flux reached a certain upper limit, a "dry burning" phenomenon appeared in the evaporation chamber. "Dry burning" refers to the state wherein the working fluid at the inlet and outlet of the evaporation chamber begins to evaporate while the evaporation chamber is continuously heated, finally resulting in the vaporization of all the liquid working fluid in the evaporation chamber. This occurs because the liquid replenishment rate is less than the phase change rate of the working fluid, which means that the liquid cannot be replenished, the liquid film on the wall of the silicon-based microchannel is evaporated quickly, and the gas–liquid two-phase interface regresses until the liquid working fluid in the evaporation chamber completely changes phase. At this time, the single gas-phase state does not have the conditions for capillary force generation and cannot be recycled. As presented in Figure 13, the s-UTLHP filled with deionized water was dried at a heat flux of 44.8 W/cm^2. Another reason for the drying is that the pressure of the non-condensable gas bubble itself increased due to continuous heating and combined with the vapor. The gas phase flowed back to the evaporation chamber and then dried up instantly; after that, the gas–liquid interface at the gas phase pipeline retrogressed. Because the capillary limit of s-UTLHP was not been reached, the heat flux could further improved if the influence of non-condensable gas was excluded.

In summary, by analyzing the operation state of the s-UTLHP filled with deionized water, we concluded that it was suitable for heat dissipation with high heat flux and environments that are sensitive to temperature fluctuations. In addition, the s-UTLHP failed because of insufficient liquid refrigerant supply after reaching the limit heat flux and because of non-condensable gas reflux.

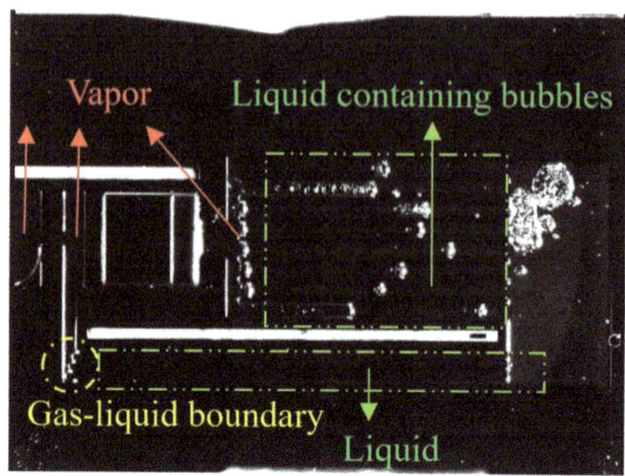

Figure 13. Imaging of s-UTLHP during drying.

4. Conclusions

In this study, an s-UTLHP with a size of 40 mm in length, 20 mm in width, and 1.5 mm in thickness was designed. The width of the parallel microchannel was 30 μm, and the aspect ratio was 6:1. The design could achieve high capillary force and high mass flow. Furthermore, the longitudinal section of the microchannel with a high aspect ratio could fully extend the liquid film and realize heat exchange with heat flux of 37.5575 W/cm^2.

Deionized water was selected as the working fluid. In the charging experiment, the s-UTLHP was pumped and charged, and the non-condensable gas was eliminated by combining with visualization. After the s-UTLHP was separately charged with deionized water, it could run stably for a long time. When the heat flux exceeded the limit, the "dry burning" phenomenon was observed, which caused the failure of the s-UTLHP. At low heat flux, the evaporation chamber presented the coexistence of bubbles and droplets, that is, the dynamic equilibrium of evaporation and condensation. At high heat flux, there was pure gas in the evaporation chamber, and the evaporation rate was much higher than the condensation rate, so an efficient evaporation replenishment mechanism was carried out.

This study explored a self-circulating heat dissipation scheme suitable for microchip integration, realized phase-change heat transfer on the surface of a silicon substrate, and proved the feasibility of the scheme for high-heat-flux heat dissipation in a narrow space. Moreover, the regular structure of the high-aspect-ratio microchannel array in the core of the evaporation chamber means that these arrays can be mass-produced by a 3D IC manufacturing process, which could help with popularization and application. In addition, this research adopted a visual method to conduct macroscopic and microscopic observational studies of evaporation and condensation, respectively, and to further explore the boiling flow and condensation cycle characteristics of the working fluid in the microchannel. It also provided references for the improvement of fine structure and surface morphology and for the selection of a working fluid, which can enhance the heat transfer capacity and the stability of the cyclic heat transfer mechanism. Therefore, this study has important reference value in the field of micro-scale heat dissipation for electronic chips.

Author Contributions: Conceptualization, C.G.; data curation, W.S.; formal analysis, W.S.; funding acquisition, C.G.; investigation, H.L.; methodology, Y.X.; project administration, C.G.; software, L.X.; supervision, C.G.; validation, Y.X.; visualization, W.S. and Y.X.; writing—original draft, W.S.; writing—review and editing, W.S., C.G., and Y.X. All authors have read and agreed to the published version of the manuscript.

Funding: This work was supported by the Focus on Research and Development Plan in Shandong Province (Grant No. 2018GGX104011) and the Natural Science Foundation of Shandong Province (Grant No. ZR2017BEE012).

Data Availability Statement: The data that support the findings of this study are available from the authors upon reasonable request. The data are not publicly available due to that they are planned to be used in our following experiments.

Conflicts of Interest: The authors declare no conflict of interest.

Nomenclature

T	temperature, °C
U	voltage, V
I	current, A
J	serial number of the repeated experiment
i	serial number of the temperature point tested
A	nominal error
V_{wf}	volume filled with liquid working fluid, m^3
V_{sys}	volume of the s-UTLHP system, m^3
t	time, s
	Greek symbols
Φ	charging rate

References

1. Cho, J.; Goodson, K.E. Thermal transport: Cool electronics. *Nat. Mater.* **2015**, *14*, 136–137. [CrossRef] [PubMed]
2. Tang, H.; Tang, Y.; Wan, Z.; Li, J.; Yuan, W.; Lu, L.; Li, Y.; Tang, K. Review of applications and developments of ultra-thin micro heat pipes for electronic cooling. *Appl. Energy* **2018**, *223*, 383–400. [CrossRef]
3. Ji, H.; Zhang, Y.; Wang, F.; Zhang, J. Thermal Performance Simulation of Novel Micro Heat Pipe. *J. Phys. Conf. Ser.* **2021**, *1936*, 012025. [CrossRef]
4. Li, J.; Lv, L.; Zhou, G.; Li, X. Mechanism of a microscale flat plate heat pipe with extremely high nominal thermal conductivity for cooling high-end smartphone chips. *Energy Convers. Manag.* **2019**, *201*, 112202. [CrossRef]
5. Liu, F.F.; Lan, F.C.; Chen, J.Q.; Li, Y.G. Experimental Investigation on Cooling/Heating Characteristics of Ultra-Thin Micro Heat Pipe for Electric Vehicle Battery Thermal Management. *Chin. J. Mech. Eng.* **2018**, *31*, 1–10. [CrossRef]
6. Lim, J.; Kim, S.J. A channel layout of a micro pulsating heat pipe for an excessively localized heating condition. *Appl. Therm. Eng.* **2021**, *196*, 117266. [CrossRef]
7. Kim, J.; Kim, S.J. Experimental investigation on working fluid selection in a micro pulsating heat pipe. *Energy Convers. Manag.* **2020**, *205*, 112462. [CrossRef]
8. Narayanasamy, M.P.; Gurusamy, S.; Sivan, S.; Senthilkumar, A.P. Heat transfer analysis of looped micro heat pipes with graphene oxide nanofluid for Li-ion battery. *Therm. Sci.* **2021**, *25*, 395–405. [CrossRef]
9. Wei, L.; Cai, J.; Liang, J. Experimental performance of ammonia-charged micro heat pipe array for energy saving. *Appl. Therm. Eng.* **2021**, *186*, 116525.
10. Chen, G.; Tang, Y.; Wan, Z.; Zhong, G.; Tang, H.; Zeng, J. Heat transfer characteristic of an ultra-thin flat plate heat pipe with surface-functional wicks for cooling electronics. *Int. Commun. Heat Mass Transf.* **2019**, *100*, 12–19. [CrossRef]
11. Wang, G.; Quan, Z.; Zhao, Y.; Wang, H. Performance of a flat-plate micro heat pipe at different filling ratios and working fluids. *Appl. Therm. Eng.* **2019**, *146*, 459–468. [CrossRef]
12. Kamijima, C.; Yoshimoto, Y.; Abe, Y.; Takagi, S.; Kinefuchi, I. Relating the thermal properties of a micro pulsating heat pipe to the internal flow characteristics via experiments, image recognition of flow patterns and heat transfer simulations. *Int. J. Heat Mass Transf.* **2020**, *163*, 120415. [CrossRef]
13. Kim, Y.B.; Kim, H.; Sung, J. Investigation of thermal enhancement factor in micro pulsating heat exchanger using LIF visualization technique. *Int. J. Heat Mass Transf.* **2020**, *159*, 120121. [CrossRef]
14. Kim, Y.B.; Song, H.W.; Sung, J. Flow behavior of rapid thermal oscillation inside an asymmetric micro pulsating heat exchanger. *Int. J. Heat Mass Transf.* **2018**, *120*, 923–929. [CrossRef]

15. Xue, Y.; Guo, C.; Gu, X.; Xu, Y.; Xue, L.; Lin, H. Study on Flow Characteristics of Working Medium in Microchannel Simulated by Porous Media Model. *Micromachines* **2020**, *12*, 18. [CrossRef]
16. Kuznetsov, V.V. Capillary hydrodynamics and heat transfer in two-phase gas-liquid microscale systems. *J. Phys. Conf. Ser.* **2020**, *1677*, 012068. [CrossRef]
17. Fukagata, K.; Kasagi, N.; Ua-arayaporn, P.; Himeno, T. Numerical simulation of gas–liquid two-phase flow and convective heat transfer in a micro tube. *Int. J. Heat Fluid Flow* **2007**, *28*, 72–82. [CrossRef]
18. Liao, X.; Jian, Q.; Zu, S.; Li, D.; Huang, Z. Visualization study and analysis on the heat transfer performance of an ultra-thin flat-plate heat pipe. *Int. Commun. Heat Mass Transf.* **2021**, *126*, 105464. [CrossRef]
19. Taft, B.S.; Williams, A.D.; Drolen, B.L. Review of Pulsating Heat Pipe Working Fluid Selection. *J. Thermophys. Heat Transf.* **2012**, *26*, 651–656. [CrossRef]
20. Lee, S.; Phelan, P.E.; Dai, L.; Prasher, R.; Gunawan, A.; Taylor, R.A. Experimental investigation of the latent heat of vaporization in aqueous nanofluids. *Appl. Phys. Lett.* **2014**, *104*, 151908. [CrossRef]
21. Wan, Z.-P.; Wang, X.-W.; Tang, Y. Condenser design optimization and operation characteristics of a novel miniature loop heat pipe. *Energy Convers. Manag.* **2012**, *64*, 35–42. [CrossRef]
22. Setyawan, I.; Putra, N.; Hakim, I.I. Experimental investigation of the operating characteristics of a hybrid loop heat pipe using pump assistance. *Appl. Therm. Eng.* **2018**, *130*, 10–16. [CrossRef]
23. Jiang, P.X.; Fan, M.H.; Si, G.S.; Ren, Z.P. Thermal–hydraulic performance of small scale micro-channel and porous-media heat-exchangers. *Int. J. Heat Mass Transf.* **2001**, *44*, 1039–1051. [CrossRef]

 *micromachines*

Article

Effects of Microscopic Properties on Macroscopic Thermal Conductivity for Convective Heat Transfer in Porous Materials

Mayssaa Jbeili and Junfeng Zhang *

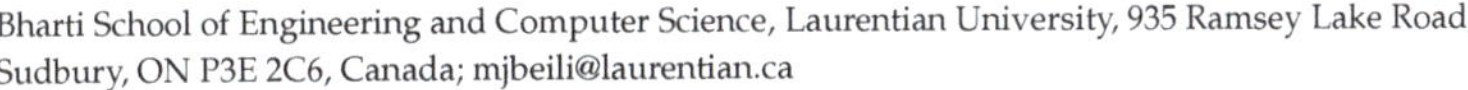

Bharti School of Engineering and Computer Science, Laurentian University, 935 Ramsey Lake Road, Sudbury, ON P3E 2C6, Canada; mjbeili@laurentian.ca
* Correspondence: jzhang@laurentian.ca; Tel.: +1-705-675-1151 (ext. 2248)

Abstract: Porous materials are widely used in many heat transfer applications. Modeling porous materials at the microscopic level can accurately incorporate the detailed structure and substance parameters and thus provides valuable information for the complex heat transfer processes in such media. In this study, we use the generalized periodic boundary condition for pore-scale simulations of thermal flows in porous materials. A two-dimensional porous model consisting of circular solid domains is considered, and comprehensive simulations are performed to study the influences on macroscopic thermal conductivity from several microscopic system parameters, including the porosity, Reynolds number, and periodic unit aspect ratio and the thermal conductance at the solid–fluid interface. Our results show that, even at the same porosity and Reynolds number, the aspect ratio of the periodic unit and the interfacial thermal conductance can significantly affect the macroscopic thermal behaviors of porous materials. Qualitative analysis is also provided to relate the apparent thermal conductivity to the complex flow and temperature distributions in the microscopic porous structure. The method, findings and discussions presented in this paper could be useful for fundamental studies, material development, and engineering applications of porous thermal flow systems.

Keywords: heat transfer; porous media; pore-scale modeling; boundary condition; thermal conductivity; porosity; conjugate interface; aspect ratio

Citation: Jbeili, M.; Zhang, J. Effects of Microscopic Properties on Macroscopic Thermal Conductivity for Convective Heat Transfer in Porous Materials. *Micromachines* **2021**, *12*, 1369. https://doi.org/10.3390/mi12111369

Academic Editor: Kwang-Yong Kim

Received: 11 October 2021
Accepted: 5 November 2021
Published: 7 November 2021

Publisher's Note: MDPI stays neutral with regard to jurisdictional claims in published maps and institutional affiliations.

1. Introduction

Fluid flows and the associated heat transfer in porous materials have attracted great interest over the decades for their scientific and practical importance. With the large contact area between the solid matrix and the coolant fluid, heat transfer performance is significantly enhanced in such porous materials. Relevant applications can be found in various industrial areas, including catalytic beds for chemical reactions, water treatment, nanofluids, heat exchangers, heat sinks and thermal energy storage systems [1–8]. Extensive experimental and theoretical investigations have been conducted to study the heat transfer performance in porous materials [9–11]. In addition to the high cost and long time duration, experimental studies are not able to reveal the flow and temperature distributions in the complex microscopic porous geometry, while such information is crucial for understanding the mechanism and relationships among various factors involved. On the other hand, theoretical analysis is limited to simple geometry structures and idealized flow situations, and such conditions can hardly be satisfied in realistic systems. Fortunately, with the advances in computer and software technologies, numerical simulations have been proved to be useful for various complex systems, including flows and heat transfer in porous media [12–16]. In the literature, there exist two typical approaches for modeling the flow and heat transfer in porous media: the continuum and the pore-scale approaches. The first method treats the porous materials as continuum media and solves the macroscopic flow and energy equations with apparent parameters [17,18]. Despite the model's simplicity and computational efficiency, this approach relies on the accuracy of those apparent parameters and also cannot incorporate the specific microscopic structure and material thermophysical properties in the porous media. On the other hand, the pore-scale method solves the flow and energy equations with the microscopic structure and properties of the matrix material considered explicitly [19,20]. Pore-scale simulations can provide detailed flow and thermal fields at the microscopic level, and apparent properties can be obtained from these microscopic distributions for macroscopic analysis [21–23]. In recent years, the lattice Boltzmann method (LBM) has become especially popular in such simulations, mainly for its convenience in dealing with complex boundary geometry [22–25].

With the microscopic porous structure modeled explicitly, the physical scale of pore-scale simulations is limited to small sizes due to the large computation demand. For this reason, pore-scale simulations typically work with a small unit and assume that such identical units are repeated in space. To solve the flow and energy equations in a periodic unit, appropriate boundary conditions are necessary on the side surfaces of this computational unit. Such side surfaces are simply for computational convenience and they are not physical surfaces or boundaries; therefore, definite conditions on such surfaces are not available. To create a thermal gradient, previous studies are usually assigned constant temperatures on two opposite unit surfaces and assumed adiabatic (no heat flux) conditions on other side walls [24–27]. This treatment has been examined recently and significant concerns have been raised in terms of its validity and accuracy [28]. Few studies treated the unit surface conditions more reasonably by considering the periodic relations of temperature, however, the temperature or heat flux on the fluid-solid interface was fixed at constant values [29–31]. Clearly this is physically not true: the thermal condition (temperature and heat flux) at the microscopic fluid-solid interface is determined by the particular local flow and thermal situations and it cannot be specified in advance. The interfacial thermal resistance may also exist at the fluid-solid interface in porous materials [32] or the constituent interface in composite materials [33], and this feature has typically been neglected in previous studies. Furthermore, previous studies usually only considered situations with the flow along the temperature gradient direction [22,25,30], whereas in many experimental setups and industrial applications the temperature gradient is mainly in the orthogonal plane to the fluid flow [34–36].

In this paper, we study the convective heat transfer in porous materials using the recent generalized periodic boundary method for thermal flows by Jbeili and Zhang [28]. This boundary method was established with rigorous mathematical justifications and validated by carefully designed numerical tests. Using this method, extensive simulations are conducted based on a two-dimensional (2D) porous model of circular solid particles. Unlike previous studies usually focusing on the porosity and Reynolds number effects, we also investigate the influences of the unit aspect ratio and the interface conductance on the macroscopic thermal performances of porous flows. Our results show that, even with the same porosity and Reynolds number, the unit aspect ratio can greatly affect the apparent thermal conductivity at the microscopic level. It is also interesting to observe that a weaker interfacial conductance can reduce the tortuosity conductivity but increase the dispersion conductivity, and that the dispersion conductivity can respond to the porosity change in a non-monotonic fashion. In addition, qualitative discussions are provided to relate the macroscopic conductivity coefficients to the microscopic flow and thermal fields for a better understanding of the complex nature of porous thermal flows.

2. Governing Equations and Boundary Conditions for Periodic Unit

For simplicity and clarity, in this study we consider the 2D porous model with circular solid domains as shown in Figure 1a. The fluid flows vertically from the bottom to the top, while the thermal gradient is imposed in the horizontal direction. The governing equations for this thermal flow system include the continuity equation, Equation (1), and the Navier–Stokes equation Equation (2) for the fluid domain Ω_0:

$$\nabla \cdot \mathbf{u} = 0, \tag{1}$$

$$\rho\left(\frac{\partial \mathbf{u}}{\partial t} + \mathbf{u} \cdot \nabla \mathbf{u}\right) = -\nabla p + \mu \nabla^2 \mathbf{u} + \mathbf{f}; \tag{2}$$

and the energy equation

$$\frac{\partial T_f}{\partial t} + \mathbf{u} \cdot \nabla T_f = \alpha_f \nabla^2 T_f, \quad \text{in } \Omega_0,$$

$$\frac{\partial T_s}{\partial t} = \alpha_s \nabla^2 T_s, \quad \text{in } \Omega_1, \tag{3}$$

for the fluid (Ω_0) and solid (Ω_1) domains, respectively [37]. In these equations, ρ represents the fluid density, $\mathbf{u}$ is the fluid velocity, p is the pressure, μ is the dynamic viscosity of fluid, $\mathbf{f}$ is the body force on the fluid, T is the temperature, and α is the thermal diffusivity. The subscripts f and s for T and α denote the fluid and solid domains respectively. On the fluid-solid interface Γ, the no-slip boundary condition is applied for the fluid flow

$$\mathbf{u} = \mathbf{0}, \quad \text{on } \Gamma, \tag{4}$$

to incorporate the possible thermal resistance and therefore the temperature discontinuity at the solid–fluid interface Γ, a general conjugate condition for temperature is given as [38–40]:

$$q = C\left(T_f - T_s\right) = K_s \frac{\partial T_s}{\partial n} = K_f \frac{\partial T_f}{\partial n}, \quad \text{on } \Gamma. \tag{5}$$

Here, q is the heat flux along the local normal direction $\mathbf{n}$, which points from Domain Ω_1 toward Domain Ω_0 (Figure 1). C is the interface conductance, and K_f and K_s are the thermal conductivities of the fluid and solid substances, respectively.

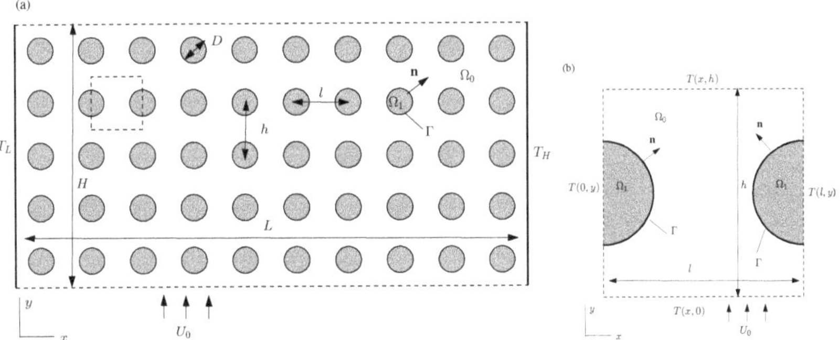

Figure 1. Schematic representations of the 2D porous model used in this study (**a**) and one periodic unit considered in pore-scale simulations (the dashed box in (**a**), and with more details in (**b**). The key parameters involved in the model description are also labeled, and more details can be found in the text.

In the pore-scale approach, only one periodic unit of the porous material is considered in the simulation (Figure 1). To solve the governing equations in this periodic unit, appropriate boundary conditions are required for the side boundaries of the periodic unit. These side boundaries are virtual but not real surfaces, and thus physical constraints on such boundaries are not directly available. For flows in such periodic structures with negligible fluid property changes, it has been well recognized that the flow field would be identical in each periodic unit [21,41–43]:

$$\mathbf{u}(x + ml, y + nh) = \mathbf{u}(x, y), \tag{6}$$

where l and h are the unit length and height (Figure 1), and m and n are arbitrary integers. Based on previous practices in periodic thermal flow simulations [28,41–45], the following relationship is proposed for the temperature field among periodic units in Figure 1a:

$$T(x + ml, y + nh) = T(x, y) + mlT_g'. \tag{7}$$

Here, T_g' is the global thermal gradient in the horizontal direction, and it can be obtained from the temperature difference $T_H - T_L$ and the distance L between the locations where these two temperatures T_L and T_H are imposed (see Figure 1a) as $T_g' = (T_H - T_L)/L$.

Assume $\mathbf{u}(x, y)$ and $T(x, y)$ are the correct flow and temperature solutions in one periodic unit ($0 \leq x \leq l$ and $0 \leq y \leq h$), it can be readily shown that the velocity from Equation (6) and temperature from Equation (7) satisfy all the governing equations and boundary conditions given in Equations (1)–(5), and thus they are the correct solutions for the unit of $ml \leq x \leq (m+1)l$ and $nh \leq y \leq (n+1)h$. Therefore, in pore-scale simulations, one can simulate the flow and thermal fields in one periodic unit as shown in Figure 1b, and the solutions can be extended to other units according to Equations (6) and (7). These relations can then be used to establish correct boundary conditions for the side surfaces of the periodic unit. For example, the right boundary $x = l$ for the unit shown in Figure 1b actually is also the left boundary of the next unit on its right. According to Equations (6) and (7) with $m = 1$ and $n = 0$, we have the left-right boundary relations for the periodic unit as:

$$\mathbf{u}(l, y) = \mathbf{u}(0, y), \quad T(l, y) = T(0, y) + lT_g'. \tag{8}$$

Similarly, the top-bottom boundary relations for flow and temperature are expressed as:

$$\mathbf{u}(x, h) = \mathbf{u}(x, 0), \quad T(x, h) = T(x, 0). \tag{9}$$

These relations are similar to those used by Kuwahara et al. [21]; however, no mathematical justifications and numerical validations were provided there. In addition to these boundary relations, due to the linearity and homogeneity of the governing equations and boundary conditions, extra anchoring conditions are necessary to ensure the numerical convergence [39,41,42]. For our current system in Figure 1a, the following conditions are adopted in our next simulations:

$$\frac{1}{l} \int_0^l u(x, 0)dx = U_0, \quad \frac{1}{h} \int_0^h T(0, y)dy = T_0; \tag{10}$$

where U_0 is the mean velocity through the porous medium and T_0 is the mean temperature at the left unit boundary. The mean velocity U_0 is related to the Reynolds number Re of the system (to be defined later), while the mean temperature T_0 can be set at an arbitrary value and it has no impact on the result analysis and interpretation.

3. Numerical Validation of the Periodic Relations among Units

To further verify the flow and temperature relations among units given in Equations (6) and (7), we conduct a direct numerical simulation for the system in Figure 1a. This simulation is also helpful to illustrate our concerns with the periodic unit boundary treatments used in previous studies. The diameter of the solid domains is $D = 50$ and they are arranged as a square array with $l = h = 2D$. The porosity of this model medium is thus $\epsilon = \pi/16$. The rectangular simulation domain has a length of $L = 20D$ in the horizontal direction and a hight of $H = 10D$ in the vertical direction, and thus the simulation domain consists $10 \times 5 = 50$ identical periodic units as shown in Figure 1b. Constant temperatures are imposed on the left and right boundaries: $T_L = 0$ at $x = 0$ and $T_H = 1$ at $x = L$, resulting a global thermal gradient of $T_g' = (T_H - T_L)/L = 10^{-3}$. The classical periodic condition is imposed on the top and bottom boundaries for both flow velocity and temperature, and on the left and right boundaries for the fluid flow:

$\mathbf{u}(0,y) = \mathbf{u}(L,y)$, $\mathbf{u}(x,0) = \mathbf{u}(x,H)$, and $T(x,0) = T(x,H)$. Please note that the temperature relation Equation (7) is not involved in this validation simulation. A body force $\mathbf{f}$ in the upward direction is utilized to generate the flow, and its magnitude is adjusted to obtain the Reynolds number $Re = \rho U_0 \sqrt{lh}/\mu = 50$. Here we use the geometric mean value of the periodic unit length l and height h for the Reynolds number Re, and this choice is made for our convenience in examining the effect of the aspect ratio $\beta = h/l$ in Section 4.1. The Prandtl number of the fluid is 0.7. The thermal conductivities are set as $K_f = 0.2$ for fluid and $K_s = 10K_f = 2$ for solid. A relatively large interface conductance value $C = 10$ is adopted to minimize the temperature discontinuity across the interface. The lattice Boltzmann method (LBM) with the D2Q9 (2D and with nine lattice velocities) lattice structure is used to solve the flow and thermal fields [44,46–48] for all calculations in this paper, and the counter-extrapolation method [49] is adopted for the conjugate thermal condition on domain interfaces. As in general LBM studies [16,22,31,48–50], all quantities provided above and in the later result presentation are non-dimensional based on LBM simulation units (for example, length in the lattice grid resolution, and time in the simulation time step).

The calculated flow and temperature fields from this direct simulation are displayed in Figure 2. The flow pattern in Figure 2a appears identical in each periodic unit. For the temperature in Figure 2b, it can be seen that the temperature increases in general along the horizontal direction, agreeing with the global thermal gradient generated from the two boundary temperatures $T_L = 0$ at the left and $T_H = 1$ at the right. In Figure 2b, neither the temperature nor the heat flux are constant along the solid–fluid interface (edges of the circular patches), meaning that the constant temperature or flux assumptions for the interfaces in previous studies [22,30,31] are not valid. The wider gaps between isotherm contours in the solid domain imply smaller temperature gradients there, and this is consistent with the larger solid conductivity used in this simulation. With a relatively larger conductance value for the solid–fluid interface, no apparent discontinuity in temperature can be observed across the interfaces.

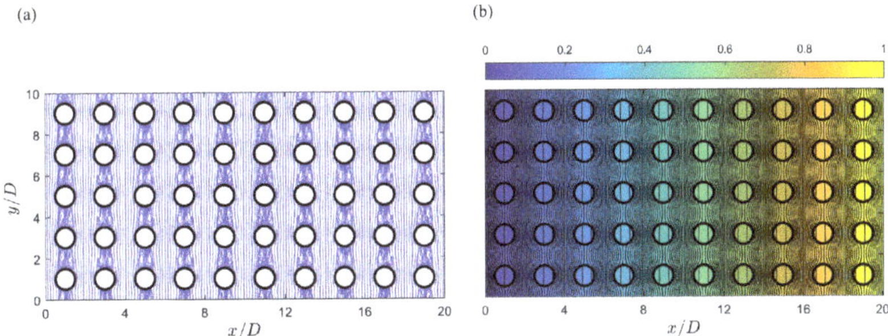

Figure 2. Flow (a) and temperature (b) distributions obtained from the direct simulation for the 2D porous model system in Figure 1a.

The apparently linear increase in temperature and the similarity in local isotherm contours in Figure 2b appear rational according to the temperature relation Equation (7). Moreover, $T \sim y$ profiles at six horizontal locations are plotted in Figure 3, one in each panel from the left and with the x positions labeled on top. These horizontal positions are selected at the same relative locations to the nearby solid columns ($0.4D$ to the left of the patch centers). The flat segments on these curves occur in the solid domains, and they are due to the vertical isotherm lines there. The strong variations in these profiles clearly show that in general one should not assign constant temperatures on the periodic unit boundaries, as done in [24–26]. According to Equation (7), these temperature profiles should have the same variation features along the vertical direction, except different offset values. This is apparently true by looking at these individual profiles. For a more quantitative confirmation, we then shift each profile by its mean temperature T_m, and the six shifted profiles are plotted in the last panel on the right in Figure 3. These six shifted profiles overlap with each other perfectly and we only see one single curve there. These identical shifted profiles precisely confirm that the temperature values at same relative positions in individual periodic units are only different in respective mean temperatures and the variation fashion is identical. Furthermore, the mean temperature values calculated along these six profiles are listed in Table 1. According to Equation (7), these mean temperatures can be related to the mean value at $x/D = 0.6$ by

$$T_m(x_i) = T_m(x = 0.6D) + (x_i - 0.6D)T'_g, \tag{11}$$

with x_i = 0.6D, 2.6D, 6.6D, 10.6D, 14.6D and 18.6D for the six profiles in Figure 3 (from left to right). The predicted values from this equation are also provided in Table 1 for comparison, and the excellent agreement convincingly confirms the validity of the temperature relation Equation (7) for such porous thermal flows.

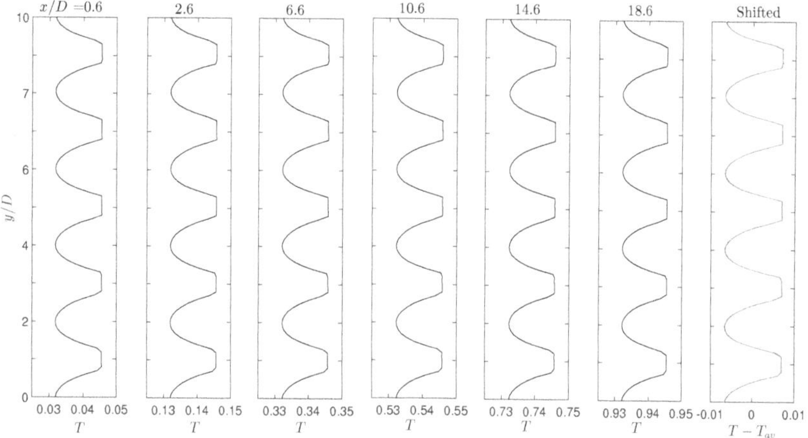

Figure 3. Temperature profiles along the vertical direction at different horizontal positions (labeled on top) from the direct simulation results in Figure 2b. The last panel on the right collects all the temperature profiles in other panels, however shifted by their individual mean temperatures. These six shifted profiles become identical and completely overlap each other, confirming the thermal relation Equation (7) among periodic units along the horizontal direction.

Table 1. Comparison of the mean temperature values at different horizontal positions obtained from the simulation and predicted by Equation (11).

x/D	Calculated from Simulation Results	Predicted from Equation (11)
0.6	0.03847	-
2.6	0.13857	0.13847
6.6	0.33872	0.33847
10.6	0.53881	0.53847
14.6	0.73878	0.73847
18.6	0.93860	0.93847

4. Simulation Results and Discussion

The enhanced heat transfer performance in porous flow systems, in principle, benefits from the large solid–fluid contact area and the long, twisted flow path as the fluid passes through the microscopic porous structure. At the macroscopic, practical level, the apparent thermal conductivity is often used to quantify the overall thermal behaviors. The volume averaging analysis [21,51–53] can be applied to obtain the effective conductive tensor from the microscopic flow and thermal fields. Kuwahara et al. [21] further decomposed this conductivity tensor $\mathbf{K}$ into three parts

$$\mathbf{K} = K_e\mathbf{I} + \mathbf{K}_{tor} + \mathbf{K}_{dis}, \tag{12}$$

where K_e is the stagnant conductivity, which is simply the volume average of the fluid and solid conductivities based on the porosity ϵ:

$$K_e = \epsilon K_f + (1 - \epsilon)K_s. \tag{13}$$

Please note that the expression for K_e in Equation (13) is simply the coefficient for the isotropic part of the conductivity matrix $\mathbf{K}$ from the volume average analysis [21,51–53]; and that there are no assumptions involved on the microscopic porous structures. $\mathbf{K}_{tor}$ and $\mathbf{K}_{dis}$ are called the tortuosity and dispersion conductivity tensors, respectively. When the mean flow direction is along one of the coordinate directions, only diagonal elements are nonzero; and obviously the primary concerns are the diagonal elements in the global thermal gradient direction [21,53,54]. For the setup in Figure 1, we simply use K_{tor} and K_{dis} to denote the xx components of tensors $\mathbf{K}_{tor}$ and $\mathbf{K}_{dis}$; and they can be calculated from the simulated flow and temperature distributions in one periodic unit by [21,53]

$$K_{tor} = -\frac{K_s - K_f}{hlT_g'} \int_\Gamma n_x T_f d\Gamma, \tag{14}$$

$$K_{dis} = -\frac{\rho c_{pf}}{hlT_g'} \int_{\Omega_0} (T - \bar{T})(\mathbf{u} - \bar{\mathbf{u}}_f)_x d\Omega; \tag{15}$$

where c_{pf} is the fluid specific heat, and the volume average temperature $\bar{T}$ and the intrinsic average velocity $\bar{\mathbf{u}}_f$ are given by:

$$\bar{T} = \frac{1}{lh}\left(\int_{\Omega_0} T_f d\Omega + \int_{\Omega_1} T_s d\Omega\right), \quad \bar{\mathbf{u}}_f = \frac{1}{\epsilon lh} \int_{\Omega_0} \mathbf{u} d\Omega. \tag{16}$$

The subscript x denotes the x component of the corresponding vector.

In this section, we apply the boundary conditions in Section 2 to the periodic unit shown in Figure 1b, and examine the effects on the effective macroscopic conductivity coefficients K_{tor} and K_{dis} from several microscopic parameters, including the aspect ratio $\beta = h/l$, the Reynolds number $Re = \rho\sqrt{hl}U_0/\mu$, the interfacial thermal conductance C, and the porosity $\epsilon = 1 - \pi D^2/4hl$. We keep the unit area $lh = 90,000$ constant in our next simulations, so that the Reynolds number Re is directly proportional to the mean flow velocity U_0, and the porosity ϵ depends purely on the solid diameter D and not affected by the aspect ratio β. The mean temperature at $x = 0$ is set as $T_0 = -0.05$ and the global thermal gradient is $T_g' = 0.1/l$. Please note that, due to the linearity and homogeneity of the energy equation Equation (3) and the associated thermal boundary conditions, the particular values of T_0 and T_g' used in the simulations do not affect the calculated conductivities K_{tor} and K_{dis} from Equations (14) and (15). To make this point clear, we consider a periodic unit under two situations: Situation (a) with mean inlet temperature $T_{0,a}$ and thermal gradient $T_{g,a}'$; and Situation (b) with mean inlet temperature $T_{0,b}$ and thermal gradient $T_{g,b}'$. If $T_a(x,y)$ is the solution in the periodic unit with Situation (a), the temperature field for Situation (b) should be $T_b(x,y) = T_{0,b} + \frac{T_{g,b}'}{T_{g,a}'}[T_a(x,y) - T_{0,a}]$. Clearly, $T_b(x,y)$ satisfies the energy equation Equation (3) and all temperature conditions in Equations (5), (8), (9), (10), . Substituting this expression of $T_b(x,y)$ in Equations (14) and (15) and considering $\int_\Gamma u_x d\Gamma = 0$ for periodic structures, one can find that the K_{tor} and K_{dis} values remain the same for Situations (a) and (b). Other simulation parameters are kept the same as given in Section 3, unless otherwise mentioned. As for the initial state, we start simulations with a linear transition from T_0 at the inlet to $T_0 + T'_g l$ at the outlet for the temperature, and zero velocity for the flow. The results are taken when steady or quasi-steady states are established. The computer code for this work has been developed based on our programs used in previous publications [28,38,44,49,55,56], and all important elements involved in our simulations, including the LBM algorithms for flow and heat transfer, the no-slip boundary and conjugate interface treatments, as well as the mesh resolution selection, have been validated and confirmed in these previous studies.

4.1. Effects of the Aspect Ratio β and Reynolds Number Re

We start with simulations to study the effect of the aspect ratio β on the macroscopic thermal conductivity, which has not been well addressed in previous investigations. Here we set the porosity $\epsilon = 0.85$, and accordingly the solid domain diameter is $D = 131.11$. Following previous studies [29,57,58], two representative Reynolds number values, $Re = 50$ and 100, are considered in this section. Higher Reynolds numbers are possible for gas flows through porous media [59,60]. Our simulations cover a range of 0.25~4 for the aspect ratio β; further increasing (>4) or decreasing (<0.25) of the β value will make the gap between the solid surfaces too small for accuracy and stable computations. The flows are always steady for all tested β values at $Re = 50$ and for $\beta \leq 1$ at $Re = 100$; however, the flow becomes unsteady for $\beta > 1$ and $Re = 100$. This transition is reasonable, since for a larger β, the gap between the solid surfaces becomes narrower, and the fluid velocity through this gap increases accordingly. This situation is similar to a flow jet coming out from a small opening shooting into a relatively large space. Compared to the condition with the same mean flow velocity (same Reynolds number) but a smaller aspect ratio β, the reduction in flow passage width is less significant (a wider gap between solid surfaces) and the fluid space after the gap is relatively limited. Figure 4 shows the streamlines and temperature distributions for two representative aspect ratios, $\beta = 0.44$ and 1.44 at $Re = 50$ and 100. For steady flows in Figure 4a–c, we see smooth and symmetric streamlines and isotherm contours. On the other hand, when the flow become unsteady in Figure 4d, these lines are distorted and less organized. Both for the steady and unsteady cases, thermal features discussed in Section 3 are noticed as well, including the relatively uniform temperature distributions inside the solid domains (due to the high solid conductivity K_s) and the smooth temperature transition across the solid–fluid interface (due to the large interface conductance C). Moreover, we see significant temperature variations along the unit boundaries for the unsteady system in Figure 4d. The variation fashions are similar on opposite edges. More specific, the temperature variations along the top and bottom boundaries appear identical, whereas the temperature is low on the left and high on the right. All these observations are consistent with the thermal boundary conditions in Equations (8) and (9).

For unsteady flow situations, the mean flow velocity U_0 and Reynolds number Re both vary with time, and it is difficult to achieve the time-averaged Reynolds number exactly at 100. For such situations, we accept the simulation results when the Re variation is limited in the range of 95 $\sim$ 105. The unsteady flow certainly affects the thermal field, and accordingly, the tortuosity and dispersion conductivities K_{tor} and K_{dis} calculated from Equations (14) and (15) also vary with time. Figure 5 displays the temporal oscillations of Re, K_{tor} and K_{dis} for $Re = 100$ and $\beta = 1.44$. The oscillation periods for K_{tor} and K_{dis} appear to be identical, as twice that for Re. In our next analysis, we use the average values over a variation period for the tortuosity and dispersion conductivities K_{tor} and K_{dis}.

Figure 6 plots the apparent conductivities K_{tor} and K_{dis} changing with the aspect ratio β and Reynolds number Re. Clearly, even under the same porosity $\epsilon = 0.85$, the macroscopic conductivities can be greatly affected by the aspect ratio β, and the influence can be enhanced with a higher Reynolds number Re. The tortuosity conductivity K_{tor} decreases with β in Figure 6a, which can be explained by looking at the K_{tor} definition Equation (14) and the temperature distributions in Figure 4. Equation (14) can be interpreted as a weighted sum of the fluid temperature over the solid–fluid interface Γ, with n_x, the x-component of the normal vector **n**, as the weighting factor. For the system considered here, the left surface has $n_x > 0$ and the right surface has $n_x < 0$. On each semicircular surface, the portion around the horizontal centerline (the surface is approximately aligned in the vertical direction and thus it has a larger $|n_x|$ value) plays a more determinant role than the portions near the unit edges (the surface is approximately aligned in the horizontal and thus $|n_x| \approx 0$). Therefore, by comparing the temperatures on the parts around the horizontal centerline of the two semicircular surfaces in Figure 4, we can have a qualitative understanding of the K_{tor} dependence on β: For a large β, the two semicircular surfaces are closer and the temperature difference on the closest portions becomes smaller, and this smaller temperature difference results in a smaller tortuosity conductivity, as observed in Figure 6a. The Reynolds number Re appears to be less influential on K_{tor}, except for the highly unsteady case with $\beta = 4$. The weak influence of Re on K_{tor} is consistent with that observed in Ref. [21].

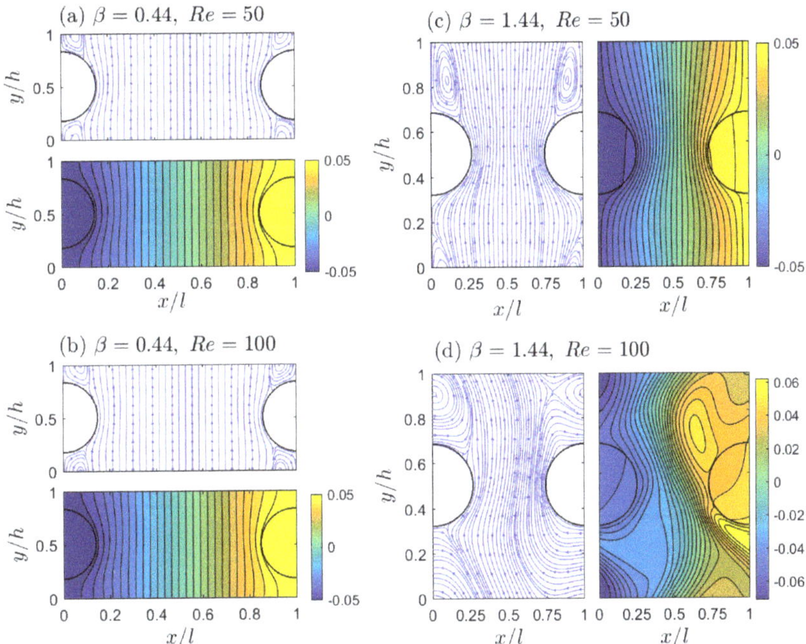

Figure 4. Simulation results of the flow streamlines and temperature distributions for $\beta = 0.44$ (**a**,**b**) and $\beta = 1.44$ (**c**,**d**) at $Re = 50$ (**a**,**c**) and 100 (**b**,**d**).

Unlike the tortuosity conductivity K_{tor}, the dispersion conductivity K_{dis} increases by orders with β in Figure 6b. Similarly, we attempt to interpret this observation by examining the K_{dis} definition Equation (15) and flow and thermal fields in Figure 4. Clearly, Equation (15) represents the disorder level of the flow and thermal distributions in the periodic unit, since the integrand is simply the product of the temperature fluctuation $T - \bar{T}$ and the flow fluctuation $(\mathbf{u} - \bar{\mathbf{u}}_f)_x$. Therefore, the more disordered the flow and temperature distributions in the periodic unit, the larger K_{dis} value will be obtained. This analysis is consistent with our general understanding of the thermal dispersion process and it explains the increasing trend of K_{dis} with β and Re in Figure 6b as well. The similarity in flow and thermal fields between $Re = 50$ and 100 for $\beta \leq 1$ steady systems suggests that K_{dis} does not change much with Re there; however, for $\beta > 1$, the $Re = 100$ systems become unsteady and the flow and temperature distributions become significantly disordered, resulting in an abrupt increase in K_{dis}. We also notice that only the x-component of the flow fluctuation is involved in Equation (15). For the steady flows at $\beta \leq 1$, the nonzero x fluid velocity is limited in the small circulation areas near the four unit corners, whereas for large β values, the circulation region is large. The high Reynolds number $Re = 100$ is also helpful in increasing the circulation size, and it can even make the x velocity component nonzero over almost the entire fluid domain for the unsteady flows with $\beta > 1$. This is the reason that we see the K_{dis} increase with β and Re by orders for $\beta > 1$ in Figure 6b.

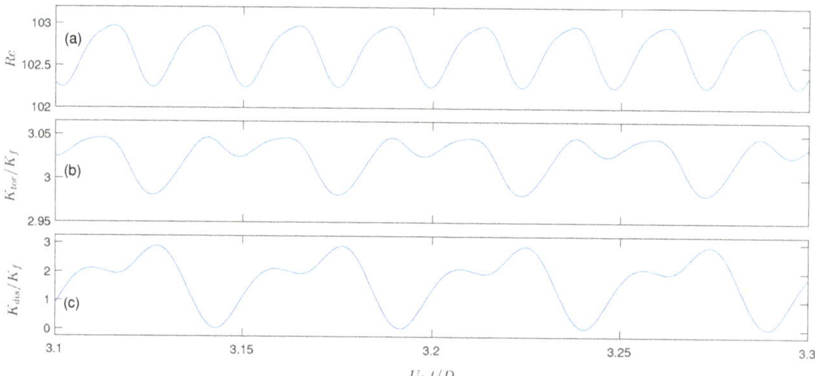

Figure 5. Variations of Re (**a**), K_{tor} (**b**) and K_{dis} (**c**) with time for the unsteady system with $\beta = 1.44$ and $Re \approx 100$. The simulation time is normalized based on the mean velocity U_0 and solid cylinder diameter D.

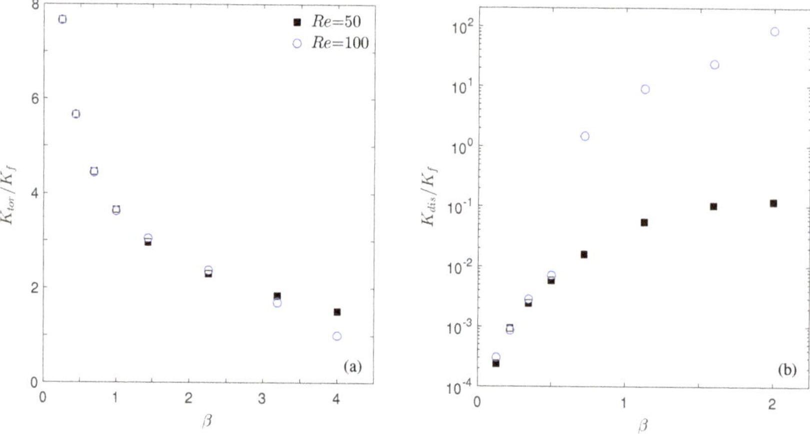

Figure 6. Plots of K_{tor} (**a**) and K_{dis} (**b**) changing with the aspect ratio β at two Reynolds numbers $Re = 50$ (black squares) and 100 (blue circles).

4.2. Effect of the Interfacial Conductance C

Next we shift our attention to the influence of the interfacial thermal conductance C on the apparent conductivities. Again this is an important topic [32,33], however, it has not been investigated adequately. In this part, we fix the porosity $\epsilon = 0.85$ and the Reynolds number $Re = 50$ for simplicity, and test three interfacial conductance values $C = 10$ (virtually no thermal resistance at the solid–fluid interface as observed in Figures 2b and 4), 5×10^{-4}, and 5×10^{-5}. Figure 7 displays the flow and thermal distributions at these three conductance values with two aspect ratios $\beta = 0.44$ and 1.44; and Figure 8 collects the calculated K_{tor} and K_{dis} conductivities in our simulations. The analysis in the previous section on the relationships between macroscopic conductivities and microscopic flow and thermal situations can still be applied here. Please note the change in interfacial conductance C does not affect the flow field and thus we can focus on the temperature response to different C values in Figure 7. As conductance C decreases, the solid domains become more insulated, resulting in a nearly constant temperature in each solid patch (i.e., the same colors and no isotherm contours in solid domains). With the solid part being insulated and thus its high-conductivity influence reduced, the fluid temperature around the solid domains exhibits a faster change along the thermal gradient direction. For example, the fluid temperatures near the solid surfaces at the narrowest gap location in Figure 7a are -0.0471 (left) vs. 0.0473 (right) for $C = 10$, -0.0364 vs. 0.0366 for $C = 5 \times 10^{-4}$, and -0.0251 vs. 0.0254 for $C = 5 \times 10^{-5}$. As discussed in the previous section, the large temperature difference at $C = 10$ introduces a larger tortuosity conductivity K_{tor}, and vise versa, as shown in Figure 8a. On the other hand, the dispersion conductivity K_{dis} increases as we reduce the interfacial conductance C, and the C influence becomes negligible for large aspect ratios. This change should be attributed to the temperature change in circulation areas; however, due to the complexity in flow structures and temperature changes with C in Figure 7, we are not able to provide a direct qualitative explanation here. Overall, the influence of the interface conductance C on the apparent conductivities is less dramatic in Figure 8 than that from the unit aspect ratio β in Figure 6, but it cannot be neglected for accurate analysis.

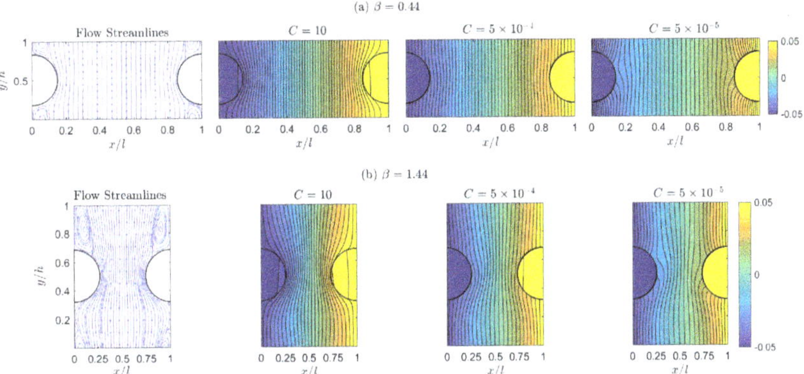

Figure 7. Simulation results of the flow streamlines (first column) and temperature distributions for different interfacial conductances $C = 10$ (second column), 5×10^{-4} (third column), and 5×10^{-5} (last column) with the aspect ratios $\beta = 0.44$ (**(a)** in the top row) and 1.44 (**(b)** in the bottom row) at $Re = 50$.

4.3. Effect of the Porosity ϵ

Unlike the effects of aspect ratio β and the interfacial conductance C, extensive studies have been conducted for the influence of the porosity ϵ on the macroscopic thermal behaviors of porous materials [21,61,62]. As mentioned in Section 1, previous pore-scale simulations usually set isothermal conditions on a pair of unit boundaries to create the thermal gradient and applied adiabatic conditions on other unit side surfaces [24,27]. Such artificial, unphysical treatments interfere with the natural heat transfer processes among microscopic periodic units, and the results from such boundary methods could be inaccurate or misleading [28]. In this section, we use the generalized periodic condition Equation (7) to examine the porosity effect on macroscopic conductivities K_{tor} and K_{dis}. Our next simulations cover a range of porosity of $\epsilon = 0.5\sim0.99$ for aspect ratios $\beta = 0.69$ and 1, and $\epsilon = 0.65\sim0.99$ for and $\beta = 1.44$. The Reynolds number is set at $Re = 50$ and the interfacial conductance is fixed at $C = 10$. Further reducing ϵ will increase the solid cylinder diameter D and thus make the gap between the solid surfaces very narrow (Figure 9). As discussed in Section 4.1, such a narrow gap may turn the flow unsteady for the large aspect ratio $\beta = 1.44$, and also a finer spacial resolution is required to accurately capture the flow and thermal variations in the gap regions.

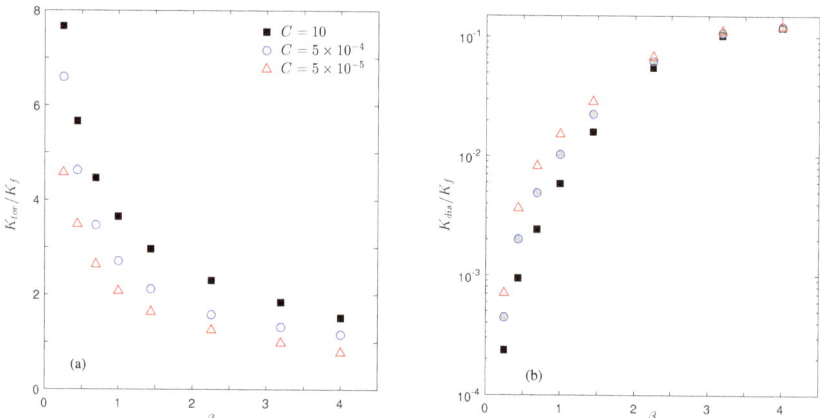

Figure 8. Plots of K_{tor} (**a**) and K_{dis} (**b**) changing with the aspect ratio β at three interfacial conductance values: $C = 10$ (black squares), 5×10^{-4} (blue circles), and 5×10^{-5} (red triangles). The Reynolds number is $Re = 50$.

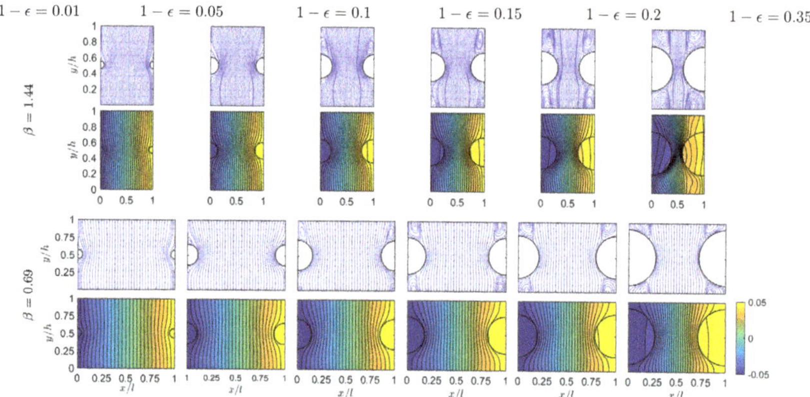

Figure 9. Simulation results of the flow streamlines (first and third rows) and temperature distributions (second and last rows) for different porosity values (labeled on top of each column) with interfacial conductance $C = 10$ and Reynolds number $Re = 50$.

Results from this set of simulations are collected in Figure 9 for the flow and temperature distributions and Figure 10 for the calculated conductivities K_{tor} and K_{dis} at different porosity values. Please note that these figures are plotted in terms of the solid fraction $1 - \epsilon$ as in Ref. [21]. With the solid diameter D increase (from left to right in Figure 9), a pair of circulation vortices are developed in the wake region above the solid cylinder, and the vortex size grows gradually till it completely fills the vertical space between two cylinders. Meanwhile, the original relatively organized temperature field is gradually distorted by the increasing size of the solid domain. Such changes in the flow and thermal patterns therefore affect the macroscopic thermal behaviors as characterized by the tortuosity (K_{tor}) and dispersion (K_{dis}) conductivities (Figure 10). The continuous increase in K_{tor} with the solid fraction $1 - \epsilon$ is mainly because of the larger area (length in our 2D system) of the solid–fluid interface Γ (see Equation (14)). This trend is similar to that observed in Ref. [21]; however, the K_{tor} magnitude is smaller here. This can be attributed to the different solid domain shapes: For the same porosity, compared to the circular shape in this work, the square solid domain shape in Ref. [21] yields a longer interface length, and more profoundly, half of the interface has the local normal direction aligned the x direction. All these are favorable for a larger K_{tor} according to its definition in Equation (14). As for the dispersion conductivity K_{dis} in Figure 10b, in general, K_{dis} increases with the solid fraction $1 - \epsilon$; however, unlike the monotonic growth in Ref. [21], local maximum and minimum states are observed here. Due to the complexity in flow and temperature fields as well as their involvement in the K_{dis} calculation, it is difficult to provide detailed insights and mechanism for the K_{dis} variations. Nevertheless, this could be an interesting topic to explore in the future. The general increasing trend can be qualitatively related to the increasing size of the circulation region, which is the major contributor to K_{dis} via the large x velocity component in this region. A similar analysis can be applied for the gentle variation and slow recovery of K_{dis} for $1 - \epsilon = 0.2 \sim 0.5$ at $\beta = 0.69$ in Figure 10b, since the flow pattern remains almost unchanged in these systems (see the subplots for $1 - \epsilon = 0.2$ and 0.35 at $\beta = 0.69$ in Figure 9). Finally, for the three curves with different aspect ratios, we see K_{tor} is smaller but K_{dis} is larger for a higher aspect ratio β. This agrees well with our findings and discussions for the aspect ratio effect on conductivity in Section 4.1 (see Figure 6).

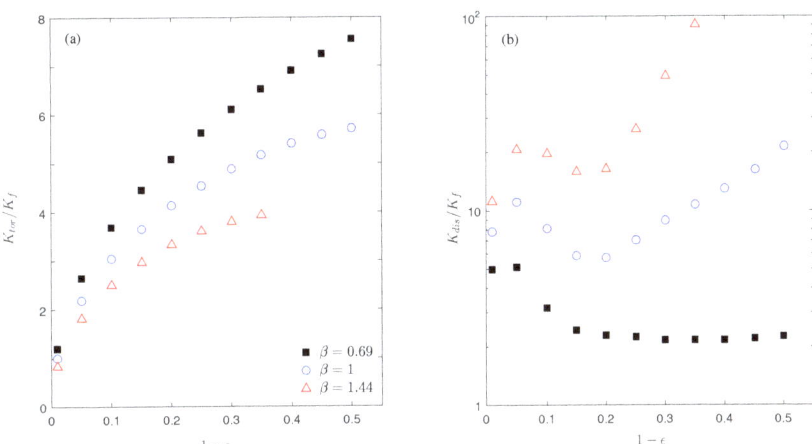

Figure 10. Plots of K_{tor} (**a**) and K_{dis} (**b**) changing with the solid fraction $1 - \epsilon$ at three aspect ratio values: $\beta = 0.69$ (black squares), 1 (blue circles), and 1.44 (red triangles). The Reynolds number is $Re = 50$.

5. Summary

In this study, we have first justified and validated the generalized periodic condition [21,28] for pore-scale simulations of thermal flows in porous media. Extensive simulations have then been carried out using a 2D porous model to investigate the influence of several microscopic parameters on macroscopic thermal conductivity. Among them, the effects of the aspect ratio of the periodic unit and the thermal conductance at the pore surface have not been addressed adequately in previous studies. Our results show that these microscopic properties can dramatically change the flow and thermal fields in the microscopic porous structure, and affect the apparent thermal performances of the porous materials at the macroscopic level. Therefore, these microscopic factors need to be considered carefully for more accurate and reliable simulation results, which are crucial for both fundamental research and practical applications. In addition, thorough discussions are attempted to qualitatively explore the relationship between the apparent conductivity at the macroscopic level and the complex thermal flow situations in the microscopic porous structure, and our analysis and findings could be helpful for a better understanding of the underlying thermal processes.

We are aware that several serious limitations exist for this study and one should not over interpret the results obtained from a specific system. The simple 2D porous model and the relatively low Reynolds numbers considered here may appear less realistic for practical systems; however, they are helpful for understanding the micro–macro relations and fundamental mechanisms involved in the complex thermal flow processes in porous materials. We have fixed the solid and fluid substance properties (conductivity, diffusivity, and the Prandtl number), which can certainly affect the apparent thermal behaviors as well. The 2D unit geometry adopted in this work is also symmetric along both the flow and thermal gradient directions; and the anisotropic effects could be an interesting topic for future research. The periodic conditions Equations (8) and (9) utilized in these calculations require the flow and thermal fields to be fully developed and thus they are not applicable to the entrance and development regions [63,64]. Moreover, in our simulations, we have assumed that the material properties are constant in the flow and heat transfer processes and thus steady or quasi-steady states can be established. In some situations, the microscopic porous structure may undergo a dynamic change due to swelling and erosion [65]; and caution must be taken for applying results from this study to such systems. Nevertheless, the boundary method, simulation results, and analysis discussions can be useful for the research and applications of porous thermal flows. The generalized periodic boundary condition, although presented in 2D, can be readily applied to three-dimensional pore-scale thermal flow simulations.

Author Contributions: Conceptualization, M.J. and J.Z.; methodology, M.J. and J.Z.; data curation, M.J.; formal analysis, M.J. and J.Z.; writing, M.J. and J.Z. All authors have read and agreed to the published version of the manuscript.

Funding: This work was supported by the Natural Science and Engineering Research Council of Canada (NSERC).

Institutional Review Board Statement: Not applicable.

Informed Consent Statement: Not applicable.

Acknowledgments: M.J. acknowledges the financial support from the Queen Elizabeth II Graduate Scholarship in Science and Technology (QEII-GSST) at Laurentian University. The calculations have been enabled by the use of computing resources provided by Compute Canada (www.computecanada.ca).

Conflicts of Interest: The authors declare no conflict of interest.

References

1. Ingham, D.B.; Pop, I. *Transport Phenomena in Porous Media III*; Elsevier: Amsterdam, The Netherlands, 2005.
2. Ottenhall, A.; Illergard, J.; Ek, M. Water Purification Using Functionalized Cellulosic Fibers with Nonleaching Bacteria Adsorbing Properties. *Environ. Sci. Technol.* **2017**, *51*, 7616–7623. [CrossRef]
3. Boules, D.; Sharqawy, M.H.; Ahmed, W.H. Enhancement of heat transfer from a horizontal cylinder wrapped with whole and segmented layers of metal foam. *Int. J. Heat Mass Transf.* **2021**, *165*, 120675. [CrossRef]
4. Bianco, N.; Busiello, S.; Iasiello, M.; Mauro, G.M. Finned heat sinks with phase change materials and metal foams: Pareto optimization to address cost and operation time. *Appl. Therm. Eng.* **2021**, *197*, 117436. [CrossRef]
5. Li, Y.; Gong, L.; Ding, B.; Xu, M.; Joshi, Y. Thermal management of power electronics with liquid cooled metal foam heat sink. *Int. J. Therm. Sci.* **2021**, *163*, 106796. [CrossRef]
6. Iasiello, M.; Mameli, M.; Filippeschi, S.; Bianco, S. Simulations of paraffine melting inside metal foams at different gravity levels with preliminary experimental validation. *J. Phys. Conf. Ser.* **2020**, *1599*, 012008. [CrossRef]
7. Fu, X.; Viskanta, R.; Gore, J. Measurement and correlation of volumetric heat transfer coeffcients of cellular ceramics. *Exp. Therm. Fluid Sci.* **1998**, *17*, 285–293.
8. Wang, G.; Zhang, Z.; Wang, R.; Zhu, Z. A Review on Heat Transfer of Nanofluids by Applied Electric Field or Magnetic Field. *Nanomaterials* **2020**, *10*, 2386. [CrossRef]
9. Chuan, L.; Wang, X.D.; Wang, T.H.; Yan, W.M. Fluid flow and heat transfer in microchannel heat sink based on porous fin design concept. *Int. Commun. Heat Mass Transf.* **2015**, *65*, 52–57. [CrossRef]

10. Behrang, A.; Taheri, S.; Kantzas, A. A hybrid approach on predicting the effective thermal conductivity of porous and nanoporous media. *Int. J. Heat Mass Transf.* **2016**, *98*, 52–59. [CrossRef]

11. Gong, L.; Wang, Y.; Cheng, X.; Zhang, R.; Zhang, H. A novel effective medium theory for modelling the thermal conductivity of porous materials. *Int. J. Heat Mass Transf.* **2014**, *68*, 295–298. [CrossRef]

12. Song, Y.S.; Youn, J.R. Evaluation of effective thermal conductivity for carbon nanotube/polymer composites using control volume finite element method. *Carbon* **2006**, *44*, 710–717. [CrossRef]

13. Ye, C.; Huang, H.; Fan, J.; Sum, W. Numerical Study of Heat and Moisture Transfer in Textile Materials by a Finite Volume Method. *Commun. Comput. Phys.* **2008**, *4*, 929–948.

14. Chiappini, D.; Festuccia, A.; Bella, G. Coupled lattice Boltzmann finite volume method for conjugate heat transfer in porous media. *Numer. Heat Transf. Part A Appl.* **2018**, *73*, 291–306. [CrossRef]

15. Guo, Z.; Zhao, T.S. A lattice Boltzmann model for convection heat transfer in porous media. *Numer. Heat Transf. Part B Fundam.* **2005**, *47*, 157–177. [CrossRef]

16. Liu, Q.; He, Y.L. Lattice Boltzmann simulations of convection heat transfer in porous media. *Phys. A Stat. Mech. Its Appl.* **2017**, *465*, 742–753. [CrossRef]

17. Azadi, P.; Farnood, R.; Yan, N. FEM-CDEM modeling of thermal conductivity of porous pigmented coatings. *Comput. Mater. Sci.* **2010**, *49*, 392–399. [CrossRef]

18. Duan, Y.; He, S.; Ganesan, P.; Gotts, J. Analysis of the horizontal flow in the advanced gas-cooled reactor. *Nucl. Eng. Des.* **2014**, *272*, 53–64. [CrossRef]

19. Zhou, Y.; Yan, C.; Tang, A.M.; Duan, C.; Shengshi, D. Mesoscopic prediction on the effective thermal conductivity of unsaturated clayey soils with double porosity system. *Int. J. Heat Mass Transf.* **2019**, *130*, 747–756. [CrossRef]

20. Huang, P.; Zhao, Y.; Niu, Y.; Ren, X.; Chang, M.; Sun, Y. Mesoscopic Finite Element Method of the Effective Thermal Conductivity of Concrete with Arbitrary Gradation. *Adv. Mater. Sci. Eng.* **2018**, *2018*, 2352864. [CrossRef]

21. Kuwahara, F.; Nakayama, A.; Koyama, H. A Numerical Study of Thermal Dispersion in Porous Media. *J. Heat Transf.* **1996**, *118*, 756–761. [CrossRef]

22. Yang, P.; Wen, Z.; Dou, R.; Liu, X. Effect of random structure on permeability and heat transfer characteristics for flow in 2D porous medium based on MRT lattice Boltzmann method. *Phys. Lett. A* **2016**, *380*, 2902–2911. [CrossRef]

23. Jeong, N.; Choi, D.H. Estimation of the thermal dispersion in a porous medium of complex structures using a lattice Boltzmann method. *Int. J. Heat Mass Transf.* **2011**, *54*, 4389–4399. [CrossRef]

24. Zhao, C.; Dai, L.; Tang, G.; Qu, Z.; Li, Z. Numerical study of natural convection in porous media (metals) using Lattice Boltzmann Methods (LBM). *Int. J. Heat Fluid Flow* **2010**, *31*, 925–1934. [CrossRef]

25. Liu, Z.; Wu, H. Pore-scale study on flow and heat transfer in 3D reconstructed porous media using micro-tomography images. *Appl. Therm. Eng.* **2016**, *100*, 602–610. [CrossRef]

26. Liu, Q.; He, Y.L.; Li, Q.; Tao, W.Q. A multiple-relaxation-time lattice Boltzmann model for convection heat transfer in porous media. *Int. J. Heat Mass Transf.* **2014**, *73*, 761–775. [CrossRef]

27. Ren, Q.; Chan, C.L. Natural convection with an array of solid obstacles in an enclosure by lattice Boltzmann method on a CUDA computation platform. *Int. J. Heat Mass Transf.* **2016**, *93*, 273–285. [CrossRef]

28. Jbeili, M.; Zhang, J. The Generalized Periodic Boundary Conditions for Microscopic Simulations of Heat Transfer in Heterogeneous Materials. *Int. J. Heat Mass Transf.* **2021**, *173*, 121200. [CrossRef]

29. Ozgumus, T.; Mobedi, M. Effect of Pore to Throat Size Ratio on Interfacial Heat Transfer Coefficient of Porous Media. *J. Heat Transf.* **2015**, *137*, 012602. [CrossRef]

30. Yang, P.; Wen, Z.; Dou, R.; Liu, X. Heat transfer characteristics in random porous media based on the 3D lattice Boltzmann method. *Int. J. Heat Mass Transf.* **2017**, *109*, 647–656. [CrossRef]

31. Grucelski, A.; Pozorski, J. Lattice Boltzmann simulations of heat transfer in flow past a cylinder and in simple porous media. *Int. J. Heat Mass Transf.* **2015**, *86*, 139–148. [CrossRef]

32. Smith, D.S.; Alzina, A.; Bourret, J.; Nait-Ali, B. Thermal conductivity of porous materials. *J. Mater. Res.* **2013**, *28*, 2260–2272. [CrossRef]

33. Fang, W.Z.; Gou, J.; Zhang, H.; Kang, Q.; Tao, W.Q. Numerical predictions of the effective thermal conductivity for needled C/C-SiC composite materials. *Numer. Heat Transf. Part A Appl.* **2016**, *70*, 1101–1117. [CrossRef]

34. Jiang, P.X.; Li, M.; Lu, T.J.; Yu, L.; Ren, Z.P. Experimental research on convection heat transfer in sintered porous plate channels. *Int. J. Heat Mass Transf.* **2004**, *47*, 2085–2096. [CrossRef]

35. Wang, S.L.; Li, X.Y.; Wang, X.D.; Lu, G. Flow and heat transfer characteristics in double-layered microchannel heat sinks with porous fins. *Int. Commun. Heat Mass Transf.* **2018**, *93*, 41–47. [CrossRef]

36. Lu, X.; Zhao, Y. Effect of flow regime on convective heat transfer in porous copper manufactured by lost carbonate sintering. *Int. J. Heat Fluid Flow* **2019**, *80*, 108482. [CrossRef]

37. Cengel, Y. *Heat and Mass Transfer: Fundamentals and Applications*; McGraw-Hill: Hoboken, NJ, USA, 2019.

38. Le, G.; Oulaid, O.; Zhang, J. Counter-extrapolation method for conjugate interfaces in computational heat and mass transfer. *Phys. Rev. E* **2015**, *91*, 033306. [CrossRef]

39. Jbeili, M.; Zhang, J. The Temperature Decomposition Method for Periodic Thermal Flows with Conjugate Heat Transfer. *Int. J. Heat Mass Transf.* **2020**, *150*, 119288. [CrossRef]

40. Guo, K.; L, L.; Xao, G.; AuYeung, N.; Mei, R. Lattice Boltzmann method for conjugate heat and mass transfer with interfacial jump conditions. *Int. J. Heat Mass Transf.* **2015**, *88*, 306–322. [CrossRef]
41. Patankar, S.V.; Liu, C.H.; Sparrow, E.M. Fully Developed Flow and Heat Transfer in Ducts Having Streamwise-Periodic Variations of Cross-Sectional Area. *ASME J. Heat Transf.* **1977**, *99*, 180–186. [CrossRef]
42. Stalio, E.; Piller, M. Direct Numerical Simulation of Heat Transfer in Converging-Diverging Wavy Channels. *J. Heat Transf.* **2007**, *129*, 769. [CrossRef]
43. Harikrishnan, S.; Tiwari, S. Unsteady Flow and Heat Transfer Characteristics of Primary and Secondary Corrugated Channels. *J. Heat Transf.* **2020**, *142*, 031803. [CrossRef]
44. Wang, Z.; Shang, H.; Zhang, J. Lattice Boltzmann simulations of heat transfer in fully developed periodic incompressible flows. *Phys. Rev. E* **2017**, *95*, 063309. [CrossRef]
45. Li, P.; Zhang, J. Simulating Heat Transfer through Periodic Structures with Different Wall Temperatures: A Temperature Decomposition Method. *ASME J. Heat Transf.* **2018**, *140*, 112002. [CrossRef]
46. He, X.; Chen, S.; Doolen, G.D. A Novel Thermal Model for the Lattice Boltzmann Method in Incompressible Limit. *J. Comput. Phys.* **1998**, *146*, 282–300. [CrossRef]
47. Guo, Z.; Shu, C. *Lattice Boltzmann Method and Its Applications in Engineering*; World Scientific Publishing: Singapore, 2013.
48. Zhang, J. Lattice Boltzmann method for microfluidics: Models and applications. *Microfluid. Nanofluid.* **2010**, *10*, 1–28. [CrossRef]
49. Wang, Z.; Colin, F.; Le, G.; Zhang, J. Counter-Extrapolation Method for Conjugate Heat and Mass Transfer with Interfacial Discontinuity. *Int. J. Numer. Methods Heat Fluid Flow* **2017**, *27*, 2231–2258. [CrossRef]
50. Succi, S. *The Lattice Boltzmann Equation*; Oxford University Press: Oxford, UK, 2001.
51. Whitaker, S. Diffusion and dispersion in porous media. *AIChE J.* **1967**, *13*, 420–427. [CrossRef]
52. Whitaker, S. Advances in theory of fluid motion in porous media. *Ind. Eng. Chem.* **1969**, *61*, 14–28. [CrossRef]
53. d'Hueppe, A. Heat Transfer Modeling at an Interface between a Porous Medium and a Free Region. Ph.D. Thesis, Ecole Centrale Paris, Paris, France, 2011.
54. de Lemos, M.J.S. *Turbulence in Porous Media: Modeling and Applications*; Elsevier: Amsterdam, The Netherlands, 2012.
55. Yin, X.; Zhang, J. An Improved Bounce-Back Scheme for Complex Boundary Conditions in Lattice Boltzmann Method. *J. Comput. Phys.* **2012**, *231*, 4295–4303. [CrossRef]
56. Chen, Q.; Zhang, X.; Zhang, J. Effects of Reynolds and Prandtl Numbers on Heat Transfer Around a Circular Cylinder by the Simplified Thermal Lattice Boltzmann Model. *Commun. Comput. Phys.* **2015**, *17*, 937–959. [CrossRef]
57. Dyga, R.; Placzek, M. Efficiency of heat transfer in heat exchangers with wire mesh packing. *Int. J. Heat Mass Transf.* **2010**, *53*, 5499–5508. [CrossRef]
58. Alshare, A.A.; Strykowski, P.J.; Simon, T.W. Modeling of unsteady and steady fluid flow, heat transfer and dispersion in porous media using unit cell scale. *Int. J. Heat Mass Transf.* **2010**, *53*, 2294–2310. [CrossRef]
59. He, J.; Zhang, J.; Shang, H. Two-Phase Dynamic Modelling and Simulation of Transport and Reaction in Catalytic Sulfur Dioxide Converters. *Ind. Eng. Chem. Res.* **2019**, *58*, 10963–10974. [CrossRef]
60. Bonnet, J.P.; Topin, F.; Tadrist, L. Flow Laws in Metal Foams: Compressibility and Pore Size Effects. *Transp. Porous Media* **2008**, *73*, 233–254. [CrossRef]
61. Sumirat, I.; Ando, Y.; Shimamura, S. Theoretical consideration of the effect of porosity on thermal conductivity of porous materials. *J. Porous Mater.* **2006**, *13*, 439–443. [CrossRef]
62. Zhao, C.Y. Review on thermal transport in high porosity cellular metal foams with open cells. *Int. J. Heat Mass Transf.* **2012**, *55*, 3618–3632. [CrossRef]
63. Iasiello, M.; Cunsolo, S.; Bianco, N.; Chiu, W.K.S.; Naso, V. Developing thermal flow in open-cell foams. *Int. J. Therm. Sci.* **2017**, *111*, 129–137. [CrossRef]
64. Suleiman, A.S.; Dukhan, N. Forced convection inside metal foam: simulation over a long domain and analytical validation. *Int. J. Therm. Sci.* **2014**, *86*, 104–114. [CrossRef]
65. Matias, A.F.V.; Coelho, R.C.V.; Andrade, J.S., Jr.; Araujo, N.A.M. Flow through time–evolving porous media: Swelling and erosion. *J. Comput. Sci.* **2021**, *53*, 101360. [CrossRef]

 *micromachines*

Article

Significance of Rosseland's Radiative Process on Reactive Maxwell Nanofluid Flows over an Isothermally Heated Stretching Sheet in the Presence of Darcy–Forchheimer and Lorentz Forces: Towards a New Perspective on Buongiorno's Model

Ghulam Rasool [1,*], Anum Shafiq [2,*], Sajjad Hussain [3], Mostafa Zaydan [4], Abderrahim Wakif [4], Ali J. Chamkha [5] and Muhammad Shoaib Bhutta [1]

[1] College of International Students, Wuxi University, Wuxi 214105, China; cqu2012170@cqu.edu.cn
[2] School of Mathematics and Statistics, Nanjing University of Information Science and Technology, Nanjing 210044, China
[3] Department of Mathematics, Quaid-i-Azam University, Islamabad 44000, Pakistan; shussain@math.qau.edu.pk
[4] Laboratory of Mechanics, Faculty of Sciences Aïn Chock, Hassan II University of Casablanca, Casablanca 20000, Morocco; na.zidane@gmail.com (M.Z.); wakif.abderrahim@gmail.com (A.W.)
[5] Faculty of Engineering, Kuwait College of Science and Technology, Kuwait City 35004, Kuwait; a.chamkha@kcst.edu.kw
* Correspondence: grasool@zju.edu.cn (G.R.); anum_shafiq@nuist.edu.cn (A.S.)

Citation: Rasool, G.; Shafiq, A.; Hussain, S.; Zaydan, M.; Wakif, A.; Chamkha, A.J.; Bhutta, M.S. Significance of Rosseland's Radiative Process on Reactive Maxwell Nanofluid Flows over an Isothermally Heated Stretching Sheet in the Presence of Darcy–Forchheimer and Lorentz Forces: Towards a New Perspective on Buongiorno's Model. *Micromachines* **2022**, *13*, 368. https://doi.org/10.3390/mi13030368

Academic Editors: Junfeng Zhang and Ruijin Wang

Received: 9 January 2022
Accepted: 21 February 2022
Published: 26 February 2022

Publisher's Note: MDPI stays neutral with regard to jurisdictional claims in published maps and institutional affiliations.

Abstract: This study aimed to investigate the consequences of the Darcy–Forchheimer medium and thermal radiation in the magnetohydrodynamic (MHD) Maxwell nanofluid flow subject to a stretching surface. The involvement of the Maxwell model provided more relaxation time to the momentum boundary layer formulation. The thermal radiation appearing from the famous Rosseland approximation was involved in the energy equation. The significant features arising from Buongiorno's model, i.e., thermophoresis and Brownian diffusion, were retained. Governing equations, the two-dimensional partial differential equations based on symmetric components of non-Newtonian fluids in the Navier–Stokes model, were converted into one-dimensional ordinary differential equations using transformations. For fixed values of physical parameters, the solutions of the governing ODEs were obtained using the homotopy analysis method. The appearance of non-dimensional coefficients in velocity, temperature, and concentration were physical parameters. The critical parameters included thermal radiation, chemical reaction, the porosity factor, the Forchheimer number, the Deborah number, the Prandtl number, thermophoresis, and Brownian diffusion. Results were plotted in graphical form. The variation in boundary layers and corresponding profiles was discussed, followed by the concluding remarks. A comparison of the Nusselt number (heat flux rate) was also framed in graphical form for convective and non-convective/simple boundary conditions at the surface. The outcomes indicated that the thermal radiation increased the temperature profile, whereas the chemical reaction showed a reduction in the concentration profile. The drag force (skin friction) showed sufficient enhancement for the augmented values of the porosity factor. The rates of heat and mass flux also fluctuated for various values of the physical parameters. The results can help model oil reservoirs, geothermal engineering, groundwater management systems, and many others.

Keywords: Maxwell nanofluid; Darcy–Forchheimer model; thermal radiation; chemical reaction; Brownian diffusion

1. Introduction

The concept of nanofluid and nanotechnology in fluid flow phenomena has received enormous attraction in the community of researchers working in fluid mechanics. For sure, the thermophysical properties of fluids in the pure state have many variations, and mostly fluids face a lack of conductivity in the purely natural form. However, a mixture based on nanoparticles of some highly conductive metals in base fluids categorically reduces this deficiency, and the performance of the fluid increases drastically. Notable improvements have appeared in industry due to this innovation in nanotechnology. Specifically, the industries based on drug delivery, paints, ceramics, coatings, and similar products have been quite famous in the recent past. Similarly, enhancing the nanofluids' heat absorption properties due to a mixture of metallic nanoparticles is another significant achievement in this regard. Nanofluids are considered top coolants in fluid flow. Therefore, the performance rate increases sufficiently. The pioneering study introducing the concept of nanofluids was reported by Choi [1], which received much attention, and after that, millions of studies have been reported so far to testify to the properties of nanofluid formulations; however, experimental evidence was provided by Buongiorno [2] in the form of the two-phase model by introducing the terms of the Brownian diffusion and thermophoresis in nanofluid flow. Benos et al. [3] reported an analytic investigation on magnetohydrodynamic naturally convective nanofluids via a horizontal cavity with local heat generation capacity. Bhattacharyya et al. [4] analyzed the impact of the hybrid structure in nanofluids and applied a statistical approach to judge the characteristics of graphene- and copper-type nanoparticles in nanofluids. Gowda et al. [5] analyzed the significance of the activation energy and second-order chemical reaction in the context of heat and mass flux rates in non-Newtonian Marangoni-driven nanofluid flow. Hussain et al. [6] disclosed the features of thermal enhancement in nanofluid flow using an embrittled cone as the core surface. Benos et al. [7] reported some crucial effects of aggregations focusing on CNT–water-based nanofluid flow subject to magnetohydrodynamic (MHD) impact. Yusuf et al. [8] considered Williamson nanofluid using an inclined surface involved with gyrotactic microorganisms to analyze the implications of MHD and bio-convection together with entropy optimization. Benos et al. [9] reported a theoretical investigation on the natural convection of CNT–water-based nanofluid flow subject to MHD under the umbrella of the revised Hamilton–Crosser theory. Apostolos et al. [10] analyzed the flow of Al_2O_3–water-based nanofluid considering the printed circuit heat exchangers to see the impact of the interfacial layer in the context of heat flux. Some further relevant studies can be found in [11–14] and the references cited therein.

The concept of fluid flow in a porous medium is quite natural. The flow-through a rocky surface, the natural flow of fluids through sand and dusty areas, etc., are the very realistic situations around us since the creation of the universe. In the modern world, this concept is now widely used in the manufacturing industry for multiple purposes, especially in modeling oil reservoirs, geothermal engineering, groundwater management systems, and many similar aspects [15–17]. Attention was received by classical Darcy law, which was valid under exceptional circumstances in situations where the porosity factor remains low. However, the classic law fails to accept the higher rate of fluid momentum through the porous medium. Thus, an improvement was needed in the classical Darcy law to enhance its applicability. Therefore, Forchheimer [18] added the squared velocity term in the momentum equation of the governing model for the classical Darcy law to tackle the higher porosity rates. Later on, Muskat [19] named this term the Forchheimer term, and the governing model was called the Darcy–Forchheimer model of fluid flow. Pal and Mondal [20] reported interesting findings on convective diffusion of the species using a non-uniform heat sink/source under the umbrella of the Darcy–Forchheimer model. Hayat et al. [21] presented the variable thermal-conductivity-based Darcy–Forchheimer flow of nanofluids using the Cattaneo–Christov model. Eid and Mabood [22] implemented the Darcy–Forchheimer model using the two-phase cross nanofluid flow. The consequence of Arrhenius activation was disclosed. Furthermore, entropy optimization was analyzed in this study. Shankaralingappa et al. [23] configured the impact of the Cattaneo–Christov theory of double diffusion on an Oldroyd-B-type fluid using a stretching surface considering the thermophoresis deposition of particles and the chemical reaction. Liquids incorporating the MHD and thermal radiation effects have been hot spots recently. The involvement of fluids with the MHD impact is relatively high in many industrial processes such as gastric medications, wound treatments, sterilized devices, medical sciences, X-ray technology, and many others. Numerous studies have been reported so far mentioning the impact of these variables in industrial applications of nanofluids. The significance of MHD is also very relatable in fluid flow analysis because it helps to control the fluid motion and thermal state of the fluid. The sudden bumps created by the magnetic field in the fluid flow phenomena are remarkably used to tackle various abnormal situations in fluid flow. Numerous related articles mention multiple parameters such as MHD, thermal radiation, chemical reaction, and many others in nanofluid flow analysis. Jamshed et al. [24] reported second-grade nanofluidic flow considering the radiation impact in a single-phase model via a flat porous surface. Sheikholeslami et al. [25] reported an analytic investigation of the MHD-type nanofluid flow subject to a semi-permeable channel. Kumar et al. [26] explored the influence of a magnetic dipole in radiative nanofluidic flow via the stretching surface using the KKL model. Kumar et al. [27] modeled Casson-type nanofluid flow using a curved stretching surface to highlight the impact of MHD and the chemical reaction. Hayat et al. [28] reported essential findings on radiative chemically reactive three-dimensional flow. Sarada et al. [29] reported the impact of magnetohydrodynamics on the heat flux rate in non-Newtonian fluids flowing over a stretching surface subject to local non-equilibrium thermal conditions. Sheikholeslami [30] analyzed the effect of thermal radiation and MHD in nanofluid flow. Charakopoulos et al. [31] examined the influence of magnetohydrodynamics in a channel flow using complex network analysis. Hamid et al. [32] reported critical data in axisymmetric nanomaterial flow towards a radiative shrinking disk. Furthermore, the impact of the Darcy–Forchheimer model, together with various parameters, including thermal radiation, chemical reaction, activation energy, and many others, were already reported (see, for example, [33–39]). Wakif et al. [40] reported a novel approach to the MHD analysis of Casson fluids over the horizontal surface (stretching) using the impact of thermal conductivity and temperature-dependent viscosity. Ramesh and Joshi [41] reported the MHD analysis of Jeffrey-type fluid flow between two parallel plates through a porous medium using an unsteady flow model.

The above-mentioned studies motivated the authors to look for a model comprising the given flow constraints and physical parameters. Herein, we considered the Darcy–Forchheimer flow model together with the Maxwell nanofluid boundary layer assumptions to examine the influence of thermal radiation and a porous medium using Buongiorno's model of Brownian diffusion and thermophoresis phenomena. The governing system of equations was subject to the homotopy analysis method (HAM) (see, for example, [42–46]), which is a highly efficient and frequently used analytic approach to solve highly nonlinear governing equations providing the freedom of choice for choosing the linear auxiliary operators and base functions. The results were plotted graphically, and data of the skin friction and the Nusselt and Sherwood numbers are given in tables. The study concludes with a discussion on the results and concluding remarks. A comparison of results for the Nusselt number for convective boundary and the non-convective boundary is provided. The results can help model oil reservoirs, geothermal engineering, groundwater management systems, and many others.

2. Formulation of the Problem

Assume the two-dimensional flowchart based on a Maxwell nanofluid. The thermal radiation appearing from the famous Rosseland approximation is involved in the energy equation. In addition, the Darcy–Forchheimer model was adopted to saturate the fluid in a certain porous boundary. Furthermore, thermophoresis, Brownian diffusion, and the first-order chemical reaction were retained. The surface that generates the fluid flow was assumed to stretch linearly. Uniform magnetic impact directly influences the flow model with a term in the momentum equation. However, considering a small Reynolds number helps dismiss the magnetic field induction. The fluid was assumed to proceed alongside the x-axis, while no velocity was considered alongside the y-axis. At the initial condition, the velocity is the same as the stretching rate of the sheet, and it becomes zero as the distance approaches the free surface from the solid sheet towards the y-axis. The temperature and concentration terms are typically considered T and C, respectively having wall conditions (T_w, C_w) at the surface and ambient conditions (T_∞, C_∞) at the free surface. The physical scenario can be visualized in Figure 1. The governing equations (see, for example, [42,43,47,48]) are as follows.

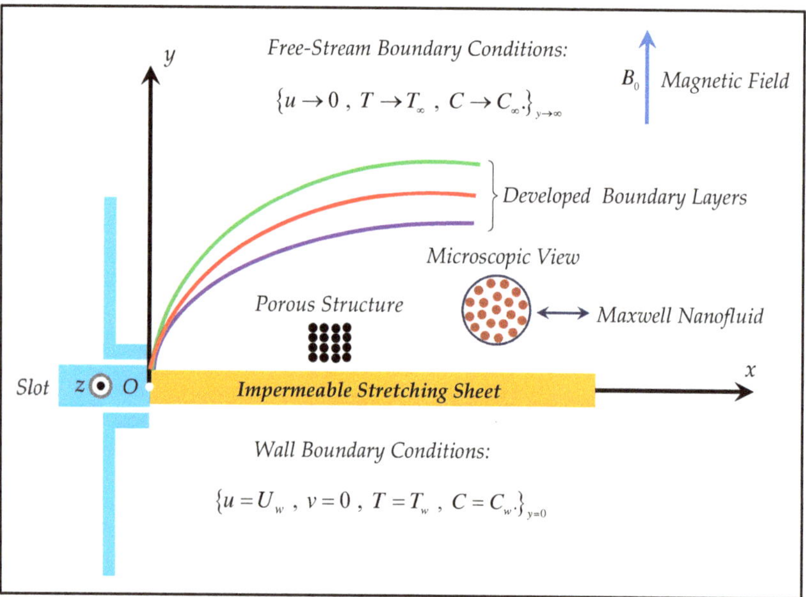

Figure 1. Geometry of the nanofluid flow.

$$\frac{\partial u}{\partial x} + \frac{\partial v}{\partial y} = 0, \tag{1}$$

$$u\frac{\partial u}{\partial x} + v\frac{\partial u}{\partial y} + \lambda_1\left(u^2\frac{\partial}{\partial x}\left(\frac{\partial u}{\partial x}\right) + v^2\frac{\partial}{\partial y}\left(\frac{\partial u}{\partial y}\right) + 2vu\frac{\partial}{\partial x}\left(\frac{\partial u}{\partial y}\right)\right)$$
$$= v\frac{\partial}{\partial y}\left(\frac{\partial u}{\partial y}\right) - \frac{\sigma B_0^2 u}{\rho} - \frac{\sigma \lambda_1 B_0^2}{\rho}\left(v\frac{\partial u}{\partial y}\right) - \left(\frac{v}{K_1} + \frac{C_b}{x\sqrt{K_1}}u\right)u, \tag{2}$$

$$u\frac{\partial T}{\partial x} + v\frac{\partial T}{\partial y} = \alpha\frac{\partial}{\partial y}\left(\frac{\partial T}{\partial y}\right) + \frac{\tau D_B}{\delta_C}\frac{\partial T}{\partial y}\frac{\partial C}{\partial y} + \frac{\tau D_T}{T_\infty}\left(\frac{\partial T}{\partial y}\right)^2 - \frac{1}{\rho c}\frac{\partial q_r}{\partial y}, \tag{3}$$

$$u\frac{\partial C}{\partial x} + v\frac{\partial C}{\partial y} = D_B\frac{\partial}{\partial y}\left(\frac{\partial C}{\partial y}\right) + \frac{\delta_C D_T}{T_\infty}\frac{\partial}{\partial y}\left(\frac{\partial T}{\partial y}\right) - Cr(C - C_\infty). \tag{4}$$

$$u = U_w = ex, \quad v = 0, \quad T = T_w \quad C = C_w \quad \text{at y} = 0, \tag{5}$$

$$u \to 0, \quad T \to T_\infty, \quad C \to C_\infty \quad \text{as y} \to \infty. \tag{6}$$

In the last term in Equation (3), the quantity q_r actually represents the radiative heat flux appearing from the famous Rosseland's approximation (see, for example, [49]). Mathematically,

$$q_r = -\frac{4}{3}\frac{\sigma_{SB}}{k_{ABS}}\frac{\partial T^4}{\partial y}, \tag{7}$$

where σ_{SB} is called the Stefan–Boltzmann constant and k_{ABS} is known as the mean absorption factor. Using the Taylor series expansion on T^4 and neglecting the second- and higher-order terms in $(T - T_\infty)$, one can write:

$$\frac{\partial q_r}{\partial y} = -\frac{16}{3}\frac{\sigma_{SB}T_\infty^3}{k_{ABS}}\left(\frac{\partial^2 T}{\partial y^2}\right). \tag{8}$$

Therefore, Equation (3) re-appears as follows:

$$u\frac{\partial T}{\partial x} + v\frac{\partial T}{\partial y} = \alpha\frac{\partial}{\partial y}\left(\frac{\partial T}{\partial y}\right) + \frac{\tau D_B}{\delta_C}\frac{\partial T}{\partial y}\frac{\partial C}{\partial y} + \frac{\tau D_T}{T_\infty}\left(\frac{\partial T}{\partial y}\right)^2 + \frac{16}{3(\rho c)}\frac{\sigma_{SB}T_\infty^3}{k_{ABS}}\frac{\partial^2 T}{\partial y^2}. \tag{9}$$

Furthermore, in Equation (4), Cr represents the first-order chemical reaction. Define (see, for example, [42,50]):

$$u = exf'(\eta), \qquad v = -(ev)^{1/2}f(\eta), \qquad \theta(\eta) = \frac{(T-T_\infty)}{(T_w-T_\infty)}, \qquad \phi(\eta) = \frac{(C-C_\infty)}{(C_w-C_\infty)}, \tag{10}$$

$$\eta = \left(\frac{e}{v}\right)^{1/2}y. \tag{11}$$

The application of Equations (10) and (11) in Equations (1), (2), (4) and (9) results in the following non-dimensional ODEs:

$$f''' + \left(1 + M^2\gamma\right)ff'' + 2\gamma ff'f'' - \gamma f^2 f''' - \left(\lambda + M^2\right)f' - (F_r + 1)f'^2 = 0, \tag{12}$$

$$\left(1 + \frac{4}{3}Rd\right)\theta'' + Prf\theta' + PrN_b\theta'\phi' + PrN_t\theta'^2 = 0, \tag{13}$$

$$\phi'' + PrLef\phi' + \frac{N_t}{N_b}\theta'' - K\phi = 0, \tag{14}$$

$$f(0) = 0, \qquad f'(0) = 1, \qquad \theta(0) = 1 \qquad \phi(0) = 1, \tag{15}$$
$$f'(\infty) \to 0, \qquad \theta(\infty) \to 0, \qquad \phi(\infty) \to 0. \tag{16}$$

In the process of non-dimensionalization, the quantities that appeared as the coefficient of f, θ, ϕ are called various physical parameters involved in the problem model. These quantities are mathematically defined as follows:

$$M^2 = \frac{\sigma}{e}\frac{B_0^2}{\rho}, \qquad \gamma = \lambda_1 e, \qquad \lambda = \frac{v}{eK_1}, \qquad K = \frac{Cr}{e}, \tag{17}$$

$$F_r = \frac{C_b}{\sqrt{K_1}}, \qquad Pr = \frac{v}{\alpha}, \qquad Le = \frac{\alpha}{D_B}, \tag{18}$$

$$N_t = \tau D_T\frac{(T_w - T_\infty)}{vT_\infty}, \; N_b = \tau D_B\frac{(C_w - C_\infty)}{v\delta_C}, \; Rd = \frac{4\sigma_{SB}T_\infty^3}{k_f k_{ABS}}. \tag{19}$$

The quantities appearing in Equation (17) are the magnetic parameter, the Deborah number, the porosity parameter, and the first-order chemical reaction. The quantities appearing in Equation (18) are the Forchheimer number, the Prandtl number, and the Lewis factor. The first two quantities in Equation (19) are the two significant nanofluids known as thermophoresis and Brownian diffusion. The last quantity represents thermal radiation.

In natural fluid flow phenomena, some critical factors affect fluid motion and the thermal state. The three crucial factors are called the drag force (skin friction), the heat flux rate (Nusselt number), and the mass flux rate (Sherwood number). The final representation of these three quantities in non-dimensional form is given below:

$$Re_{fx}^{1/2}C_f x = f'', \quad at \quad \eta = 0, \tag{20}$$

$$Re_x^{-1/2}Nu_x = -\theta', at \quad \eta = 0, \tag{21}$$

$$Re_x^{-1/2}Sh_x = -\phi', at \quad \eta = 0. \tag{22}$$

where Re_x is known as the local Reynolds number.

3. Solution Methodology

The homotopy analysis method was implemented to obtain the convergent series solutions. The graphs were prepared using Mathematica 9.0. Let,

$$f_0 = 1 - e^{-\eta}, \; \theta_0 = e^{-\eta}, \; \phi_0 = e^{-\eta}, \tag{23}$$

$$\hat{\mathscr{I}}_f = f''' - f', \; \hat{\mathscr{I}}_\theta = \theta'' - \theta, \; \hat{\mathscr{I}}_\phi = \phi'' - \phi, \tag{24}$$

with the following hypothesis,

$$\hat{\mathscr{I}}_f\left[L_1 e^{-\eta} + L_2 e^{\eta} + L_3\right] = 0, \; \hat{\mathscr{I}}_\theta\left[L_4 e^{-\eta} + L_5 e^{\eta}\right] = 0, \; \hat{\mathscr{I}}_\phi\left[L_6 e^{-\eta} + L_7 e^{\eta}\right] = 0, \tag{25}$$

where L_i, $i = 1, 2, \cdots, 7$, are constant numbers. Subsequently, the zeroth-order equations of deformation can be symbolized as $\mathscr{Q}_f\left[\hat{f}\right]$ for the momentum equations, $\mathscr{Q}_\theta\left[\hat{f}, \hat{\theta}, \hat{\phi}\right]$ for the energy equation, and $\mathscr{Q}_\phi\left[\hat{f}, \hat{\theta}, \hat{\phi}\right]$ for the concentration equations given in Equations (12)–(14), such that:

$$(1 - e)\,\hat{\mathscr{I}}_f\left[\hat{f}(\eta, e) - f_0(\eta)\right] = e\hat{h}_f\mathscr{Q}_f[\hat{f}],$$

$$(1 - e)\,\hat{\mathscr{I}}_\theta\left[\hat{\theta}(\eta, e) - \theta_0(\eta)\right] = e\hat{h}_\theta\mathscr{Q}_\theta[\hat{f}, \hat{\theta}, \hat{\phi}], \tag{26}$$

$$(1 - e)\,\hat{\mathscr{I}}_\phi\left[\hat{\phi}(\eta, e) - \phi_0(\eta)\right] = e\hat{h}_\phi\mathscr{Q}_\phi[\hat{f}, \hat{\theta}, \hat{\phi}].$$

with the transformed boundary conditions given in (15–16). It is important to mention here that $\hat{h}_f$ is the auxiliary function corresponding to the velocity equation, $\hat{h}_\theta$ is the auxiliary function corresponding to the energy equation, and $\hat{h}_\phi$ is the auxiliary function corresponding to the concentration equation. Furthermore, $e \in [0, 1]$ is called the embedding. $\mathscr{Q}_{\hat{f}}, \mathscr{Q}_{\hat{\theta}}$, and $\mathscr{Q}_{\hat{\phi}}$ are named the non-linear operators. The Taylor series implementation results in the following equations:

$$\hat{f} = \sum_{i=0}^{\infty} f_i(\eta)e^i, \; \hat{\theta} = \sum_{i=0}^{\infty} \theta_i(\eta)e^i, \; \hat{\phi} = \sum_{i=0}^{\infty} \phi_i(\eta)e^i, \tag{27}$$

where $\mathscr{E}_i(\eta) = \frac{1}{i!}\frac{\partial^i \mathscr{E}}{\partial e^i}\Big|_{e=0}$ for $\mathscr{E} = \hat{f}, \hat{\theta},$ or $\hat{\phi}$. The efficient and smoothly convergent results are strictly dependent on the numerical choice of $\hat{h}$. The values of e fluctuate between $e = 0, 1$. General solutions are given as follows,

$$f_i = L_1 + L_2 e^{\eta} + L_3 e^{-\eta} + f_i^{\star}(\eta),$$

$$\theta_i = L_4 e^{\eta} + L_5 e^{-\eta} + \theta_i^{\star}(\eta), \tag{28}$$

$$\phi_i = L_6 e^{\eta} + L_7 e^{-\eta} + \phi_i^{\star}(\eta),$$

where the functions with $\star$ represent the special solutions.

4. Results and Discussion

In this paper, we investigated the consequences of the Darcy–Forchheimer medium and thermal radiation in the magnetohydrodynamic (MHD) Maxwell nanofluid flow subject to a stretching surface confined within the simple boundary conditions. Physical parameters such as thermal radiation, the chemical reaction, the porosity factor, the Forchheimer number, the Deborah number, the Prandtl number, thermophoresis, and Brownian diffusion and their impact on the fluid profiles are discussed in the following lines. Figures 2 and 3 represent the behavior of the velocity profiles for the variation in the Deborah number and porosity factor. Specifically, Figure 2 shows the behavior of the velocity profile for altered values of the Deborah number. The constitution of the Deborah number is based on the relaxation time parameter, which in the physical context means providing more time to the nanoparticles to be diluted in the base fluid. The higher the Deborah number is, the lower the fluid velocity, and the consequent boundary layer drops to a certain extent. This appearance of the velocity profile was obtained fixing the other three physical parameters involved in the momentum equation. Figure 3 represents the variations in the velocity profile subject to the altered values of the porosity factor. Physically, the presence of the porous medium is itself a reason for the increase in the frictional retardation force offered to the fluid in motion. The higher the porous ratio in the medium, the more retardation is offered to the fluid. Therefore, the velocity profile shows a reduction in its trend for incremental values of λ. As for the energy equations, the final non-dimensional ODE involves several physical parameters already defined in the previous section. To see their impact on the temperature profile, we plotted the data in graphs given in Figures 2–8. In particular, Figure 4 represents the consequent impact of the Deborah number on the temperature profile. The profile apprises the behavior of the altered, augmented values of the Deborah number. Here again, the justification of this behavior relates to the relaxation time provided to the model by the Maxwell model. The more is the relaxation time, the more is the temperature profile and vice versa. The continuous offering of more friction to the fluid in motion is the main property of the porous medium, which is mathematically involved in the model using the two important factors, the porosity factor and the Forchheimer number. The appearance of these parameters in the energy equation has a high impact on the temperature profile. Figure 5 gives this impact in the porosity factor versus the temperature profile. The higher the rate of resistance provided to the system, the higher is the system's temperature due to the high rate of collisions between the molecules of the base fluid and the nanoparticles diluted in it. This behavior gives rise to another important aspect of fluid flow analysis, i.e., thermal radiation. The inside-out thermal radiation is another important source of raising the temperature profile. A dominant rising trend in the temperature profile due to thermal radiation can be seen in Figure 6. The result shows that even with a slight variation in the thermal radiation, a very high variation is noted in the corresponding boundary layer formulation of the temperature profile. The impact of Brownian diffusion is given in Figure 7, which is physically related to the predicted movement of nanoparticles and collisions. The higher the value of Nb, the higher is the temperature profile and vice versa. However, an inverse trend was found in the case of the Prandtl number. The higher values of the Prandtl number, as given in Figure 8, result in a reduction in the temperature profile. The constituent term of the Prandtl number involves kinematic viscosity inversely related to the thermal diffusivity. Higher values of the Prandtl number result in a reduction of thermal diffusivity and an increment in the kinematic viscosity, which results in the reduction of the temperature profile. The behavior of both the temperature and concentration profiles towards the Deborah number is quite similar. In both cases, a rise in the values of the Deborah number results in the increment of the respective profile. Figure 9 shows the same thing discussed in the above lines. Higher values of the Deborah number mean more convenience is provided to the nanoparticles to adjust and be diluted in the base fluid.

The concentration profile rises after that. The porosity factor, when increased, provides more space for the nanoparticles to be spaced in the base fluid, and therefore, the consequence is shown in Figure 10. The concentration behavior in response to the altered values of thermal radiation is given in Figure 11. The interval of rising and lowering is very short. Thus, a shorter, but incremental trend is noticed for higher values of the radiation factor. The impact of Lewis's number is quite dominant and prominent. From the Lewis number given in Equation (18), we see that the inverse relation of Brownian diffusion and thermal diffusivity is called the Lewis number. Thus, Brownian diffusion and Le are inversely proportional to each other. The higher the values of Le, the lower the diffusion will be and, therefore, the lower the concentration of nanoparticles in the base fluid, as displayed in Figure 12. Figure 13 gives the impact of Nb (Brownian motion parameter) over the concentration profile. Higher values result in a low concentration and vice versa. The Prandtl number decreases the concentration profile given in Figure 14 with the same justification as given in Figure 8 because the diffusivity is linked to the Prandtl number. The first-order chemical reaction is a source of the reduction in the concentration profile. The fluid faces a descending concentration near the surface at a lower intensity of the reactive material. However, the concentration reduces sufficiently with higher values of K, as shown in Figure 15. Table 1 provides the numerical data of the mass flux, heat flux, and wall drag (also known as skin friction) for fluctuating values of various parameters. In particular, the porosity, the Forchheimer number, and the Deborah number increase the drag force. The Nusselt number reduces for the thermal radiation factor; however, the same parameter enhances the Sherwood number. The mass flux rate is enhanced for larger values of the Lewis number. Figures 16 and 17 are based on the two comparative results, i.e., with and without the convective boundary, setting the radiation factor and chemical reaction factor equal to zero. The convective boundary has a significant variation in the values of the Nusselt number as compared to a simple boundary. In both cases, the trend of the Nusselt number is the same, but the rates are different at the same values of Pr and M for both cases of different boundary conditions.

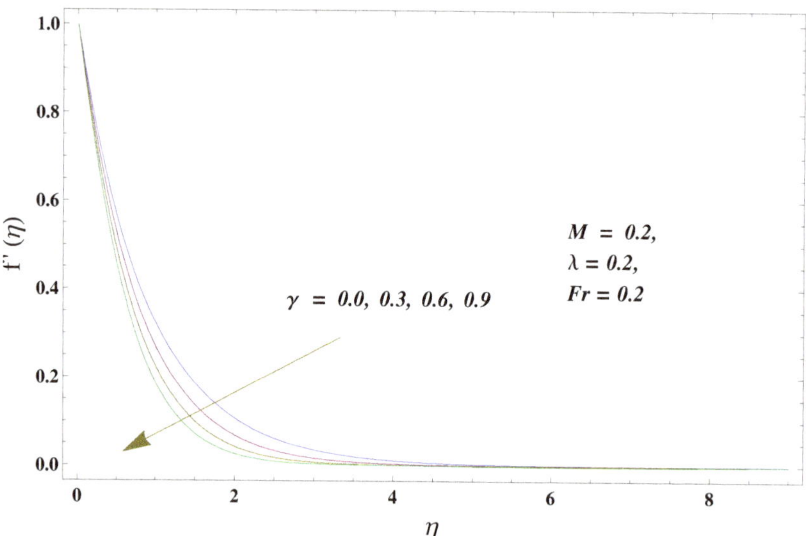

Figure 2. Deborah number and its impact on the velocity profile.

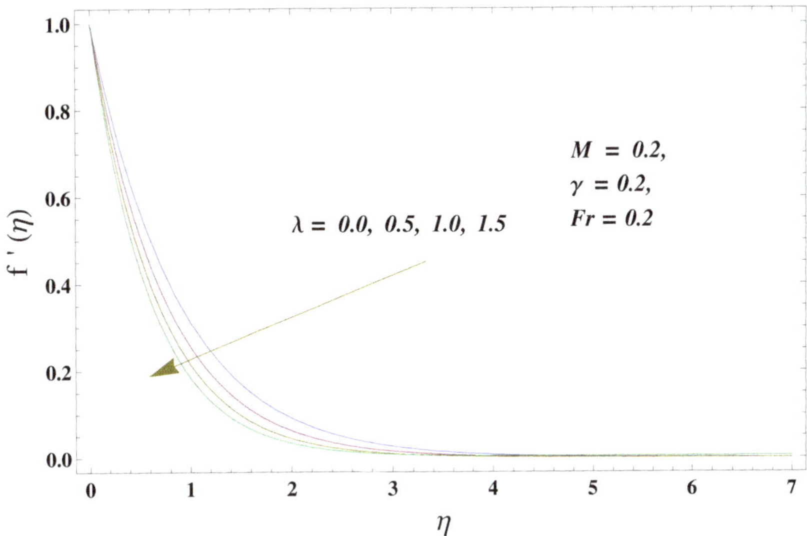

Figure 3. Porosity number and its impact on the velocity profile.

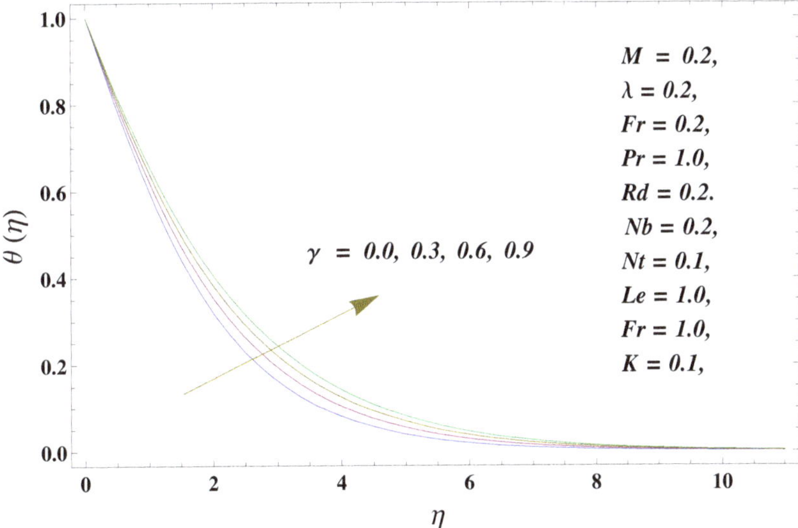

Figure 4. Deborah number and its impact on the temperature profile.

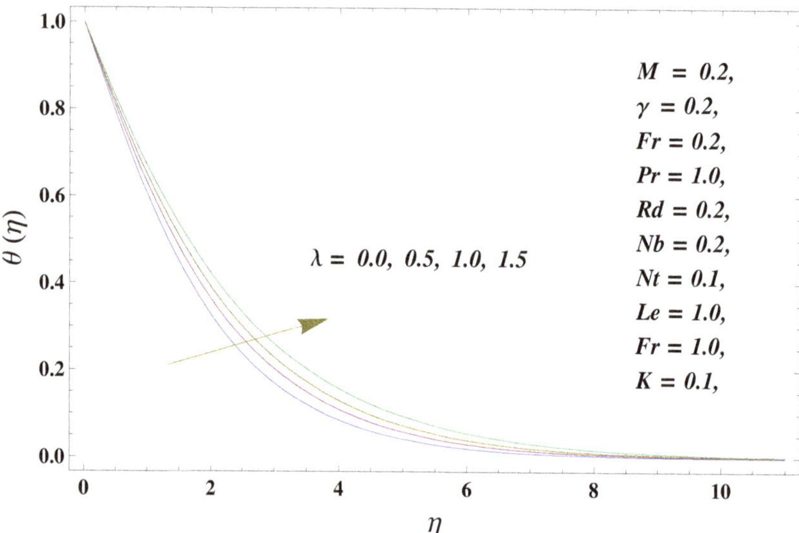

Figure 5. Porosity number and its impact on the velocity profile.

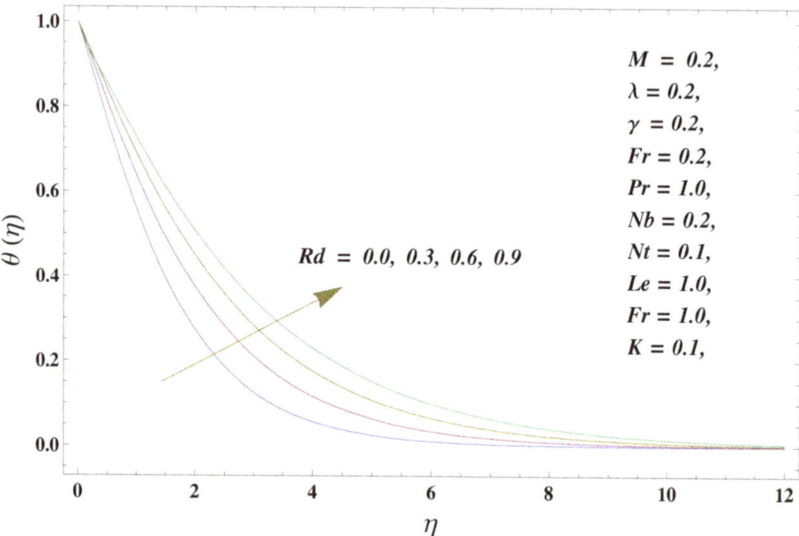

Figure 6. Thermal radiation and its impact on the temperature profile.

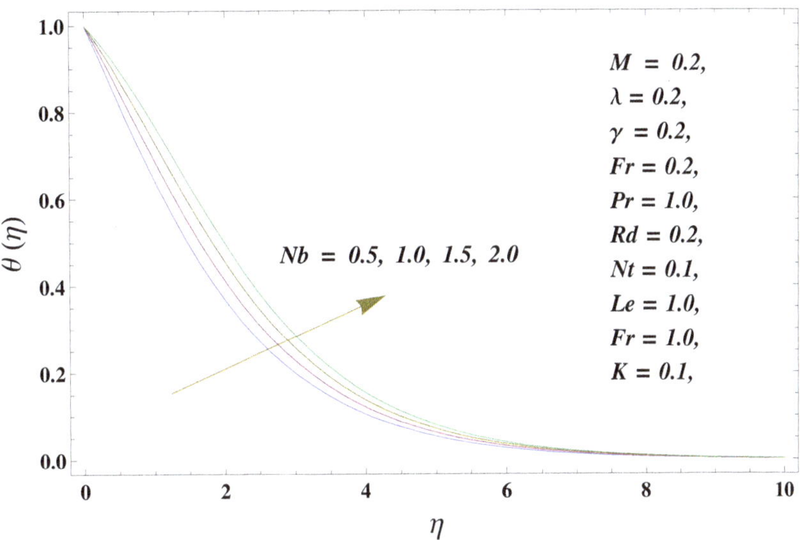

Figure 7. Brownian diffusion and its impact on the temperature profile.

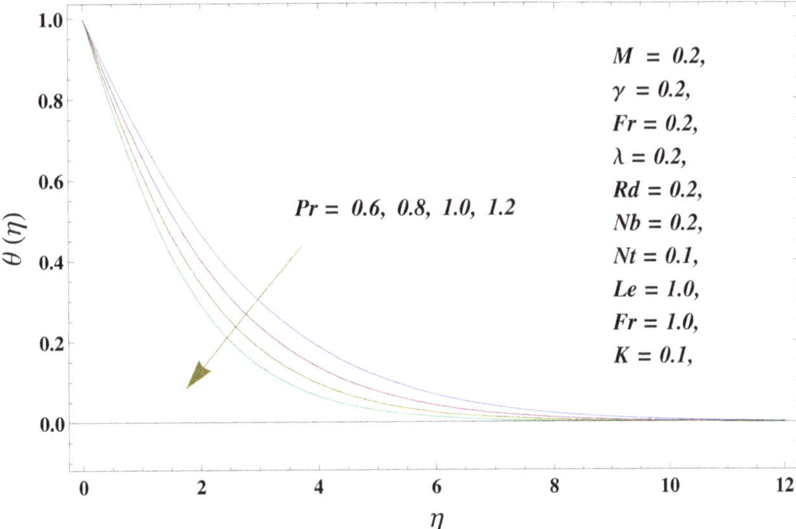

Figure 8. Prandtl number and its impact on the temperature profile.

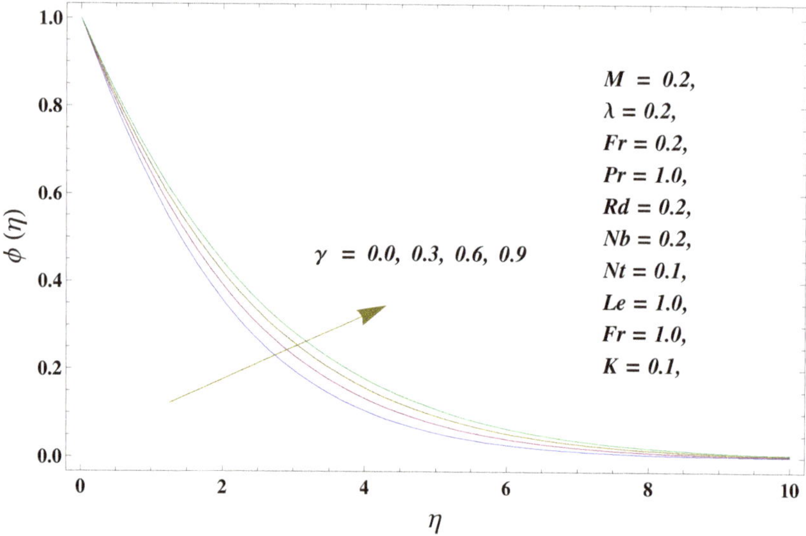

Figure 9. Deborah number and its impact on the concentration profile.

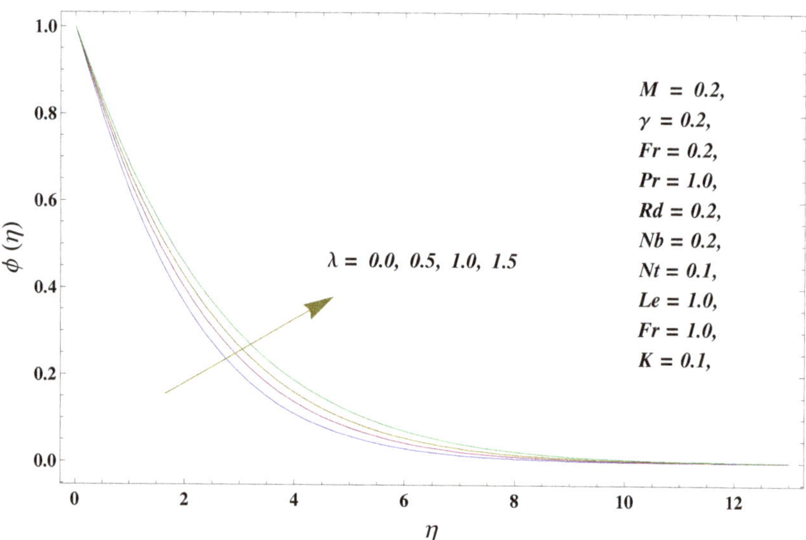

Figure 10. Porosity number and its impact on the concentration profile.

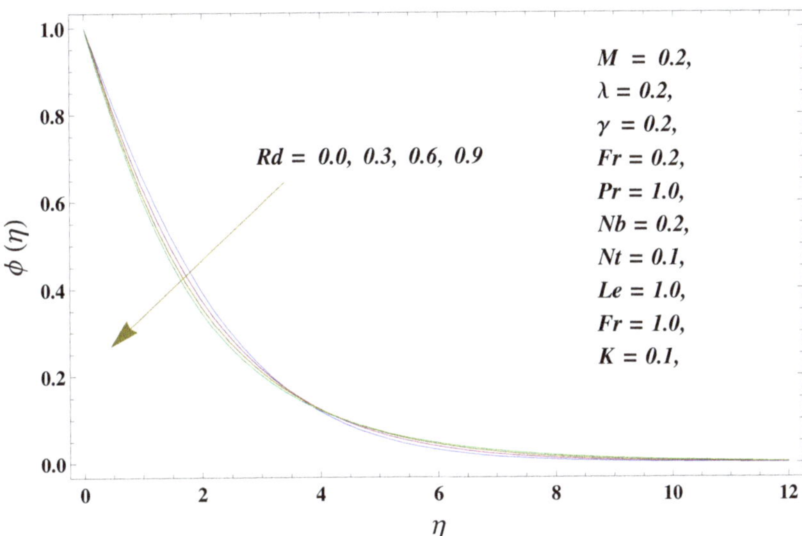

Figure 11. Thermal radiation and its impact on the concentration profile.

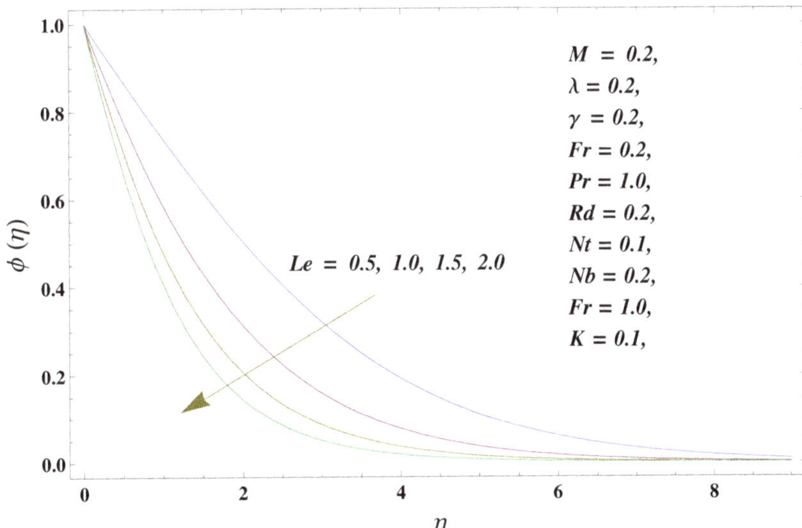

Figure 12. Lewis number and its impact on the concentration profile.

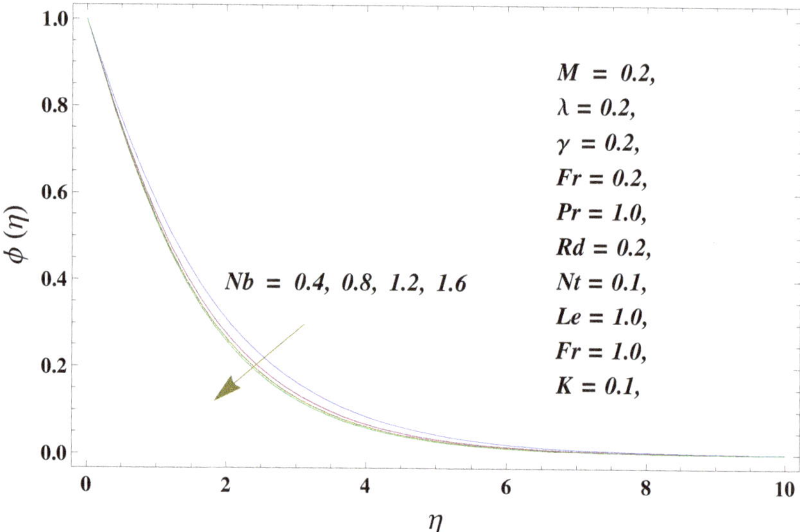

Figure 13. Brownian diffusion and its impact on the concentration profile.

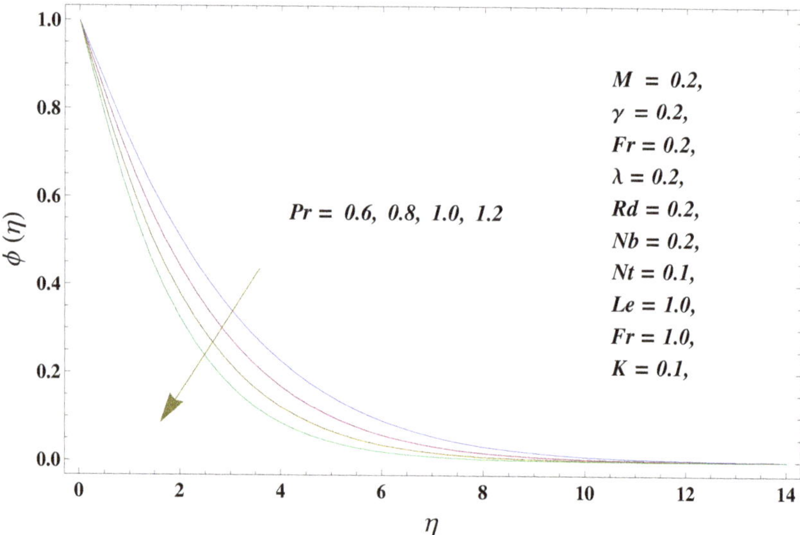

Figure 14. Prandtl number and its impact on the concentration profile.

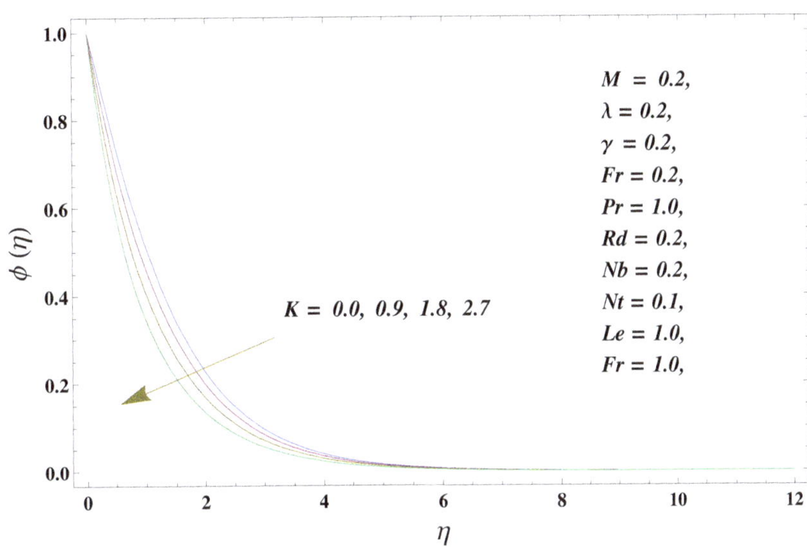

Figure 15. Chemical reaction and its impact on the concentration profile.

Table 1. Numerical results of the skin friction (wall drag) and Nusselt number (heat flux). The default values are: $\lambda = 2/10$, $M = 2/10$, $Fr = 1$, $\gamma = 2/10$, $K = 2/10$, $Le = 1$, $Nb = 2/10$, $Nt = 1/10$, $Pr = 1$, $Rd = 2/10$.

λ	Fr	γ	K	Le	M	Nb	Nt	Pr	Rd	$-Re_x^{1/2}C_{fx}$	$-Re_x^{-1/2}Nu_x$	$-Re_x^{-1/2}Sh_x$
0.0										1.1199	0.4129	0.5306
0.3										1.2456	0.3962	0.5175
0.6										1.3601	0.3822	0.5067
	0.0									1.1786	0.5115	0.4085
	0.3									1.3065	0.4027	0.3957
	0.6									1.4242	0.3981	0.3933
		0.0								1.3514	0.4172	0.5341
		0.3								1.4492	0.3941	0.5159
		0.6								1.5233	0.3744	0.5007
			0.0							1.4235	0.4050	0.4003
			0.3							1.4235	0.3969	0.7075
			0.6							1.4235	0.3931	0.9137
				0.4						$--$	0.4097	0.3353
				0.8						$--$	0.4040	0.4581
				1.2						$--$	0.3991	0.5846
					0.0					1.4089	0.4039	0.5236
					0.3					1.4429	0.3984	0.5193
					0.6					1.5983	0.3838	0.5079
						0.4				$--$	0.3650	0.5840
						0.8				$--$	0.2999	0.6142
						1.2				$--$	0.2445	0.6236
							0.0			$--$	0.4123	0.6312
							0.3			$--$	0.3807	0.3263
							0.6			$--$	0.3520	0.0868

Table 1. *Cont.*

0.6		— —	0.3071	0.4292
1.2		— —	0.4445	0.5693
2.0		— —	0.5784	0.7616
	0.0	— —	0.4649	0.4990
	0.5	— —	0.3369	0.5445
	1.0	— —	0.1025	0.8359

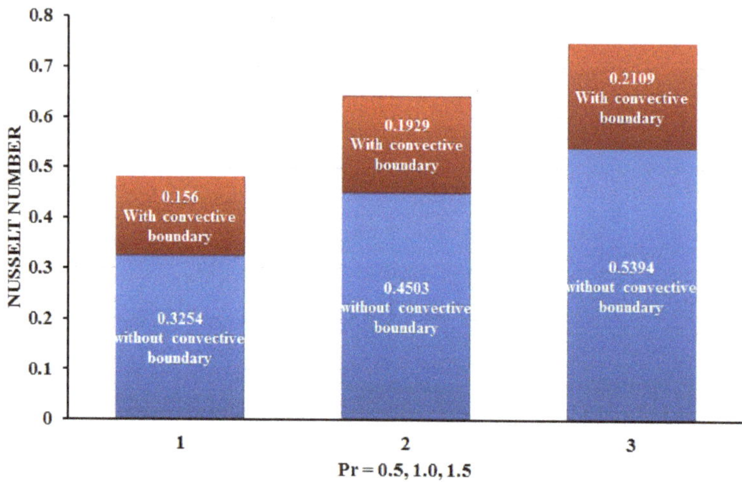

Figure 16. Nusselt number at Rd = 0 = K corresponding to Pr = 0.5, 1.0, 1.5.

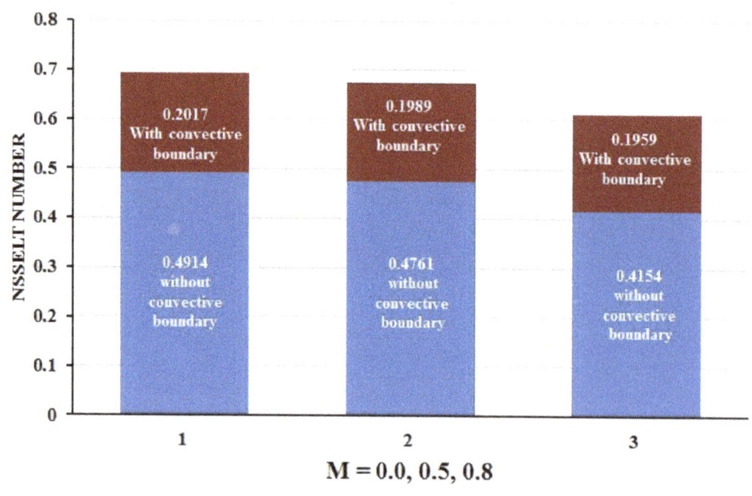

Figure 17. Nusselt number at Rd = 0 = K corresponding to M = 0.0, 0.5, 0.8.

5. Conclusions

We considered the Darcy–Forchheimer medium and thermal radiation in the MHD Maxwell nanofluid flow subject to a stretching surface. The significant features appearing from Buongiorno's model, i.e., thermophoresis and Brownian diffusion, were retained. The governing equations after conversion into ODEs were solved for convergent series solutions using the HAM. Graphs were plotted for the three profiles of the flow model against various physical parameters such as thermal radiation, the chemical reaction, the porosity factor, the Forchheimer number, the Deborah number, the Prandtl number, thermophoresis, and Brownian diffusion. The following salient features were the critical points in this investigation:

- The involvement of the Maxwell model in nanofluid flow provided more relaxation time for the diffusion and dilution of nanoparticles in the base fluid;
- The presence of the porosity factor was a significant source of increment in the drag force and the reduction in fluid flow along the horizontal axis;
- The impact of thermal radiation was prominent in the case of the temperature profile; however, it impacted the other two profiles as well;
- The Deborah number coming from the relaxation time provided in the Maxwell model enhanced the concentration and temperature profiles and decelerated the fluid;
- Brownian diffusion enhanced the temperature profile and reduced the concentration profile;
- The chemical reaction appeared to be a reducing factor for the concentration of nanoparticles in the base fluid;
- The trio of the porosity, the Forchheimer number, and the Deborah number increased the drag force;
- The heat flux reduced for the thermal radiation factor; however, the same parameter enhanced the mass flux rate;
- A rise was noted in the mass flux rate for augmented values of the Lewis number;
- Despite a difference in the numerical data of the Nusselt number for the convective and non-convective boundary, the trend of the increase and decrease of the flux rate was identical for both types of boundary conditions.

Author Contributions: Conceptualization, G.R., A.W. and A.S.; methodology, G.R. and S.H.; software, G.R., A.S. and S.H.; validation, G.R., A.W., M.Z. and M.S.B.; formal analysis, A.J.C., M.Z. and M.S.B.; investigation, G.R., A.W. and A.S.; resources, G.R. and A.S.; data curation, S.H., M.Z. and A.J.C.; writing—original draft preparation, G.R. and A.W.; writing—review and editing, G.R., A.S., A.W., S.H., M.Z., A.J.C. and M.S.B.; visualization, G.R.; supervision, A.S. and A.J.C.; project administration, G.R.; funding acquisition, G.R. and A.S. All authors have read and agreed to the published version of the manuscript.

Funding: This research received no external funding.

Institutional Review Board Statement: Not applicable.

Data Availability Statement: All data is available within the manuscript.

Acknowledgments: The authors are thankful to reviewers for their valuable suggestions and feedback on paper.

Conflicts of Interest: The authors have no conflict of interest.

Nomenclature

The following nomenclature is used in this manuscript:

x, y	Cartesian coordinates$/m$
$u(x, y), v(x, y)$	Coordinates of the velocity vector $(m \cdot s^{-1})$
v	Nanofluid viscosity (kinematic) $(m^2 \cdot s^{-1})$
μ	Nanofluid viscosity (dynamic) $(Kg \cdot m^{-1} \cdot s^{-1})$
C_b	Inertial coefficient (m)
σ	Electric conductivity $((\Omega \cdot m)^{-1})$
B_0	Magnetic impact $(A \cdot m^{-1})$

K_1	Permeability (m^2)
ρ	Density of the nanofluid ($Kg \cdot m^{-3}$)
σ_{SB}	Stefan–Boltzmann constant ($W \cdot K^{-4} \cdot m^{-2}$)
α	Thermal diffusivity ($m^2 \cdot s^{-1}$)
k_{ABS}	Mean absorption factor (m^{-1})
T	Temperature (K)
C	Concentration ($mol \cdot m^{-3}$)
T_w	Temperature at the wall (K)
C_w	Concentration of nanoparticles at the wall ($mol \cdot m^{-3}$)
(ρc)	Nanofluid's productive heat capacity
T_∞	Temperature far away from the surface (boundary condition)
C_∞	Concentration of nanoparticles far away from the surface (boundary condition)
Cr	Chemical reaction (s^{-1})
$(\rho c)_p$	Nanoparticles' productive heat capacity ($J \cdot m^{-3} \cdot K^{-1}$)
D_T	Thermophoretic effect ($m^2 \cdot s^{-1}$)
D_B	Brownian diffusion factor ($m^2 \cdot s^{-1}$)
F_r	Local inertia
M	Magnetic parameter
e	Positive constant number (s^{-1})
Le	Lewis factor
Nb	Brownian diffusion parameter
Pr	Prandtl factor
Nt	Thermophoretic parameter
Nu_x	Local Nusselt number (heat flux)
Sh_x	Local Sherwood number (mass flux)
f'	Dimensionless velocity
η	Dimensionless variable
ϕ	Dimensionless concentration of the nanoparticles
θ	Dimensionless temperature field
λ_1	Relaxation time involvement
q_r	Radiative heat flux ($W \cdot m^{-2}$)
γ	Deborah number
λ	Porosity parameter
Rd	Radiation factor
K	First-order chemical reaction
τ	Dimensionless thermal coefficient
δ_C	Corrective concentration coefficient

References

1. Choi, S.U.S. Enhancing thermal conductivity of fluids with nanoparticles. *Dev. Appl. Non-Newt. Flows* **1995**, *2*, 99–105.
2. Buongiorno, J.; Hu, W. Nanofluid coolants for advanced nuclear power plants. *Proc. ICAPP* **2005**, *2005*, 5705.
3. Benos, L.; Sarris, I.E. Analytical study of the magnetohydrodynamic natural convection of a nanofluid filled horizontal shallow cavity with internal heat generation, *Int. J. Heat Mass Transf.* **2019**, *130*, 862–873. [CrossRef]
4. Bhattacharyya, A.; Sharma, R.; Hussain, S.M.; Chamkha, A.J.; Mamatha, E. A Numerical and Statistical Approach to Capture the Flow Characteristics of Maxwell Hybrid Nanofluid Containing Copper and Graphene Nanoparticles. *Chin. J. Phys.* **2021**. [CrossRef]
5. Gowda, R.J.P.; Kumar, R.N.; Jyothi, A.M.; Prasannakumara, B.C.; Sarris, I.E. Impact of Binary Chemical Reaction and Activation Energy on Heat and Mass Transfer of Marangoni Driven Boundary Layer Flow of a Non-Newtonian Nanofluid. *Processes* **2021**, *9*, 702. [CrossRef]
6. Hussain, A.; Hassan, A.; Arshad, M.; Rehman, A.; Matoog, R.T.; Abdeljawad, T. Numerical Simulation and Thermal Enhancement of Multi-Based Nanofluid Over an Embrittled Cone. *Case Stud. Therm. Eng.* **2021**, *28*, 101614. [CrossRef]
7. Benos, L.T.; Karvelas, E.G.; Sarris, I.E. Crucial effect of aggregations in CNT-water nanofluid magnetohydrodynamic natural convection. *Therm. Sci. Eng. Prog.* **2019**, *11*, 263–271. [CrossRef]
8. Yusuf, T.A.; Mabood, F.; Prasannakumara, B.C.; Sarris, I.E. Magneto-Bioconvection Flow of Williamson Nanofluid over an Inclined Plate with Gyrotactic Microorganisms and Entropy Generation. *Fluids* **2021**, *6*, 109. [CrossRef]
9. Benos, L.T.; Karvelas, E.G.; Sarris, I.E. A theoretical model for the magnetohydrodynamic natural convection of a CNT-water nanofluid incorporating a renovated Hamilton–Crosser mode. *Int. J. Heat Mass Transf.* **2019**, *135*, 548–560. [CrossRef]

10. Apostolos, A.G.; Benos, L.T.; Sofiadis, G.N.; Sarris, I.E. A printed-circuit heat exchanger consideration by exploiting an Al2O3-water nanofluid: Effect of the nanoparticles interfacial layer on heat transfer. *Therm. Sci. Eng. Prog.* **2021**, *22*, 100818.
11. Karvelas, E.G.; Karakasidis, T.E.; Sarris, I.E. Computational analysis of paramagnetic spherical Fe_3O_4 nanoparticles under permanent magnetic fields. *Comput. Mater. Sci.* **2018**, *154*, 464–471. [CrossRef]
12. Alghamdi, M.; Wakif, A.; Thumma, T.; Khan, U.; Baleanu, D.; Rasool, G. Significance of variability in magnetic field strength and heat source on the radiative-convective motion of sodium alginate-based nanofluid within a Darcy-Brinkman porous structure bounded vertically by an irregular slender surface. *Case Stud. Therm. Eng.* **2021**, *28*, 101428. [CrossRef]
13. Rasool, G.; Shafiq, A.; Chu, Y.M.; Bhutta, M.S.; Ali, A. Optimal Homotopic Exploration of Features of Cattaneo–Christov Model in Second Grade Nanofluid Flow via Darcy–Forchheimer Medium Subject to Viscous Dissipation and Thermal Radiation. *Comb. Chem. High Throughput Screen.* **2021**, *24*. [CrossRef] [PubMed]
14. Shafiq, A.; Lone, S.A.; Sindhu, T.N.; Al-Mdallal, Q.M.; Rasool, G. Statistical modeling for bioconvective tangent hyperbolic nanofluid towards stretching surface with zero mass flux condition. *Sci Rep.* **2021**, *11*, 13869. [CrossRef]
15. Nield, D.A.; Bejan, A. *Convection in Porous Media*; Springer: New York, NY, USA, 1999.
16. Karniadakis, G.; Beskok, A. *Micro Flows*; Springer: New York, NY, USA, 2002.
17. Karniadakis, G.; Beskok, A.; Aluru, N. *Micro Flows and Nano Flows: Fundamentals and Simulation*; Springr: New York, NY, USA, 2005.
18. Forchheimer, P. Wasserbewegung durch boden. *Zeitschrift Ver. D. Ing.* **1901**, *45*, 1782–1788.
19. Muskat, M. *The Flow of Homogeneous Fluids Through Porous Media*; Edwards: Ogemaw, MI, USA 1946.
20. Pal, D.; Mondal, H. Hydromagnetic convective diffusion of species in Darcy–Forchheimer porous medium with non-uniform heat source/sink and variable viscosity. *Int. Commun. Heat Mass Transf.* **2012**, *39*, 913–917. [CrossRef]
21. Hayat, T.; Taseer, M.; Al-Mezal, S.; Liao, S.J. Darcy–Forchheimer flow with variable thermal conductivity and cattaneo-christov heat flux. *Int. Numer. Methods Heat Fluid Flow* **2016**, *26*, 2355–2369. [CrossRef]
22. Eid, M.R.; Mabood, F. Two-phase permeable non-Newtonian cross-nanomaterial flow with Arrhenius energy and entropy generation: Darcy–Forchheimer model. *Phys. Scr.* **2020**, *95*, 105209. [CrossRef]
23. Shankaralingappa, B.M.; Prasannakumara, B.C.; Gireesha, B.J.; Sarris, I.E. The Impact of Cattaneo–Christov Double Diffusion on Oldroyd-B Fluid Flow over a Stretching Sheet with Thermophoretic Particle Deposition and Relaxation Chemical Reaction. *Inventions* **2021**, *6*, 95. [CrossRef]
24. Jamshed, W.; Nisar, K.S.; Gowda, R.J.P.; Kumar, R.N.; Prasannakumara, B.C. Radiative heat transfer of second grade nanofluid flow past a porous flat surface: A single-phase mathematical model. *Phys. Scr.* **2021**, *96*, 064006. [CrossRef]
25. Sheikholeslami, M.; Hatami, M.; Ganji, D.D. Analytical investigation of MHD nanofluid flow in a semi-porous channel. *Powder Tech.* **2013**, *246*, 327–336. [CrossRef]
26. Kumar, R.N.; Suresha, S.; Gowda, R.J.; Megalamani, S.B.; Prasannakumara, B.C. Exploring the impact of magnetic dipole on the radiative nanofluid flow over a stretching sheet by means of KKL model. *Pramana J. Phys.* **2021**, *95*, 180. [CrossRef]
27. Kumar, R.S.V.; Dhananjaya, P.G.; Kumar, R.N.; Gowda R.J.P.; Prasannakumara, B.C. Modeling and theoretical investigation on Casson nanofluid flow over a curved stretching surface with the influence of magnetic field and chemical reaction. *Int. J. Comput. Methods Eng. Sci. Mech.* **2021**, *23*, 12–19. [CrossRef]
28. Hayat, T.; Taseer, M.; Shehzad, S.A.; Alsaedi, A.; Al-Solamy, F. Radiative three-dimensional flow with chemical reaction. *Int. J. Chem. Reactor Eng.* **2016**, *14*, 79–91. [CrossRef]
29. Sarada, K.; Gowda, R.J.P.; Sarris, I.E.; Kumar, R.N.; Prasannakumara, B.C. Effect of Magnetohydrodynamics on Heat Transfer Behaviour of a Non-Newtonian Fluid Flow over a Stretching Sheet under Local Thermal Non-Equilibrium Condition. *Fluids* **2021**, *6*, 264. [CrossRef]
30. Sheikholeslami, M. Magnetic field influence on nanofluid thermal radiation in a cavity with tilted elliptic inner cylinder. *J. Mol. Liq.* **2017**, *229*, 137–147. [CrossRef]
31. Charakopoulos, A.; Karakasidis, T.; Sarris, I.E. Analysis of magnetohydrodynamic channel flow through complex network analysis. *Chaos* **2021**, *31*, 043123. [CrossRef]
32. Hamid, A.; Chu, Y.M.; Khan, M.I.; Kumar, R.N.; Gowda, R.J.P.; Prasannakumara, B.C. Critical values in axisymmetric flow of magneto-Cross nanomaterial towards a radially shrinking disk. *Int. J. Mod. Phys.* **2021**, *35*, 2150105. [CrossRef]
33. Hayat, T.; Saif, R.S.; Ellahi, R.; Muhammad, T.; Ahmad, B. Numerical study for Darcy–Forchheimer flow due to a curved stretching surface with Cattaneo–Christov heat flux and homogeneous-heterogeneous reactions. *Results Phys.* **2017**, *7*, 2886–2892. [CrossRef]
34. Ali, B.; Nie, Y.; Hussain, S.; Manan, A.; Sadiq, M.T. Unsteady magneto-hydrodynamic transport of rotating maxwell nanofluid flow on a stretching sheet with Cattaneo–Christov double diffusion and activation energy. *Therm. Sci. Eng. Prog.* **2020**, *20*, 100720. [CrossRef]
35. Sultana, U.; Mushtaq, M.; Muhammad, T.; Albakri, A. On Cattaneo–Christov heat flux in carbon-water nanofluid flow due to stretchable rotating disk through porous media. *Alex. Eng. J.* **2021**, *61*, 3463–3474. [CrossRef]
36. Reddy, M.G.; Rani, M.S.; Kumar, K.G.; Prasannakumar, B.C.; Lokesh, H.J. Hybrid dusty fluid flow through a Cattaneo–Christov heat flux model. *Phys. Stat. Mech. Its Appl.* **2020**, *551*, 123975. [CrossRef]
37. Farooq, U.; Waqas, H.; Khan, M I.; Khan, S.U.; Chu, Y.M.; Kadry, S. Thermally radioactive bioconvection flow of Carreau nanofluid with modified Cattaneo–Christov expressions and exponential space-based heat source. *Alex. Eng. J.* **2021**, *60*, 3073–3086. [CrossRef]
38. Reddy, M.G.; Kumar, K.G. Cattaneo-Christof heat flux feature on carbon nanotubes filled with micropolar liquid over a melting surface: A streamline study. *Int. Commun. Heat. Mass Transf.* **2021**, *122*, 105142. [CrossRef]

39. Shah, F.; Khan, M.I.; Hayat, T.; Momani, S.; Khan, M.I. Cattaneo–Christov heat flux (CC model) in mixed convective stagnation point flow towards a Riga plate. *Comput. Methods Programs Biomed.* **2020**, *196*, 105564. [CrossRef] [PubMed]
40. Wakif, A. A novel numerical procedure for simulating steady MHD convective flows of radiative Casson fluids over a horizontal stretching sheet with irregular geometry under the combined influence of temperature-dependent viscosity and thermal conductivity. *Math. Probl. Eng.* **2020**, *2020*, 1675350. [CrossRef]
41. Ramesh, K.; Joshi, V. Numerical Solutions for Unsteady Flows of a Magnetohydrodynamic Jeffrey Fluid Between Parallel Plates Through a Porous Medium. *Int. J. Comput. Methods Eng. Sci. Mech.* **2019**, *20*, 1–13. [CrossRef]
42. Muhammad, T.; Alsaedi, A.; Shehzad, S.A.; Hayat, T. A revised model for Darcy–Forchheimer flow of Maxwell nanofluid subject to convective boundary condition. *Chin. J. Phys.* **2017**, *55*, 963–976. [CrossRef]
43. Lodhi, R. K.; Ramesh, K. Comparative study on electroosmosis modulated flow of MHD viscoelastic fluid in the presence of modified Darcy's law. *Chin. J. Phys.* **2020**, *68*, 106–120. [CrossRef]
44. Shijun, L. Homotopy analysis method: A new analytic method for nonlinear problems. *Appl. Math. Mech.* **1998**, *19*, 957–962. [CrossRef]
45. Rasool, G.; Zhang, T. Darcy–Forchheimer nanofluidic flow manifested with Cattaneo–Christov theory of heat and mass flux over non-linearly stretching surface. *PLoS ONE* **2019**, *14*, e0221302. [CrossRef] [PubMed]
46. Rasool, G.; Chamkha, A.J.; Muhammad, T.; Shafiq, A.; Khan, I. Darcy–Forchheimer relation in Casson type MHD nanofluid flow over non-linear stretching surface. *Propuls. Power Res.* **2020**, *9*, 159–168. [CrossRef]
47. Wakif, A.; Animasaun, I.L.; Khan, U.; Shah, N.A.; Thumma, T. Dynamics of radiative-reactive Walters-B fluid due to mixed convection conveying gyrotactic microorganisms, tiny particles experience haphazard motion, thermo-migration, and Lorentz force. *Phys. Scr.* **2021**, *96*, 125239. [CrossRef]
48. Dawar, A.; Wakif, A.; Thumma, T.; Shah, N.A. Towards a new MHD non-homogeneous convective nanofluid flow model for simulating a rotating inclined thin layer of sodium alginate-based Iron oxide exposed to incident solar energy. *Int. Commun. Heat Mass Transf.* **2022**, *130*, 105800. [CrossRef]
49. Modest, M.F. *Radiative Heat Transfer*, 2nd ed.; Academic Press: San Diego, CA, USA, 2021.
50. Pakdemirli, M.; Yurusoy, M. Similarity Transformations for Partial Differential Equations. *SIAM Rev.* **1998**, *40*, 96–101. [CrossRef]

 *micromachines*

Article

A One-Dollar, Disposable, Paper-Based Microfluidic Chip for Real-Time Monitoring of Sweat Rate

Hongcheng Wang [1,*], Kai Xu [1], Haihao Xu [1], Along Huang [1], Zecong Fang [2,3], Yifan Zhang [1], Ze'en Wang [1], Kai Lu [1], Fei Wan [1], Zihao Bai [1], Qiao Wang [4], Linan Zhang [1] and Liqun Wu [1]

[1] School of Mechanical Engineering, Hangzhou Dianzi University, Hangzhou 310018, China; xukai192010065@hdu.edu.cn (K.X.); xhh761335698@hdu.edu.cn (H.X.); huangalong@hdu.edu.cn (A.H.); 202010139@hdu.edu.cn (Y.Z.); 202010053@hdu.edu.cn (Z.W.); lukai@hdu.edu.cn (K.L.); 211010006@hdu.edu.cn (F.W.); 212010101@hdu.edu.cn (Z.B.); zln@hdu.edu.cn (L.Z.); wuliqun@hdu.edu.cn (L.W.)

[2] Institute of Biomedical and Health Engineering, Shenzhen Institute of Advanced Technology, Chinese Academy of Sciences, Shenzhen 518055, China; zc.fang@siat.ac.cn

[3] Shenzhen Engineering Laboratory of Single-Molecule Detection and Instrument Development, Shenzhen 518055, China

[4] School of Pharmacy, Hangzhou Medical College, Hangzhou 310059, China; wangqiao-1@163.com

* Correspondence: wanghc@hdu.edu.cn; Tel.: +86-1595-8149-841

Citation: Wang, H.; Xu, K.; Xu, H.; Huang, A.; Fang, Z.; Zhang, Y.; Wang, Z.; Lu, K.; Wan, F.; Bai, Z.; et al. A One-Dollar, Disposable, Paper-Based Microfluidic Chip for Real-Time Monitoring of Sweat Rate. *Micromachines* **2022**, *13*, 414. https://doi.org/10.3390/mi13030414

Academic Editor: Khashayar Khoshmanesh

Received: 25 January 2022
Accepted: 4 March 2022
Published: 6 March 2022

Publisher's Note: MDPI stays neutral with regard to jurisdictional claims in published maps and institutional affiliations.

Abstract: Collecting sweat and monitoring its rate is important for determining body condition and further sweat analyses, as this provides vital information about physiologic status and fitness level and could become an alternative to invasive blood tests in the future. Presented here is a one-dollar, disposable, paper-based microfluidic chip for real-time monitoring of sweat rate. The chip, pasted on any part of the skin surface, consists of a skin adhesive layer, sweat-proof layer, sweat-sensing layer, and scale layer with a disk-shape from bottom to top. The sweat-sensing layer has an impressed wax micro-channel containing pre-added chromogenic agent to show displacement by sweat, and the sweat volume can be read directly by scale lines without any electronic elements. The diameter and thickness of the complete chip are 25 mm and 0.3 mm, respectively, permitting good flexibility and compactness with the skin surface. Tests of sweat flow rate monitoring on the left forearm, forehead, and nape of the neck of volunteers doing running exercise were conducted. Average sweat rate on left forearm (1156 $g \cdot m^{-2} \cdot h^{-1}$) was much lower than that on the forehead (1710 $g \cdot m^{-2} \cdot h^{-1}$) and greater than that on the nape of the neck (998 $g \cdot m^{-2} \cdot h^{-1}$), in good agreement with rates measured using existing common commercial sweat collectors. The chip, as a very low-cost and convenient wearable device, has wide application prospects in real-time monitoring of sweat loss by body builders, athletes, firefighters, etc., or for further sweat analyses.

Keywords: wearable device; microfluidic chip; sweat collecting

1. Introduction

Sweat is known to contain important information corresponding to the status of an individual's health [1]. Sweat production is related to the stimuli underlying body thermoregulation [2]. Secretion of sweat from eccrine glands on the skin surface is an essential means of heat loss in regulating body temperature and for maintaining homeostasis [3] during heat acclimation [4]. Lost body water must be replaced to maintain normal physiologic processes; this is particularly important for body builders, athletes [5], firefighters, etc. Furthermore, sweat, like saliva and tears [6], is noninvasively induced from deeper in the body and carries a diverse array of biomolecules, ranging from small electrolytes (including Na^+, K^+, and Ca^{2+}) and metabolites (such as glucose, lactate [7], and ethanol [8]) to hormones and larger proteins [9], which may provide vital information about physiological status and fitness level [10]. Sweat analysis could become an alternative to invasive blood tests in the future [11,12], as Heikenfeld et al. [13] demonstrated—for the first time in vivo—the complete correlation between continuous sweat data and blood data. Collecting and monitoring the rate of sweat production, which is the primary and key step for sweat research, are important for determining body condition and enabling further sweat analyses.

The whole-body wash-down method [14] is an early sweat-sampling technology and well-known as the gold standard for determining whole-body sweat loss, as all sweat runoff is collected. Subjects wearing minimal clothing ride a cycle ergometer in a plastic box. The subject, box, equipment, clothes, and all objects touched by the subject are thoroughly rinsed with deionized water to determine the whole-body sweat loss. However, this method is limited by the controlled laboratory setting, complexity of steps, and single mode of exercise testing; thus, it is not practical for field studies. Patches, composed of an absorbent material with a hydrophilic and porous structure [15], are used for regional skin surface collection and localized sweat sampling. This enables collection of sweat for hours positioned in a specific location [16] (e.g., forearm, thigh, back, or calf), but the collected sweat patch must be peeled off and carefully weighed to obtain the average sweat flow rate. The error rate is relatively high due to evaporation of sweat during the process, and it cannot show real-time flow rate data [17]. In addition, Zhang et al. [18] designed a microfluidic device with one-way-opening chambers and hydrophobic valves for sweat collection and analysis. Pan et al. [19] presented the first digital droplet flowmetry implemented on existing textile substrates for real-time flow rate measurement by counting the number of droplets. Eliot et al. [20] developed a flow rate sensor that easily couples to the outlet of a microfluidic channel to measure flow rate via periodic temporary shorting caused by droplets passing between two electrodes. The device was tested in a dynamic range as low as 25 nL·min^{-1} and as high as 9×10^5 nL·min^{-1}. Lindsay et al. [21] designed a skin-interfaced microfluidic system involving multilayered stacks of thin-film polymers that contain intricate microfluidic channels for personalized sweating rate and sweat chloride analytics for sports science applications.

One of the most common current commercial samplers is the Macroduct [22], which consists of a concave disk and a spiral plastic tube that collects sweat. Compared with patches, Macroducts avoid sweat leakage, contamination, and potential hydromeiosis, because the sweat is almost immediately removed from the skin. However, the device cannot be positioned on any position of the human body, so the popularity of Macroducts remains limited. A variety of modified absorbent materials, such as paper [23,24], nonwoven fabrics, textiles [25], cellulosic materials, hydrogels [26,27], and rayon pads [28] are widely used as sweat-collecting carriers. Among these materials, filter paper composed of disorderly stacked cellulose fibers with abundant hydroxyl (-OH) active groups provides high porosity, thus facilitating rapid imbibition of fluid and rendering it a very promising substrate for the immobilization of bioactive substances. Filter paper-based sweat-collecting chips are paper-based microfluidic analytical devices [29] that have generated great interest among researchers due to their portability, low cost [30], versatility, and ease of results interpretation in the analytical area due to their attractive passive movement properties (capillary phenomenon [31]) of analytes without any external forces. These chips show great promise for applications in point-of-care health systems [32], environmental monitoring [33], and food safety.

Most of the above systems, however, have complex structures or require an external collection mechanism, which is not suitable for widespread application or batch production. Paper-based microfluidic chips, which are presented in this article for the first time, can be used for real-time monitoring of sweat secretion rate. The chips, which can be pasted on any part of the skin surface, are low cost and disposable, consisting mainly of filter paper and adhesive tape with a disk shape. Sweat volume can be read directly by scale lines without any electronic elements.

2. Materials and Methods

2.1. Structure of Paper-Based Sweat Rate Monitoring Chips

Paper-based sweat rate monitoring chips (P-SRMCs) consist of a skin adhesive layer, a sweat-proof layer, a sweat-sensing layer, and a scale layer from the bottom to top, as shown in Figure 1a. The skin adhesive layer on the bottom, made of medical-grade double-sided adhesive tape (PICARO), is used to attach the chip to the skin surface. A small hole with a diameter of 2 mm in the center serves as the inlet for sweat secreted from the skin. The sweat-proof layer with a center hole the same size as that of the bottom layer hole is made of single-sided transparent adhesive tape (202102022207, DELE) and prevents sweat from penetrating through the double-sided adhesive material to the sweat-sensing layer and ensures that sweat flows through the center hole.

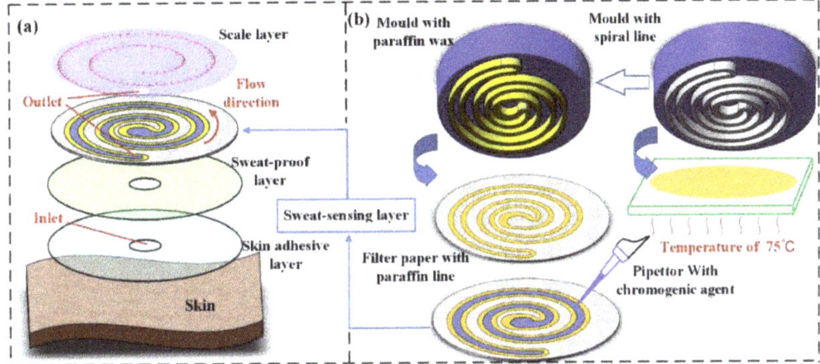

Figure 1. Structure of the paper-based sweat rate monitoring chip: (**a**) exploded view showing all layers of the whole chip, including the skin adhesive layer, sweat-proof layer, sweat-sensing layer, and scale layer; (**b**) method for fabrication of the sweat-sensing layer using a 3D-printed mold with a double-spiral structure.

The sweat-sensing layer, as the core layer, is used to determine sweat volume. A schematic illustration of its fabrication process is shown in Figure 1b. Paraffin wax (58#, Jinmen Weijia Industro Co., Ltd., Jinmen, China) is heated using a constant-temperature heating device (ET-200, ETOOL) to the melting state of 72 °C on a section of glass slide. A mold with two parallel spiral structures 1.5 mm high is printed using a 3D printer (Pro2, Raise3D) with white polylactate as the material. The double spiral structure has a smooth flow path, can form longer channel per area than other designs, and is suitable for a disc-shape sensing chip. The parallel spiral structure is dipped into melted paraffin and then transferred onto a piece of filter paper (102, Aoke, maximum void in the range of 15–20 µm) to form paraffin wax lines (marked in yellow). The two parallel spiral wax lines constitute a micro-channel through which collected sweat travels, because the wax material is incompatible with aqueous sweat and functions as a boundary. The wax-impressed, paper-based microfluidic chip is thus one of the most promising methods for future applications because it is inexpensive, easy to use, provides rapid and robust results, and is harmless to human skin [34].

Cobalt chloride solution (AR grade, Chengdu Huaze Cobalt and Nickel Material Co., Ltd., Chengdu, China) with a mass fraction of 0.157 g/mL (0.65 mol/L), which is a good chromogenic agent to H_2O molecules, is used as a precursor for the sweat chromogenic agent. The color of anhydrous $CoCl_2$ changes from blue to red when it absorbs H_2O molecules, forming $CoCl_2 \cdot 6H_2O$. Cobalt chloride solution is harmless to human skin surface and the color of $CoCl_2 \cdot 6H_2O$ turns back to blue when H_2O molecules are removed, which makes the chip possible to be reused if necessary. Cobalt chloride is slowly added to the sweat flow micro-channel using a pipettor. Then the filter paper with $CoCl_2$ solution and wax lines is dried for 90 min in a vacuum drying oven (ZKXF-1, Shanghai Shuli Yiqi Yibiao Co., Ltd., Shanghai, China) set at 45 °C to remove H_2O molecules.

The scale layer, as the top layer, is also made of single-sided transparent adhesive tape. A small hole with a diameter of 2 mm is punched and aligned to the end of the sweat flow channel in the sensing layer. The hole exposed to air is used for releasing increased air pressure caused by the entry of sweat and continuously draws sweat through the micro-channel [35]. In addition to the higher flow rate, the biggest advantage of the small hole for evaporation is that it enables convenient control of the flow rate [36]. By changing the size of the small hole, the flow rate can be easily regulated. The scale, made of red stamp ink (YY01, GSD), is impressed on the reverse side using a mold to avoid the possibility of being removed while the subject is exercising.

The diameter of all four above-mentioned layers is 25 mm and cut by a Laser Cutting Machine (3020, KETAILASER). A piece of P-SRMC is prepared according to the process shown in Figure 1, and the edge of the assembled chip is covered with paraffin wax to prevent external water molecules from affecting the test results. To make subjects more comfortable while doing exercise, the overall thickness of the microfluidic patch was limited to 0.3 mm to enable good flexibility and good compactness against the skin surface.

2.2. Sweat Micro-Channel Parameters

The ratio of the maximum volume of sweat monitored to the size of the whole chip is the most important index for a wearable device. The whole chip size is determined by the width of the wax line (w_w) and the sweat channel (w_c). If the wax line is too narrow, sweat will leak out across it in the filter paper and affect the measured result; thus, a sweat-leaking experiment was conducted to determine the minimum wax width. As to the sweat channel, if the channel is too narrow, the sweat travel velocity will be too slow. Thus, a sweat flow velocity experiment was also conducted to determine the minimum flow channel width. A chamfered fillet structure was applied on the inlet and outlet parts. To minimize the chip size, the shape of the microchannel was designed as a double spiral structure, as shown in Figure 2. In Cartesian coordinates, the equations of two spiral lines on external and internal sides are respectively obtained as:

$$\begin{cases} r = a + b(\theta/2\pi) \\ x = r\cos\theta \\ y = r\sin\theta \end{cases} \tag{1}$$

$$\begin{cases} R = a + w_w + w_c + b(\theta/2\pi) \\ x = R\cos\theta \\ y = R\sin\theta \end{cases} \tag{2}$$

where r is the radius of the spiral line on the external side, R is that on the internal side, and θ is the angle of spiral lines. The differential of R and r is ($w_w + w_c$).

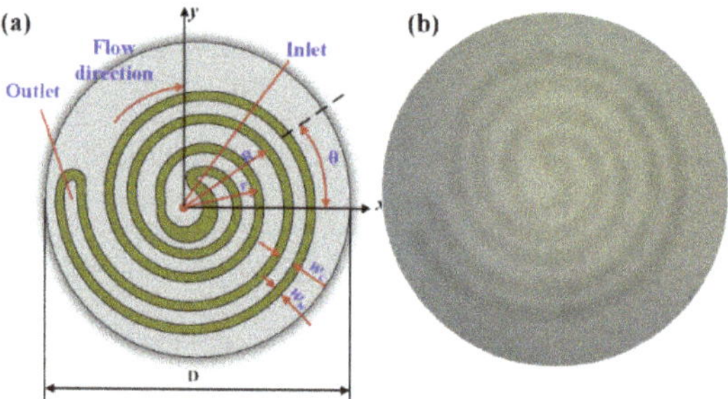

Figure 2. Sweat channel with double spiral structure: (**a**) structure sketch and (**b**) photomicrograph of the wax channel impressed on filter paper using a mold.

2.3. Chip Assembly

A complete chip fabricated according to the above process is shown in Figure 3. All the assembly steps are conducted under a microscope (3R-MSUSB401, Anyty). Scale lines should be located parallel to and on the lateral side of the wax channel to show the displacement of collected sweat traveling through the channel. The scale spacing is 1 mm, with a range of 0–90 mm. The edge of the chip appears grey in color because all the layers are shaped by a laser cutting machine, and the edge is burned to ashes. This does not affect the chip's ability to collect sweat because the whole edge is covered by paraffin wax to prevent external water molecules from affecting the test results. The assembled chip is dried for 30 min in a vacuum drying oven to thoroughly remove water molecules.

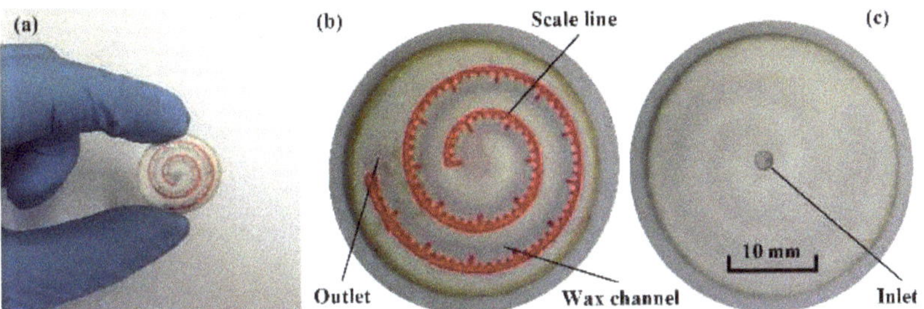

Figure 3. (**a**) Photograph of a P-SRMC chip, (**b**) top view, and (**c**) bottom view of the assembled chip.

2.4. Chip Calibration

As a type of measuring device, the chip must be calibrated before being tested on the human body. Sweat, drawn from human skin using a suction tube, is added to the chip through the inlet in the scale layer using a pipettor (volume resolution of 0.1 µL). The liquid is added drop by drop with a single drop volume of 1.0 µL. Each droplet should be added after the previous droplet has thoroughly infiltrated into the wax channel by observing the inlet area under a physical microscope. The displacement caused by sweat moving forward along the helix wax channel is recorded via the scale lines after each sweat droplet is added based on the channel with sweat molecules turning from blue to red.

2.5. Testing on Human Volunteers

Two healthy, active male and female volunteers, 24 years old, participated in indoor running sweat-collecting trials. The weight and height of male candidate are 83 kg and 175 cm, respectively and that of female are 52 kg and 162 cm, respectively. The average room temperature and humidity were 28.9 °C and 74%, respectively. The volunteers ran on a treadmill (SH-T5170) under controlled conditions with a running speed of 8.8 kph. When considering localized areas of interest, the choice of sampling area is very important, as it has been reported that sweat rate depends significantly on the sampling location [37]. The sweat-secreting rates on the forehead, nape of the neck, and left forearm are larger than the others on human body. These positions are usually exposed to air while candidates are exercising and suitable for affixing the P-SRMCs. In this study, the chips were positioned on the forehead, nape of the neck, and left forearm, as shown in Figure 4.

A mobile phone (iPhone XR with resolution of 1080p and frame rate of 240 fps), fixed on the left forearm by a designed holder and with its camera aligned with the sensing chip, was used to record video of the sweat-collection process. The displacement in chips on the forehead and nape of the neck were recorded by photos taken every 5 min and then used to calculate sweat volume using Formula (3), as the displacement value can be read from the scale lines. Scientific research shows that it takes about 30 min for an average person to have the best sports effect. So, the testing time is set as slightly longer than 30 min. Sweat rates (v_s) are expressed as $g \cdot m^{-2} \cdot h^{-1}$, for it is usually calculated in grams per square meter of body surface area per hour.

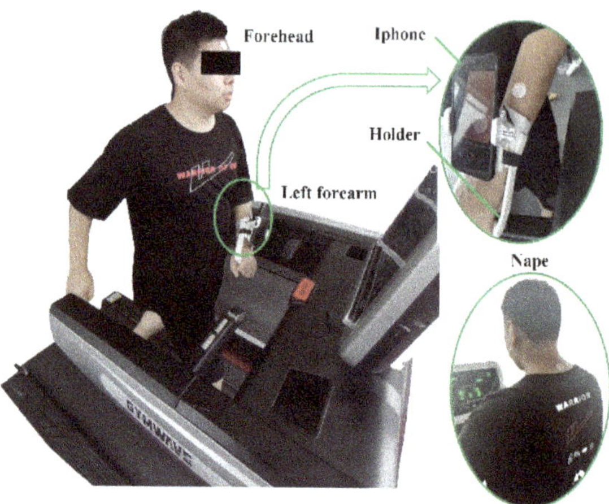

Figure 4. Experimental setup involving a male volunteer running under controlled conditions with chips located on the forehead, left forearm, and nape of the neck.

3. Results and Discussion

3.1. Sweat Micro-Channel Parameters

A sweat leakage experiment was conducted to determine the minimum wax line width to prevent sweat leaking through the filter paper. 3D-printed molds of different sizes were designed to form wax lines with different widths on the filter paper. Sweat (2.5 μL) was added on the initial site using a pipettor, and the paper was placed on a piece of horizontally situated glass slide to observe leakage using a microscope. As shown in Figure 5, when the width (w_w) of the wax lines was 0.5 mm and 0.6 mm, the $CoCl_2$ solution leaked across the wax lines. When the width was no less than 0.7 mm, the wax lines completely prevented leakage. Therefore, 0.7 mm was determined to be the minimum wax width.

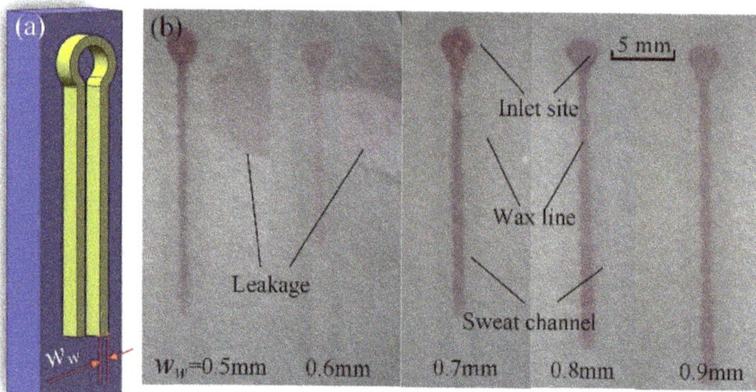

Figure 5. Sweat leakage experiment between neighboring wax lines with different widths: (**a**) mould for wax channels; (**b**) photo of channels with sweat.

The sweat channel width (w_c) is another important parameter for sweat collection, as it determines the travel rate of the sweat. Although Darcy's law, Lucas-Washburn equation [38], Modifications to Darcy's law and LW equation [39], Richards equation [40], and flow simulation or visualization tools [41] can approximately calculate the micro-fluid flow behavior in paper material, fluid imbibition into paper is a complex process governed by a highly coupled system of length and time-scaled parameters. Therefore, the sweat channel width in filter paper is typically determined by experiment, traditionally called a trial-and-error strategy. Besides that, capillary flow velocity inner filter paper is determined by pore radius of the porous media in the hydrophilic channel and the size of the channel according to Ref. [42]. The shape of channel bas little effect on flow velocity of sweat if the flow rate is relatively low. The experiment result in straight channel is suitable for spiral-shape ones.

To minimize the chip size, a width of 1.2 mm was chosen as the micro-channel parameter for the final sweat rate monitoring chip. A $CoCl_2$ solution with a concentration of 0.65 mol/L and total volume of 1.6 μL was deposited in the channel for result visualization. The flow distance was set at 10 mm to calculate the average flow velocity, $\bar{v}$. The process of sweat flow in the paper-based wax channel with a width of 1.2 mm is shown in Figure 6 and Video S1 in the Supplementary Material. Figure 7 shows the influence of channel width on sweat flow velocity in straight paper-based channels. The average flow velocity increased rapidly in channel widths set at 1.2 mm, 1.4 mm, and 1.6 mm. More than 70 s (average velocity of approximately 0.14 mm/s) was required for sweat to travel along the channel with a relatively narrow width of 1.0 mm.

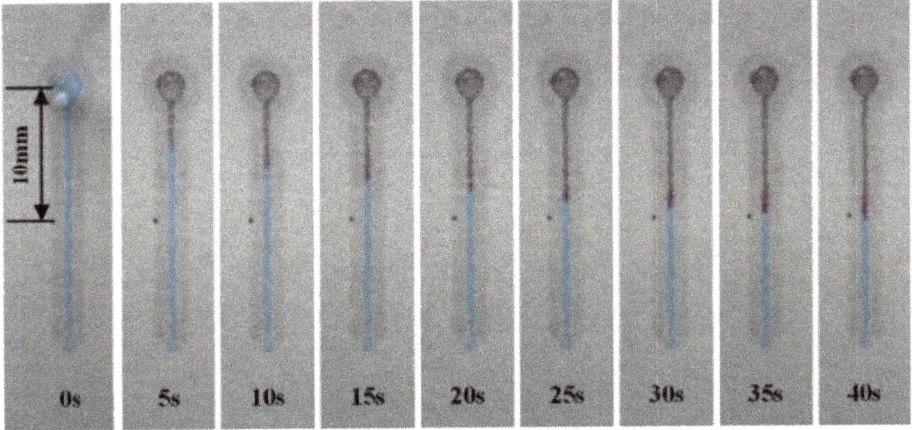

Figure 6. Recordings illustrating the process of sweat flow in a paper-based wax channel with a width of 1.2 mm and displacement of 10 mm.

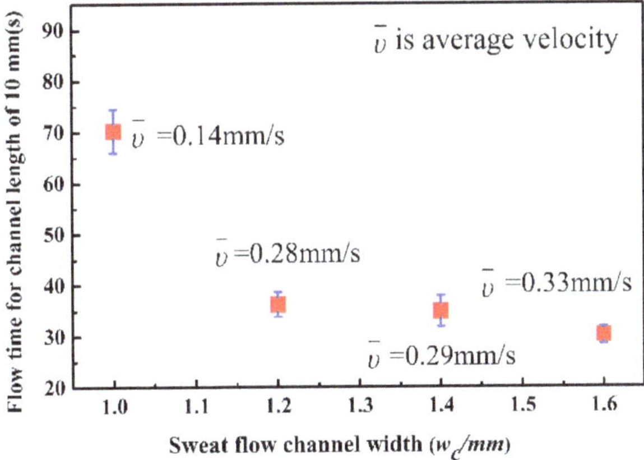

Figure 7. Variation in sweat flow velocity with width of the paper-based channel.

Figure 7 shows average sweat travel velocity data. However, the flow velocity was not constant and decreased slowly because of the increasing flow resistance downstream in the micro-channel. Therefore, flow displacement in a paper-based wax channel with a width of 1.2 mm was recorded at different times, and the results are shown in Figure 8. The velocity was relatively high near the sweat inlet site then decreased until displacement was less than approximately 7 mm and finally became approximately constant. The reason the velocity decreased in the first section is that $CoCl_2$ crystal particles were scoured downstream by sweat flow, causing an enrichment that increased the flow resistance. The experiment results show that sweat with a volume of 2.5 µL filled the paper-based channel in approximately 60 s.

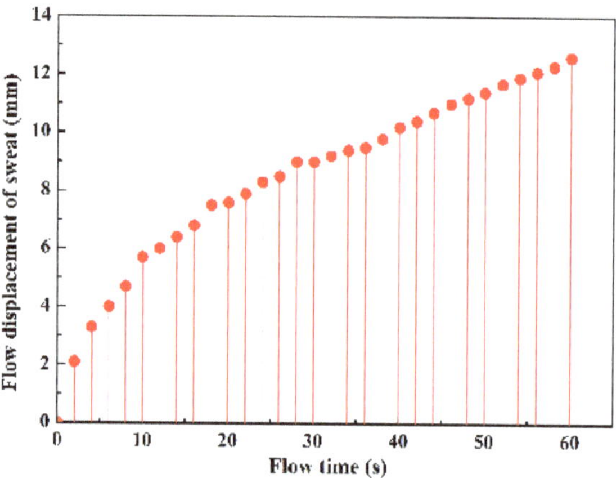

Figure 8. Record of sweat flow displacement in a paper-based wax channel with a width of 1.2 mm.

3.2. Chip Calibration Results

In chip calibration experiments, it is time-consuming to completely fill the wax channel with sweat drop by drop. A record of sweat traveling through a certain piece of P-SRMC chip during the calibration process is shown in Figure 9. The color of the channel through which the sweat travels changed from blue to red. The scale in red shows the displacement of sweat as it travels. As shown in Video S2 in the Supplementary Material, it takes each 1-µL sweat droplet approximately 5 min on average to infiltrate the wax channel completely. The time required by latter droplets is longer than that of former droplets because of the gradually increasing flow resistance as drops are added.

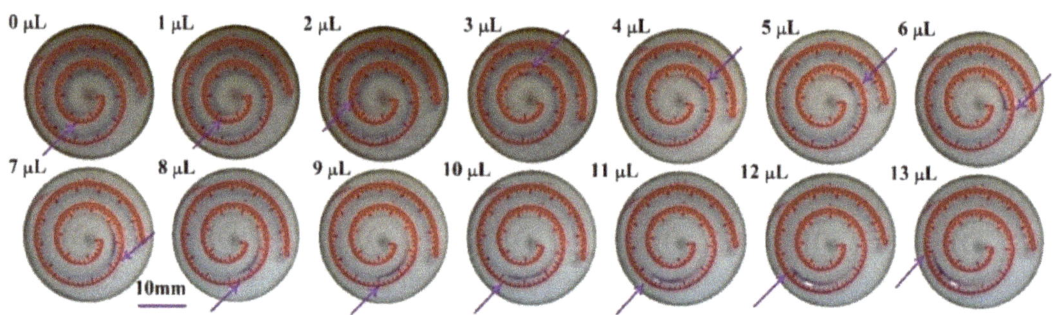

Figure 9. Calibration of a P-SRMC.

Ten samples of sweat colleting chips were tested to obtain an average result and determine their consistency. As shown in Figure 10, the displacement of sweat moving forward along the spiral wax channel was recorded using the scale lines after different volumes of sweat were added from the chip inlet.

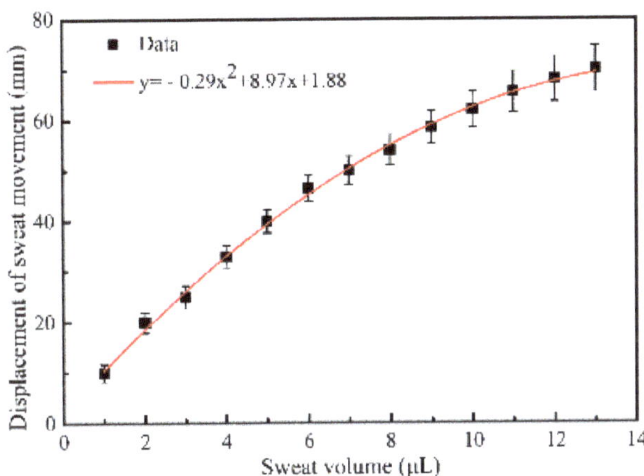

Figure 10. Second-order fitting curve for the variation in displacement with added sweat volume determined using scale lines.

A second-order fitting, as is shown in Formula (3), was carried out on the variation:

$$y = -0.29x^2 + 8.97x + 1.88 \ (0 < x < 13.4) \tag{3}$$

where y is displacement (mm) and x is the volume of sweat (μL). The relative variance was ~10%.

According to the experimental result, the maximum volume of sweat collected in a single chip is above 14 μL, and the sensitivity (s) can be calculated by:

$$s = \frac{dy}{dx} = -0.58x + 8.97 \tag{4}$$

The sensitivity decreases gradually with increasing volume of sweat collected. Figure 10 shows that it reaches a maximum value of 8.97 mm·μL^{-1}.

3.3. Testing Results Using Human Volunteers

The displacement caused by sweat moving forward was read using the scale lines on the sweat-collecting chips locating on the left forearm, forehead, and nape of the neck of the human volunteers, as shown in Table 1. The sweat secretion velocity on male candidate is faster than that on female candidate. Photographs of P-SRMC changing color on male skin surface are shown in Figure 11. The time interval for the recording was 5 min. Video of sweat traveling through the chip on the left forearm of male candidate was recorded using a mobile phone, as shown in Video S3 in the Supplementary Material. The values were consequently converted to volume of sweat collected using Formula (3). According to the shape of the above fitting function, the smaller of the two solutions for the unitary quadratic equation is the sweat volume. The conversion results for each displacement of sweat traveling through the chip are shown in Figure 12.

Table 1. Record of test results for human volunteers.

Time/min	Left Forearm/mm		Forehead/mm		Nape of Neck/mm	
	Male	Female	Mail	Female	Mail	Female
0	0	0	0	0	0	0
5	4.2	3.1	11.8	6.3	3.5	2.6
10	18.5	15.0	36.8	20.7	19.2	14.6
15	29.5	20.1	49.5	31.3	26.8	19.5
20	35.2	25.3	54.8	42.9	32.6	21.9
25	41.8	30.0	58.8	49.8	40.5	29.8
30	46.5	38.2	61.5	52.1	44.2	34.2
35	52.8	45.8	66.5	56.5	47.9	38.6

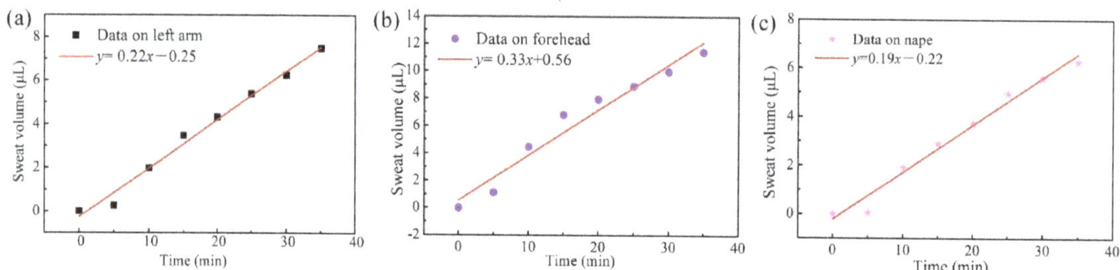

Figure 11. Photographs of P-SRMC changing color on the (**a**) left forearm, (**b**) forehead, and (**c**) nape of the neck taken every 5 min while the volunteers was running on a treadmill.

Figure 12. Flow rate of sweat collected from (**a**) the left forearm, (**b**) forehead, and (**c**) nape of the neck measured using the P-SRMC.

The test results showed that the relationship between volume of sweat secreted from the left forearm, forehead, and nape of the neck and running time was approximately linear. In other words, the sweat secretion velocity is approximately uniform. Considering that the diameter of the sweat-collecting inlet is 2 mm, the flow rate per area can be calculated. However, during the experiment, some sweat is converted into vapor around the chip, as the temperature of the skin surface at the measurement site is much higher than that of the ambient environment. The sweat vapor formed close to the inlet hole will seep into the sensing layer and cause the experimental result to be greater than the actual value. To minimize this effect, the correction coefficient δ was calculated for the collecting area and set as 2. Therefore, the collecting area ($A_c = 4\pi$ mm^2) is twice that of the inlet hole diameter.

Among the three measured sites, the forehead had the highest average sweat secreting velocity (V_f), 0.33 $\mu L \cdot min^{-1}$, which is the slope of the fitted straight line in Figure 12b. The flow velocity per area for the forehead position was calculated using Formula (5) and equaled 0.026 $\mu L \cdot mm^{-2} \cdot min^{-1}$.

$$R_f = \frac{V_f}{A_c} \tag{5}$$

Sweat rate is usually calculated in grams per square meter of body surface area per hour. Therefore, the rate of sweat secretion from the forehead position was 1734 $g \cdot m^{-2} \cdot h^{-1}$ if we assume that the density of sweat is 1.1 $g \cdot cm^{-3}$. The test result showed good agreement in order of magnitude with that measured using absorbent pads [43] and Macroduct [22], which are common commercial sweat collectors for determining sweating rate. As shown in Figure 12a,c, the sweat secretion velocity on the left forearm (V_{arm}) and nape of the neck (V_{nape}) was 0.22 $\mu L \cdot min^{-1}$ and 0.19 $\mu L \cdot min^{-1}$, respectively. The secretion rates for the above two positions were 1156 $g \cdot m^{-2} \cdot h^{-1}$ and 998 $g \cdot m^{-2} \cdot h^{-1}$, respectively. Therefore, the sweat rate of the left forearm was much lower than that of the forehead and greater than that of the nape of the neck for regional variations in human eccrine sweat gland density and local sweat secretion rates during the thermal loading in exercising individuals [44].

4. Conclusions

This paper describes a low-cost, disposable, paper-based microfluidic chip for real-time sweat secretion rate monitoring. The chip consists primarily of a skin adhesive layer, a sweat-proof layer, a sweat-sensing layer, and a scale layer with a disk-shape from the bottom to top. Double spiral wax lines impressed in a piece of filter paper using a mold serve as the micro-channel through which the collected sweat travels. The micro-channel is pre-filled with a chromogenic agent to show displacement of the sweat as it travels, so sweat volume can be read directly using scale lines without any electronic elements. The diameter and thickness of the whole chip are 25 mm and 0.3 mm, respectively, which allows for good flexibility and good compactness with the skin surface. Tests of sweat flow rate monitoring on the left forearm, forehead, and nape of the neck of human volunteers during running were conducted. The average sweat secretion rates of the three positions were 1156 $g \cdot m^{-2} \cdot h^{-1}$, 1710 $g \cdot m^{-2} \cdot h^{-1}$, and 998 $g \cdot m^{-2} \cdot h^{-1}$, respectively, in good agreement with values measured using existing common commercial sweat collectors. The P-SRMC chip, as a very low-cost, disposable, and easily fabricated wearable device, has a wide range of potential applications in real-time monitoring of sweat loss for body builders, athletes, firefighters, etc., or for further sweat analyses.

Supplementary Materials: The following are available online at https://www.mdpi.com/article/10.3390/mi13030414/s1, Video S1: Sweat flow process in paper-based channel with width of 1.2 mm; Video S2: Calibration of sweat rate real-time monitoring chip; Video S3: Test of sweat flow rate monitoring on left forearm of a male volunteer doing running exercise.

Author Contributions: H.W. conceived the idea and wrote the paper. K.X. designed the experiments, performed the experiment, and analyzed the experiment data. H.X., A.H., Y.Z., Z.W., K.L., F.W. and Z.B. performed the experiment. L.Z., L.W. and Q.W. provided scientific support and conceptual advice. Z.F. revised the paper. All authors discussed the results and commented on the manuscript. All authors have read and agreed to the published version of the manuscript.

Funding: This research was supported by the National Natural Science Foundation of China (11902107), Zhejiang Provincial Natural Science Foundation of China (LY21A020009, LY21F040005), National Natural Science Foundation of China (52175460), and Fundamental Research Funds for the Provincial Universities of Zhejiang (GK219909299001-412). ZF would like to acknowledge the support from the Shenzhen Engineering Laboratory of Single-Molecule Detection and Instrument Development (XMHT20190204002) and the Joint Research Fund for Overseas Chinese Scholars and Scholars in Hong Kong and Macao (51929501).

Institutional Review Board Statement: Not applicable.

Informed Consent Statement: Not applicable.

Data Availability Statement: The data presented in this study are available from the corresponding author, H.W, upon reasonable request.

Conflicts of Interest: The authors declare no conflict of interest.

Nomenclature

r	radius of spiral line on the external side, mm	s	sensitivity of P-SRMC sensor, mm·µL^{-1}
R	radius of spiral line on the internal side, mm	x	volume of secreting sweat, µL
w_w	width of the wax line, mm	y	is displacement, mm
w_c	width of the sweat channel, mm	A_c	sweat collecting area, mm^2
θ	angle of spiral lines, rad	V_f	sweat secreting volume velocity from forehead, µL·min^{-1}
$\bar{v}$	sweat average flow velocity in wax channel, mm/s	V_{arm}	sweat secreting volume velocity from left forearm, µL·min^{-1}
v_s	sweat secreting rate, g·m^{-S2}·h^{-1}	V_{nape}	sweat secreting volume velocity from nape of the neck, µL·min^{-1}

References

1. Harvey, C.J.; LeBouf, R.F.; Toxicol, A.B. Formulation and stability of a novel artifificial human sweat under conditions of storage and useStefaniak. *In Vitro* **2010**, *24*, 1790–1796. [CrossRef] [PubMed]
2. Gagnon, D.; Jay, O.; Kenny, G.P. The evaporative requirement for heat balance determineswhole-body sweat rate during exercise under conditions permitting full evaporation. *J. Physiol.* **2013**, *591*, 2925–2935. [CrossRef] [PubMed]
3. Lim, C.L.; Byrne, C.; Lee, J.K.W. Human Thermoregulation and Measurement of Body Temperature in Exercise and Clinical Settings. *Ann. Acad. Med. Singap.* **2008**, *37*, 347–353. [PubMed]
4. Klous, L.; Ruiter, C.D.; Alkemade, P.; Daanen, H.; Gerrett, N. Sweat rate and sweat composition during heat acclimation. *J. Therm. Biol.* **2020**, *93*, 102697. [CrossRef]
5. Salvo, P.; Pingitore, A.; Barbini, A.; Francesco, F.D. A wearable sweat rate sensor to monitor the athletes' performance during training. *Sci. Sport* **2018**, *33*, e51. [CrossRef]
6. Tseng, C.C.; Kung, C.T.; Chen, R.F.; Tsai, M.H.; Chao, H.R.; Wang, Y.N.; Fu, L.M. Recent advances in microfluidic paper-based assay devices for diagnosis of human diseases using saliva, tears and sweat samples. *Sens. Actuators B* **2021**, *342*, 130078. [CrossRef]
7. Saha, T.; Fang, J.; Mukherjee, S.; Knisely, C.T.; Dickey, M.D.; Velev, O.D. Osmotically Enabled Wearable Patch for Sweat Harvesting and Lactate Quantification. *Micromachines* **2021**, *12*, 1513. [CrossRef] [PubMed]
8. Kim, S.B.; Koo, J.; Yoon, J.; Hourlier-Fargette, A.; Lee, B.; Chen, S.; Jo, S.; Choi, J.; Oh, Y.S.; Lee, G.; et al. Soft, skin-interfaced microfluidic systems with integrated enzymatic assays for measuring the concentration of ammonia and ethanol in sweat. *Lab Chip* **2020**, *20*, 84–92. [CrossRef]
9. Mohan, A.M.V.; Rajendran, V.; Mishra, R.K.; Trends Anal Chem. Recent advances and perspectives in sweat based wearable electrochemical. *Sensors* **2020**, *131*, 116024. [CrossRef]
10. Baker, L.B.; Wolfe, A.S. Physiological mechanisms determining eccrine sweat composition. *Eur. J. Appl. Physiol.* **2020**, *120*, 719–752. [CrossRef]
11. Nyein, H.Y.Y.; Bariya, M.; Kivimäki, L.; Uusitalo, S.; Liaw, T.S.; Jansson, E.; Ahn, C.H.; Hangasky, J.A.; Zhao, J.; Lin, Y.; et al. Regional and correlative sweat analysis using high-throughput microfluidic sensing patches toward decoding sweat. *Sci. Adv.* **2019**, *5*, eaaw9906. [CrossRef] [PubMed]
12. Bandodkar, A.J.; Gutruf, P.; Choi, J.; Lee, K.; Sekine, Y.; Reeder, J.T.; Jeang, W.J.; Aranyosi, A.J.; Lee, S.P.; Model, J.B.; et al. Battery-free, skin-interfaced microfluidic/electronic systems for simultaneous electrochemical, colorimetric, and volumetric analysis of sweat. *Sci. Adv.* **2019**, *5*, eaav3294. [CrossRef] [PubMed]

13. Hauke, A.; Simmers, P.; Ojha, Y.R.; Cameron, B.D.; Ballweg, R.; T Zhang, N.T.; Brothers, M.; Gomez, E.; Heikenfeld, J. Complete validation of a continuous and blood-correlated sweat biosensing device with integrated sweat stimulation. *Lab Chip* **2018**, *18*, 3750–3759. [CrossRef] [PubMed]

14. Shirreffs, S.M.; Maughan, R.J. Whole body sweat collection in humans: An improved method with preliminary data on electrolyte content. *J. Appl. Physiol.* **1997**, *82*, 336–341. [CrossRef] [PubMed]

15. Constantinescu, M.; Hilman, B.C. The Sweat Test for Quantitation of Electrolytes. *Lab Med.* **1996**, *27*, 472–477. [CrossRef]

16. Baker, L.B.; Chavez, P.J.; Ungaro1, C.T.; Sopeña, B.; Nucciol, R.; Reimel, A.; Barnes, K. Exercise intensity effects on total sweat electrolyte losses and regional vs. whole-body sweat [Na+], [Cl−], and [K+]. *Eur. J. Appl. Physiol.* **2019**, *119*, 361–375. [CrossRef]

17. Liu, C.; T Xu, D.W.; Zhang, X. The role of sampling in wearable sweat sensors. *Talanta* **2020**, *212*, 120801. [CrossRef]

18. Zhang, Y.; Chen, Y.; Huang, J.; Liu, Y.; Peng, J.; Chen, S.; Song, K.; Ouyang, X.; Cheng, H.; Wang, X. Skin-interfaced microfluidic devices with one-opening chambers and hydrophobic valves for sweat collection and analysis. *Lab Chip* **2020**, *20*, 2635–2645. [CrossRef]

19. Yang, Y.; Xing, S.; Fang, Z.; R Li, H.K.; Pan, T. Wearable microfluidics: Fabric-based digital droplet flowmetry for perspiration analysis. *Lab Chip* **2017**, *17*, 926–935. [CrossRef]

20. Francis, J.; Stamper, I.; Heikenfeld, J.; Gomez, E.F. Digital nanoliter to milliliter flow rate sensor with in vivo demonstration for continuous sweat rate measurement. *Lab Chip* **2019**, *19*, 178–185. [CrossRef]

21. Baker, L.B.; Model, J.B.; Barnes, K.A.; Anderson, M.L.; Lee, S.P.; Lee, K.A.; Brown, S.D.; Reimel, A.J.; Roberts, T.J.; Nucciol, R.P.; et al. Skin-interfaced microfluidic system with personalized sweating rate and sweat chloride analytics for sports science applications. *Sci. Adv.* **2020**, *6*, eabe3929. [CrossRef] [PubMed]

22. Matzeu1a, G.; Fay1a, C.; Vaillant, A.; Coyle, S.; Diamond, D. A Wearable Device for Monitoring Sweat Rates via Image Analysis. *IEEE Trans. Biomed. Eng.* **2016**, *63*, 1672–1680. [CrossRef] [PubMed]

23. Liu, H.; Qing, H.; Li, Z.; Han, Y.L.; Lin, M.; Yang, H.; Li, A.; Lu, T.J.; Li, F.; Xu, F. Paper: A promising material for human-friendly functional wearable electronics. *Mat. Sci. Eng. R.* **2017**, *112*, 1–22. [CrossRef]

24. Gao, B.; Li, X.; Yang, Y.; Chu, J.; He, B. Emerging paper microflfluidic devices. *Analyst* **2019**, *144*, 6497–6511. [CrossRef] [PubMed]

25. He, J.; Xiao, G.; Chen, X.; Qiao, Y.; Xu, D.; Lu, Z. A thermoresponsive microflfluidic system integrating a shape memory polymer-modifified textile and a paper-based colorimetric sensor for the detection of glucose in human sweat. *RSC Adv.* **2019**, *9*, 23957–23963. [CrossRef]

26. Nagamine, K.; Mano, T.; Nomura, A.; Ichimura, Y.; Izawa, R.; Furusawa, H.; Matsui, H.; Kumaki, D.; Tokito, S. Noninvasive Sweat-Lactate Biosensor Emplsoying a Hydrogel-Based Touch Pad. *Sci. Rep.* **2019**, *9*, 10102. [CrossRef]

27. Zhao, F.J.; Bonmarin, M.; Chen, Z.C.; Larson, M.; D Fay, D.R.; Heikenfeld, J. Ultra-simple wearable local sweat volume monitoring patch based on swellable hydrogels. *Lab Chip* **2020**, *20*, 168–174. [CrossRef]

28. Gao, W.; Emaminejad, S.; Nyein, H.Y.Y.; Challa, S.; Chen, K.; Peck, A.; Fahad, H.M.; Ota, H.; Shiraki, H.; Kiriya, D.; et al. Fully integrated wearable sensor arrays for multiplexed in situ perspiration analysis. *Nature* **2016**, *529*, 509–514. [CrossRef] [PubMed]

29. Martinez, A.W.; Phillips, S.T.; Whitesides, G.M. Diagnostics for the Developing World: Microflfluidic Paper-Based Analytical Devices. *Anal. Chem.* **2010**, *82*, 3–10. [CrossRef]

30. Singhal, H.R.; Prabhu, A.; Nandagopal, G.; Dheivasigamani, T.; Mani, N.K. One-dollar microfluidic paper-based analytical devices: Do-It-Yourself approaches. *Microchem. J.* **2021**, *165*, 106126. [CrossRef]

31. Sriram, G.; Bhat, M.P.; Patil, P.; Uthappa, U.T.; Jung, H.; Altalhi, T.; Kumeria, T.; Aminabhavi, T.M.; Pai, R.K.; Madhuprasad; et al. Paper-based microfluidic analytical devices for colorimetric detection of toxic ions: A review. *Trends Anal. Chem.* **2017**, *93*, 212–227. [CrossRef]

32. Nishat, S.; Jafry, A.T.; Martinez, A.W.; Awan, F.R. Paper-based microfluidics: Simplified fabrication and assay methods. *Sens. Actuators B* **2021**, *336*, 129681. [CrossRef]

33. Kung, C.; Hou, C.; Wang, Y.; Fu, L. Microfluidic paper-based analytical devices for environmental analysis of soil, air, ecology and river water. *Sens. Actuators B* **2019**, *301*, 126855. [CrossRef]

34. Altundemir, S.; Uguz, A.K.; Ulgen, K. A review on wax printed microfluidic paper-based devices for international health. *Biomicrofluidics* **2017**, *11*, 041501. [CrossRef] [PubMed]

35. Nie, C.; Frijns, A.J.H.; Mandamparambil, R.; Toonder, J.M.J. A microfluidic device based on an evaporation-driven micropump. *Biomed. Microdevice* **2015**, *17*, 47–59. [CrossRef] [PubMed]

36. Chen, X.; Li, Y.; Han, D.; Zhu, H.; Xue, C.; Chui, H.; Cao, T.; Qin, K. A Capillary-Evaporation Micropump for Real-Time Sweat Rate Monitoring with an Electrochemical Sensor. *Micromachines* **2019**, *10*, 457. [CrossRef] [PubMed]

37. Havenith, G.; Fogarty, A.; Bartlett, R.; Smith, C.J.; Ventenat, V. Male and female upper body sweat distribution during running measured with technical absorbents. *Eur. J. Appl. Physiol.* **2008**, *104*, 245–255.

38. Masoodi, R.; Pillai, K.M. Darcy's Law-Based Model for Wicking in Paper-Like Swelling Porous Media. *AIChE J.* **2010**, *56*, 2257–2267. [CrossRef]

39. Engeland, C.; Haut, B.; Spreutels, L.; Sobac, B. Evaporation versus imbibition in a porous medium. *J. Colloid Interf. Sci.* **2020**, *576*, 280–290. [CrossRef]

40. Richards, L.A. Capillary conduction of liquids through porous mediums. *Physics* **1931**, *1*, 318–333. [CrossRef]

41. Mascini, A.; Cnudde, V.; Bultreys, T. Event-based contact angle measurements inside porous media using time-resolved micro-computed tomography. *J. Colloid Sci.* **2020**, *572*, 354–363. [CrossRef] [PubMed]
42. Tong, X.; Ga, L.; Zhao, R.; Ai, J. Research progress on the applications of paper chips. *RSC Adv.* **2021**, *11*, 8793. [CrossRef]
43. Smith, C.J.; Havenith, G. Body mapping of sweating patterns in male athletes in mild exercise-induced hyperthermia. *Eur. J. Appl. Physiol.* **2011**, *111*, 1391–1404. [CrossRef] [PubMed]
44. Taylor, N.; Moreira, C.M. Regional variations in transepidermal water loss, eccrine sweat gland density, sweat secretion rates and electrolyte composition in resting and exercising humans. *Extrem. Physiol. Med.* **2013**, *2*, 4. [CrossRef] [PubMed]

micromachines

Article

Microfluidics Temperature Compensating and Monitoring Based on Liquid Metal Heat Transfer

Jiyu Meng [1], Chengzhuang Yu [1], Shanshan Li [1,2,*], Chunyang Wei [1], Shijie Dai [1,*], Hui Li [1] and Junwei Li [3,*]

[1] Hebei Key Laboratory of Smart Sensing and Human-Robot Interactions, School of Mechanical Engineering, Hebei University of Technology, Tianjin 300130, China; 201521202004@stu.hebut.edu.cn (J.M.); 201811201004@stu.hebut.edu.cn (C.Y.); 201811201008@stu.hebut.edu.cn (C.W.); 202031205015@stu.hebut.edu.cn (H.L.)
[2] State Key Laboratory of Reliability and Intelligence of Electrical Equipment, Hebei University of Technology, Tianjin 300401, China
[3] School of Health Sciences and Biomedical Engineering, Hebei University of Technology, Tianjin 300401, China
* Correspondence: sli_mems@hebut.edu.cn (S.L.); dsj@hebut.edu.cn (S.D.); junwei_li@hebut.edu.cn (J.L.)

Abstract: Microfluidic devices offer excellent heat transfer, enabling the biochemical reactions to be more efficient. However, the precision of temperature sensing and control of microfluids is limited by the size effect. Here in this work, the relationship between the microfluids and the glass substrate of a typical microfluidic device is investigated. With an intelligent structure design and liquid metal, we demonstrated that a millimeter-scale industrial temperature sensor could be utilized for temperature sensing of micro-scale fluids. We proposed a heat transfer model based on this design, where the local correlations between the macro-scale temperature sensor and the micro-scale fluids were investigated. As a demonstration, a set of temperature-sensitive nucleic acid amplification tests were taken to show the precision of temperature control for micro-scale reagents. Comparisons of theoretical and experimental data further verify the effectiveness of our heat transfer model. With the presented compensation approach, the slight fluorescent intensity changes caused by isothermal amplification polymerase chain reaction (PCR) temperature could be distinguished. For instance, the probability distribution plots of fluorescent intensity are significant from each other, even if the amplification temperature has a difference of 1 °C. Thus, this method may serve as a universal approach for micro–macro interface sensing and is helpful beyond microfluidic applications.

Keywords: heat transfer; microfluidics; liquid metal; measurement; temperature monitoring; PCR

Citation: Meng, J.; Yu, C.; Li, S.; Wei, C.; Dai, S.; Li, H.; Li, J. Microfluidics Temperature Compensating and Monitoring Based on Liquid Metal Heat Transfer. *Micromachines* 2022, 13, 792. https://doi.org/10.3390/mi13050792

Academic Editors: Junfeng Zhang and Ruijin Wang

Received: 26 April 2022
Accepted: 17 May 2022
Published: 19 May 2022

Publisher's Note: MDPI stays neutral with regard to jurisdictional claims in published maps and institutional affiliations.

1. Introduction

In recent decades, advances in microfabrication techniques have led to the development of a wide variety of microfluidic devices [1–3]. For many microfluidic applications, the knowledge of the temperature field is of high technical and scientific importance [4,5]. However, it remains challenging to obtain accurate temperature data of the microfluid limited by size effects, such as manufacturing process and sensor accuracy [6,7]. Conventional high-precision types of equipment or temperature sensors suffer from expensive costs, professional operation, and poor interference immunity. With the development of micro-electro-mechanical systems (MEMS), while numbers of research on the measurement and calibration of microfluidics temperature have been conducted [8–10], a unified and systematic theory is still ambiguous. Therefore, the study on the microfluidic heat transfer characteristics and temperature calibration is kept hot [11–13].

Two main methods have been reported for the microfluidics temperature measurement [14–16]: direct contact and non-direct-contact temperature measurement. Generally, the temperature can be accurately recorded, but the inherent temperature of the targeted micro fluids will be affected via external macro instruments by contact measurement. There is no heat exchange via measurement tools of the non-contact method, which make it difficult to achieve high accuracy by utilizing the changes in the physical properties of the target object.

For contact temperature measurement, the heterojunction structure of thermocouples with micro or even nano-size can be directly applied to the thermometry of microfluid [17–19]. However, we must suffer from damage and vibration when operating such a thin and cuspate thermocouple probe, which does not meet the requirements of high reliability and ease of use. In addition, micron or sub-micron platinum film thermal resistance is also used as an essential method for microfluidic temperature measurement [20,21]. However, it must be attached to a substrate, so it is not suitable for measuring the temperature of a small volume. As active devices, their self-heating effect caused by power may influence the micro-scale temperature distribution. The emerging carbon nanotube technology measures the volume expansion of gallium in carbon nanotubes to read temperature changing [22,23], similar to a micro-nano-scale mercury thermometer. It can be applied to microfluidic temperature measurement, but the accuracy is not high.

Due to the limit of resolution, the mainstream non-contact infrared thermal imaging temperature measurement technology cannot be directly used for microfluidic temperature measurement [24]. In recent years, optical imaging technology has been a powerful means to explore the temperature distribution of microfluidics [25,26]. For example, the temperature change can be characterized by the fluorescence intensity of temperature-sensitive fluorescent dye added to the microfluid [27]. However, reference image comparison is required, and there is considerable system uncertainty, which is susceptible to the effects of cross-color in the optical path. In addition, quantum dots have become a kind of optical temperature measurement materials with development potential due to their advantages of small size, good light stability, and high plasticity [28]. However, its size and shape distribution depending on temperature will cause uneven light emission. In addition, the polymer polyacrylamide can be used as a measure of micro-scale temperature due to its structure variation with temperature [29]. The temperature sensing range can be adjusted by combining reactive polymers with fluorophores. However, polymer-based measurement is relatively slow and unsuitable for real-time temperature monitoring.

In recent years, more and more microfluidic heat transfer principles have been explored and studied to use indirect methods to invert the microfluidic temperature [30]. For example, winding copper wire around a microfluidic pipe as a temperature change resistor to measure microfluidic temperature [31], but the effect of heat transfer from the pipe material is not considered. Color-changing temperature-sensitive materials can monitor microfluidic temperature changes [32], but it is generally effective only for a specific temperature value and have a very narrow range of application. Additionally, liquid metals with low melting points have been widely studied as high thermal conductivity and easy-to-handle materials in microfluidic heat transfer [33]. Liquid metals can be used as thermal conductive media connecting the microscopic and macroscopic to achieve a coefficient of temperature conversion.

In this work, a novel indirect temperature measurement method integrating platinum resistance millimeter-scale industrial sensor and liquid metal has been demonstrated to measure and monitor microfluid temperature in a microfluidic chip channel. With this method, the close contact between the sensor and microfluid is avoided, and it is user-friendly and repeatable. First, the relationship between the chip substrate and the microfluid temperature has been studied. Numerical simulation proves that the temperature of the top surface of the glass substrate is the same as the fluid in the microchannel that contacts it. Therefore, we measure the glass surface temperature to characterize it as the targeted fluid temperature. We then punch holes at the designed position that avoid the microchannels, place temperature sensors to contact the top surface of the glass substrate, then fill liquid metal packaging. The aperture is optimized to achieve the best temperature measurement effect, and the function compensation relationship between the measurement temperature and the microfluid in the range of 30–100 °C is calibrated. In addition, a highly temperature-dependent isothermal amplification PCR experiment was presented to prove the temperature measurement validity of this method. The results show that this measurement and compensation method has great potential in microfluidic temperature monitoring.

2. Materials and Methods

2.1. Chip Design and Sensor Installation

Before conducting the numerical optimization studies, the temperature depended PCR chip is performed to validate the physical structure. Here, we present an S-shaped microchannel design that incorporates two inlets introducing two streams for reagent mixing. A through-hole is arranged on the corner, as shown in Figure 1a, which provides the design details of the proposed microfluidic chip. The microchannel consists of a rectangle cross-section channel, 100 μm wide and 30 μm deep, together with an S-shape unit at the central part, which makes the reagent mix more evenly. A circular view area is on the straight microchannel before the exit and one single outlet downstream for postprocessing.

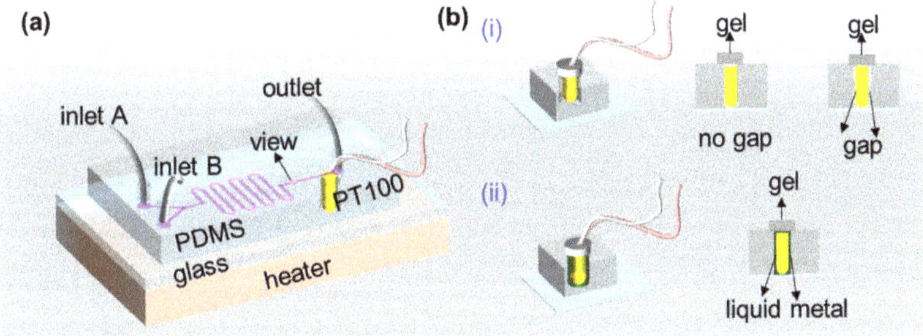

Figure 1. Illustration of the temperature depended on the PCR device, including a piece of PDMS replica bonded with a glass slider, a heater, and a PT100 sensor. (**a**) The full view of our device. The PDMS replica has 2 inlets, 1 outlet, and a through-hole. The PT100 was inserted into the through-hole to measure the temperate of the top surface of the glass slide. The heater was put under the glass substrate. (**b**) Illustration of the packing for PT100 (**i**) w/o and (**ii**) with liquid metal deposited in the through-hole.

Figure 1b shows the basic concept for sensor installation. First, a sensor is arranged to the hole to ensure its sensitive components contact with the top surface of the glass slide, as shown in Figure 1b(i). Moreover, to ensure that the sensor is in complete contact with the glass surface, the liquid metal is used to fill the gap around the sensor in the hole, as shown in Figure 1b(ii). A relative precise temperature value can be recorded by this method.

2.2. Chip Fabrication

Here, the microfluidic channel with suitable patterns is fabricated using standard soft lithography methods previously reported [34,35]. Briefly, the AutoCAD 2018 is used to design the photomasks of the electrode and channel, and then the layout is printed by a commercial high-resolution inkjet printer. Next, the SU-8 photoresist or photosensitive dry film is coated onto the silicon substrate, followed by a pre-bake, exposure, development, as well as the post-bake process to fabricate the mold master. Then, the polydimethylsiloxane (PDMS sylgard 184, Dow Corning) is cast on the mold master and cured at ~80 °C for 1~2 h. After that, the PDMS channel containing the imprint is detached from the mold master and punched to fabricate the inlets, outlets, and through-hole.

The glass slide is soaked in acetone for 20 min, rinsed with isopropyl alcohol and DI water each for 15 s, and then dried with an air gun. PDMS block mentioned above is bonded with a glass surface after plasma treatment by a plasma machine (PDC-MG, PTL Technology Co, Ltd., Shenzhen, China). The plasma treatment and bonding procedure are in four steps: a. Put PDMS block with the channel side exposed and cleaned glass slide into the plasma bonding machine. b. Vacuum the machine to a pressure value of about 100 Kpa. c. Turn on plasma high voltage discharge and last 40 s. d. Take out the PDMS block and bond it to the glass slide in time to avoid surface contamination or denaturation.

2.3. Reagent Preparation

The kit for duck-derived gene detection (including dry powder reagent tube, R buffer, B buffer, and positive gene template) was purchased from Anpu Biotech Co., Ltd. (Changzhou, China). The best amplification temperature of this reagent is 39–41 °C, and the reaction time takes 20 min. Before the experiment, it is necessary to add 45 µL R buffer, 2.5 µL positive gene template, and 2.5 µL B buffer to the dry powder reagent tube in sequence. Then, mix the reagents evenly on the rotating table. It should be noted that the entire reagent preparation process needs to be carried out in a fume hood and on an icebox to prevent air pollutants and high temperatures from affecting the reaction system. After the reagents are prepared, store them at 0–4 °C for later use, and keep the remaining reagents in a refrigerator at −20 °C.

2.4. Temperature Calibration and Solution Delivery

A transparent tempered glass heater (SAPPHIRE SC-6100, TY154, Merip Technology Co, Ltd., Nanjing, China) is used to provide a heat source for the microfluidic chip, so that the bottom surface of the microfluidic chip has a constant temperature, as shown in Figure 2, and the temperature error of the heater is 0.1 °C. The platinum resistance temperature sensor (PT 100-2mm, Sensite, Beijing, China) is encapsulated in the through-hole of the PDMS block with glue together, packaging liquid metal to measure the temperature of the upper surface of the glass slide and record the data, as shown in Figure 2. In addition, a high-precision thermocouple temperature sensor (tt-k-40-36, Omega, New York, NY, USA) is packaged inside the microchannel to measure and record the temperature data of the microfluid. Then, we make a compensation function for the temperature relationship between the two mentioned above. According to the function calculation, the glass surface temperature can be used to characterize the temperature of the microfluid, avoiding the use of precision thermocouples.

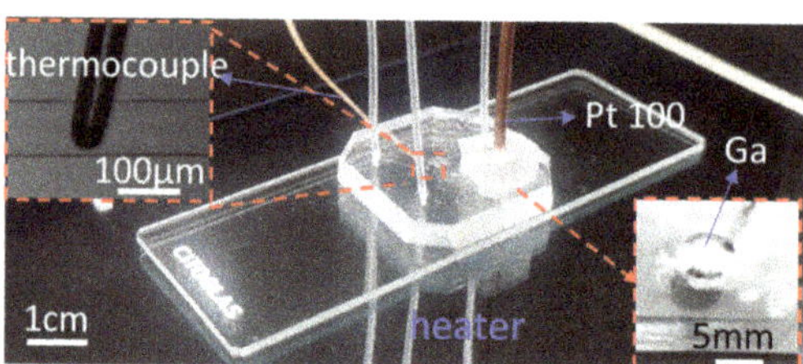

Figure 2. Photo of our microfluidic device mounted on a miniaturized heater. The high-precision smart thermocouple was used to measure the temperature of the fluid within the microchannel. The industrial PT100 sensor inserted into the through-hole was to detect the top surface of the glass slide. Gallium liquid metal was used to fill the gaps between PT100 and the through-hole.

In order to quantify the performance of the temperature measurement, a PCR amplification experiment that was highly dependent on the temperature is present. Here, two streams of aqueous reagent are pumped through the inlets with separate, independently controlled programmable flow pumps (PC1, Elveflow, Paris, France) to form a laminar flow at the very beginning of the chip. Then laminar fluids flow through the S-shaped channel, ensuring more effective reagent mixing [36]. When the fluid fills the channel, the heater is carefully adjusted to a specific temperature and lasts 20 min. Then observe the fluorescence effect by a microscope (Eclipse Ti-s, Nikon, Tokyo, Japan) in the view area downstream of the microchannel.

3. Numerical Analyses

3.1. Control Equation and Boundary Condition Settings

The physical field modules of the heat transfer in solids and fluids and events are used for simulation.

3.1.1. Heat Transfer in Solids and Fluids

Energy conservation equation:

It is assumed that it is all heat transfer between solid and fluid, the energy conservation equation is given by,

$$\rho V C_p \frac{\partial T}{\partial t} + \nabla \cdot (-V k \nabla T) = Q_0 \tag{1}$$

where ρ is the density of the applied material, V is the volume of the heated object, C_p is the capacitance measured at constant pressure, k is the thermal conductivity, Q_0 is the energy generation rate for the whole system.

In order to simulate the temperature control of the heater, a constant power p_0 and a status indicator StateHeater (set in Event interface) are selected as thermal energy source, the energy generation rate is set as,

$$Q_0 = p_0 \times \text{StateHeater} \tag{2}$$

It is assumed that the bottom of the heater is thermal insulation, the boundary condition can be set as,

$$n \cdot (-k \nabla T) = 0 \tag{3}$$

It is assumed that the outside of the system is heat dissipated by natural convection, the convective heat flux should be given by,

$$n \cdot (-k \nabla T) = h_c (T - T_0) \tag{4}$$

where h_c is the natural convection heat flux coefficient, T_0 is the ambient temperature.

The radiant heat flux of the system to the ambient can be set as,

$$n \cdot (-k \nabla T) = \varepsilon \sigma \left(T^4 - T_0^4 \right) \tag{5}$$

where ε is the emissivity of the material, $0 < \varepsilon < 1$, σ is the Stefan–Boltzmann constant.

There is contact heat loss between the heating block and the glass slide, which mainly includes contact and gap thermal resistance, which can be expressed as

$$n \cdot (-k \nabla T) = h \left(T_{down} - T_{up} \right) + r Q_0 / A_l \tag{6}$$

where r is the heat partition coefficient.

$$h = h_{contact} + h_{gap} \tag{7}$$

The heat transfer coefficient of thermal contact resistance is closely related to the surface roughness, microhardness, and contact pressure of the material. It should be expressed as [37],

$$h_{contact} = 1.25 k_{contact} \frac{m_{asp}}{\sigma_{asp}} \left(\frac{P}{H_c} \right)^{0.95} \tag{8}$$

where $k_{contact}$ is the thermal conductivity of materials in contact, m_{asp} and σ_{asp} are the asperities average slope and height surface roughness, respectively. P is contact pressure between the heater and glass slide. H_c is the microhardness of the glass.

The gap thermal resistance is related to the gas type and contact pressure between two contacting objects. Still, there is no specific expression function yet, and the value range can be checked.

3.1.2. Events

The Event interface is used to control the temperature of the heater to simulate closed-loop PID control. First, the steady-state error is defined as T_{error} = 0.1 K. First, the discrete state of the Event is defined as StateHeater = 1, and the given power $Q_0 = p_0 \times$ StateHeater, then heating starts. When heating to the upper limit of the target temperature, $T > T_{error} + T_{target}$, at this time StateHeater = 0, stop heating. When the temperature drops to $T_t < T_{targe} - T_{error}$, the heater starts working again, StateHeater = 1. The state function can be set as,

$$
\text{StateHeater} = \begin{cases} 1 & T \leq T_{target} + T_{error} & (1) \\ & \downarrow & \\ 0 & T > T_{error} + T_{target} & (2) \\ & \downarrow \quad \uparrow & \\ 1 & T < T_{target} - T_{error} & (3) \end{cases} \tag{9}
$$

After state (1) reaches its peak, states (2) and (3) act cyclically to stabilize the temperature. Feedback temperature is taken from the integral temperature of the circular sensor on the heater, as shown in Figure 3a. The simulation domain and boundary condition settings are shown in Figure 3a.

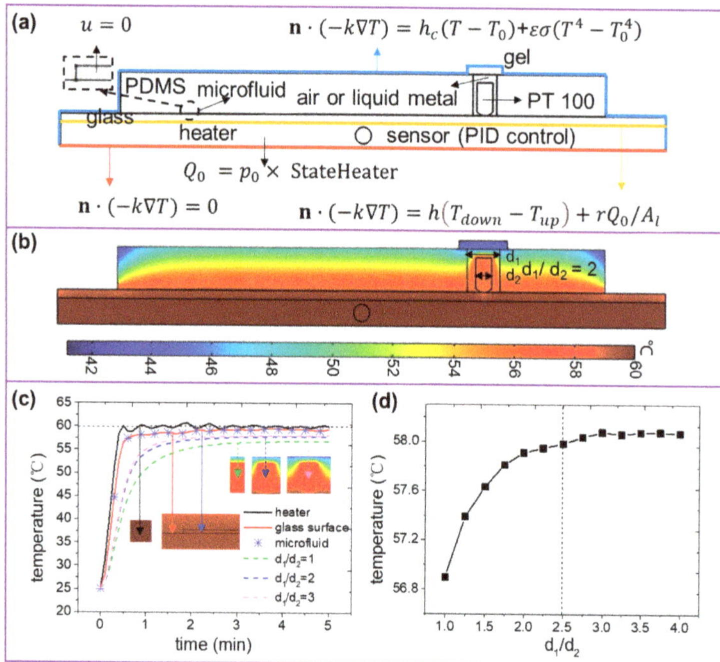

Figure 3. (**a**) Boundary condition settings of the microfluidic heating system. (**b**) The temperature field of the microfluidic heating system. Here the target temperature is 60 °C and $d_1/d_2 = 2$. (**c**) The real-time temperature of the heater, glass, and microfluid with different hole sizes. Here the heating power is 30 W. (**d**) The experimental steady-state temperature as a function of d_1/d_2.

3.2. Parameter Settings in the Simulation

Since the structure of the 3D model is relatively regular and has similar vertical sections, here we did not propose an actual 3D model to calculate the heat transfer process. For simplicity, we presented a longitudinal cross profile 2D numerical model herein. The heat transfer model is conducted using commercial finite element software COMSOL Multiphysics 5.4. The model discretization and grid independence, as well as convergence solving methods, are shown in the Supplementary Material, including a discrete grid for the simulation model (Figure S1), numbers of the domain and boundary elements in different element sizes (Table S1) and grid independence verification (Figure S2). The physical and geometry parameters in our study are set as the same as those in experiments, as listed in Table S2 in Supplementary Materials.

4. Results and Discussion

4.1. Effects of Aperture on Temperature Measurement

The temperature signal measurement by platinum (Pt) resistance sensor is contact conduction via energy. The heat transfer capacity for ambient materials placed adjacent to the sensor has a significant influence on receiving feedback signals. First, we put the Pt sensor directly into the hole, get as close as possible to contact the top surface of the glass slide, then seal it with gel. For instance, making the aperture of the hole equal to twice that of the sensor, and setting the target temperature is 60 °C. The numerical result shows the temperature distribution of the 2D microfluidic chip heat transfer system in Figure 3b. It can be seen that the temperature gradient on the heater is not visible to the naked eyes due to the excellent performance for heat transfer of aluminum products. The temperature distribution on the vertical profile of the glass slide presents a slight bottom-up temperature gradient and a lower overall temperature than the heater. Two reasons cause the loss of energy: convection heat dissipation between the top surface of the glass slide and the ambient soft-temperature air; contact heat resistance between the bottom surface and the heater. Further, the microfluidic temperature, on which visual appearance, is almost the same as that of the top surface of the glass slide. However, the temperature traversing the vertical profile of the PDMS block decrease clearly and gradually from bottom to top due to the character of low thermal conductivity of PDMS and great convective heat flux presented on the top layer. The temperature appears to be the weakest at the corners caused by concurrent heat flux on both the top and side surfaces. In addition, the temperature measured by the Pt sensor on the top surface of the glass slide is almost uniform and slightly lower than that covered by the PDMS block. The intrinsic heat conduction of the Pt sensor, air, and sealing gel results in a decrement in temperature.

The heat transfer time-dependant process can be decomposed as follows. First, the temperature of the heater rises to the steady-state status within 1 min heating by 30 W power with ± 0.1 °C error, as shown by the black solid curve in Figure 3c, which is consistent with the control method utilized in the experiment. The temperature of the top surface of the glass slide rises to the stable target later than that of the heater due to the thermodynamic hysteresis effect, as shown by the solid red curve in Figure 3c. It almost simultaneously reaches a stable status compared to microfluidic temperature with the top surface temperature of the glass, and the difference in the steady value is negligible, as shown by the blue star marks in Figure 3c. Therefore, the microfluidic temperature can be an equivalented substitute to that of the top surface of the glass slide. The Pt resistance sensor is used to sense the temperature of the top surface of the glass slide by packaging it in a cylindrical hole through the PDMS block. Actually, the feedback temperature via the Pt resistance sensor is later and lower than that of the top surface of the glass slide. The stable temperature changes regularly with the aperture, as shown by dotted curves in Figure 3c. When the aperture (d_1) is equal to the diameter of the Pt resistance sensor (d_2), that is $d_1/d_2 = 1$, the pieces of the PDMS block in contact with the Pt resistance sensor carry away the heat making the temperature measured by the sensor much less than that of the top surface of the glass slide. The feedback temperature of the Pt resistance sensor gradually increases as the aperture duo to the contact between the sensor and PDMS has been blocked by air with relatively high thermal conductivity blocks. The feedback temperature is not monotonically increasing with the continued expanding aperture but will eventually stabilize at a certain value, as shown in Figure 3d. This phenomenon is caused by the balance of heat transfer and dissipation around the sensor. Since the error value is ± 0.1 °C, the measured temperature of the sensor will reach a stable value when

$d_1/d_2 = 2.5$. The average steady-state temperature is calculated as 58.08 °C. In order to facilitate the hole processing and sensor packaging in the experiment, take $d_1/d_2 = 3$ for experimental research.

4.2. Liquid Metal Filled and Experimental Temperature Calibration

It is hard to make a Pt resistance sensor repeatedly be placed closely in a giant hole, which should affect the accuracy of the feedback signal. In order to ensure that the sensitive elements of the sensor are entirely in contact with the top surface of the glass slide, meanwhile, away from the PDMS block, liquid metal with high thermal conductivity and flexibility is used to fill the gap between the cylindrical hole and the Pt resistance sensor. In this way, the measurement accuracy of the sensor gets significant improvement. As shown in the double y-axis plot in Figure 4a, when the hole is, the temperature distribution of the sensor has become uniform with liquid metal filled in the gap, and the steady-state temperature is higher than that of air that exists. When the steady-state temperature is reached, the average measured temperature difference between the filled liquid metal and air medium is about 0.5 °C, which is five times the allowable error.

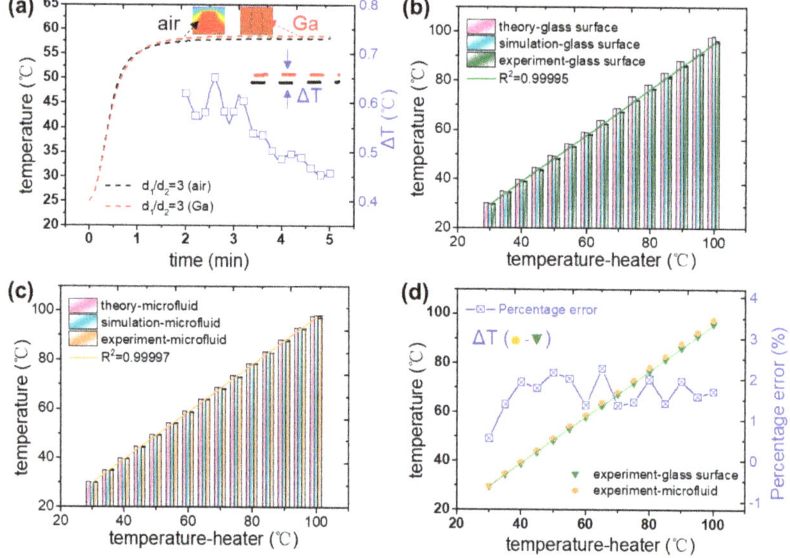

Figure 4. (**a**) The temperature (left y-axis) and temperature difference (right y-axis) as a function of time, with the condition of with and without liquid metal filled in the gap. Here $d_1/d_2 = 3$. The theoretical, simulated, and experimental temperature of glass surface (**b**) and microfluid (**c**) with the set temperature in a range of 30–100 °C. (**d**) The comparison of experimental temperature between the glass surface and microfluid. Here the average temperature error is 1.74%.

There were three times experimental measures were performed to record the steady-state temperature value of the top surface of the glass slide in five different locations within the range of 30–100 °C. The average temperature and variance are shown in Table 1.

Table 1. The steady-state temperature value of the top surface of the glass slide.

Set Temperature (°C)	Glass Surface Temperature (°C)
30	29.67 ± 0.06
35	34.40 ± 0.26
40	38.80 ± 0.17
45	43.50 ± 0.10
50	48.17 ± 0.25
55	53.03 ± 0.15
60	57.83 ± 0.12
65	62.40 ± 0.17
70	67.27 ± 0.15
75	71.97 ± 0.12
80	76.67 ± 0.21
85	81.53 ± 0.35
90	86.17 ± 0.38
95	91.03 ± 0.35
100	95.9 ± 0.4

The difference value between the top surface of the glass slide measured by Pt sensors and the set temperature gradually increases from 30 °C to 100 °C, as shown by the green column in Figure 4b. The measured temperature has a linear relationship with the set temperature of the heater (R^2 = 0.99995), within the range of 30–100 °C. The linear fitting relationship can be expressed as

$$T_{act} = 0.94T_{set} + 1.35 \tag{10}$$

where T_{act} is the measurement temperature of the top surface of the glass slide, T_{set} is the set temperature.

At the unperforated position, the temperature of the top surface of the glass slide changes with the set temperature in simulation, as shown by the cyan column in Figure 4b. The simulated results are almost the same as the theoretical values but greater than the measured by the experiment, as shown in Figure 4b. It means that under the same set temperature, the temperature of the top surface of a glass slide that is unperforated is higher than that of the sensor location. It is consistent with the law of steady-state temperature difference calculated by the simulation mentioned above when the target temperature is 60 °C. In order to further determine the relationship between the temperature of the glass surface at the perforation and that of the microfluid, the temperature of the microfluid was measured by placing a precision thermocouple in the microchannel. The average temperature value and variance of the three measurements are shown in Table 2.

The results are highly consistent with the simulated top glass surface temperature where unperforated, as shown in Figure 4c, which further illustrates the accuracy of the measurement and the rationality of the glass surface temperature characterizing the microfluidic temperature. In addition, according to the theoretical steady-state heat transfer of the multilayer medium, the heat transfer rate can be expressed as [38]

$$q = \frac{T_{set} - T_0}{[1/(h_1 A_1) + L_i/(k_i A_i) + R_{cont} + 1/(h_{i+1} A_{i+1})]} \tag{11}$$

where h_1 and h_{i+1} are the convective heat-transfer coefficient of the bottom and top, respectively. A is the surface area of the object with heat flux. L_i and k_i are the thickness and thermal conductivity of each layer (I = 2, 3, . . .), respectively. R_{cont} is the contact resistance between heater and glass slide, R_{cont} = 0.3–0.6 $\times 10^{-4}$ m^2·K/W.

Table 2. The steady-state temperature value of the microfluid.

Set Temperature (°C)	Glass Surface Temperature (°C)
30	29.83 ± 0.02
35	34.88 ± 0.06
40	39.57 ± 0.08
45	44.29 ± 0.06
50	49.24 ± 0.12
55	54.13 ± 0.07
60	58.64 ± 0.12
65	63.85 ± 0.06
70	68.19 ± 0.06
75	73.02 ± 0.18
80	78.23 ± 0.07
85	82.71 ± 0.02
90	87.88 ± 0.19
95	92.5 ± 0.17
100	97.56 ± 0.59

Assuming that each layer constructed in the microfluidic system could transfer heat uniformly in the vertical profile, and take the bottom of the heater as the zero point, the temperature of the upper layers can be expressed as:

$$T(y) = T_{set} - q\frac{y}{k_i A_i} \tag{12}$$

where y is the distance from the bottom surface of the heater.

According to Equation (12), the relationship between the temperature of the top surface of the glass slide and the set temperature can be calculated as shown by the pink column in Figure 4b. The theoretical analyses are extremely consistent with the simulation and experimental results. Therefore, we can analyze the relative temperature relationship between the top surface of the glass slide measured by the Pt 100 resistance sensor at the hole punching and the microfluid. As shown by the blue curve in Figure 4d, the difference between the microfluidic and measured temperature at each temperature is mostly concentrated between 1% and 2%. Excluding the more significant error at 30 °C, the average error is 1.74%. Therefore, the temperature function between the microfluid and the heater can be expressed as

$$T_{micro} = T_{act} \times (1 + 1.74\%) = 0.96 T_{set} + 1.37 \tag{13}$$

The microfluidic temperature value can be derived from the temperature setpoint without precision sensor measurements in this system. Even though it would be desirable to place a temperature sensor as close as possible to the location of the microfluidic, practical limits usually prevent this, such as limited space and tight bonding requirements. The emergence of precision instruments can break through the limitations of assembly space but still suffer from damage and expensive cost. On the other hand, customized MEMS sensors integrated on a microfluidic chip may lead to unexpected contaminations into liquid samples. Here, a millimeter-scale industrial grade temperature sensor is introduced for microfluidic temperature sensing, and the compensation relationship between the microfluids and the glass substrate is investigated. As a more reliable and low-cost sensor, Pt100 is also capable of providing industrial-grade precisions. It also establishes a link between micro and macro scale, which enriches the selectivity of external control devices. Although calculations are required, the optimized relational equation can be embedded inside the temperature control system to achieve a one-step operation in the future. When the substrate material is changed, the microfluidic temperature value can still be calculated from the upper surface temperature value of the substrate.

4.3. Temperature-Dependent PCR Experiment

A temperature-dependent PCR experiment is conducted to test the accuracy of the temperature measurement method described above. The reaction reagents are pumped into the microfluidic chip and then placed the chip on the heater to stay for 20 min. At a suitable temperature, the DNA template, primers, and enzymes in the reaction reagents interact to complete the unwinding and replication of the double-stranded strands. Concurrently, the fluorophores are activated so that the reacted sample exhibits a fluorescent effect. According to the fluorescence intensity, the content of the target DNA in the original sample can be calculated. The sample kit indicates that the best temperature for reagent amplification is between 39–41 °C. In order to explore the effect of temperature on the amplification effect, we set low and high temperatures relative to the optimal value for isothermal amplification of sample reagents. After the reaction is completed, the fluorescence of the reagent is observed through a microscope. The blue light is selected as the excitation light, and the fluorescence diagram of the view area is shown in Figure 5a. A 300 μm square centrally inside the view area (d = 500 μm) is cut out as the analysis area for fluorescence intensity to avoid the error caused by the edge of the microchannel. As shown in the top part of Figure 5b, the results show the fluorescence graphs of the isothermal amplification PCR with a microfluid temperature of 37–43 °C. The temperature measured by the sensor is 36.4 °C, 37.4 °C, 38.3 °C, 39.3 °C, 40.3 °C, 41.3 °C, and 42.3 °C. In order to obtain a consuming contrast, grey-level images are processed with pseudo-color, as shown in the bottom part of Figure 5b. The results show that the fluorescence intensity at 39–41 °C is higher than that at 37 °C, 38 °C, 42 °C, and 43 °C, which is consistent with the informed. The accuracy and practicability of the method mentioned above for temperature measurement are explained. The reason for the varying fluorescence intensity in the graphs is mainly due to the inevitable precipitation in the reactants, which makes the fluorescence intensity of some areas increase locally. Another reason is speculated as to the thickness error in the chip manufacturing process. Hence, we can analyze the difference in fluorescence intensity of each group through the overall surface average fluorescence data.

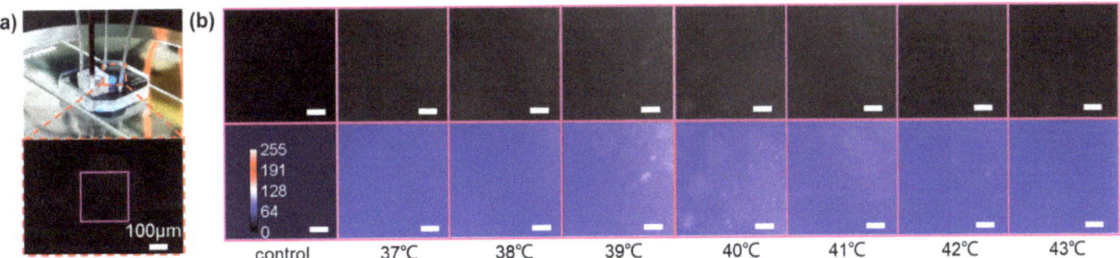

Figure 5. (**a**) The microfluidic chip for PCR tests with the microscopic image of a microchamber. (**b**) Top: fluorescent images of control and positive reagents at 37–43 °C. Bottom: blue pseudo-color images of the top. The scale bar is 50 μm.

The average surface grey level of each picture is calculated, which is used to represent the fluorescence intensity. Each picture contains 20,736 pixels. The distribution of surface grey levels (0–255) at different temperatures is shown in Figure 6a. It can be seen that the grey level of the control group is the smallest and most concentrated. The rest of the distribution is also obviously discriminative at each temperature, and it shows a trend of first increasing and then decreasing. In order to further quantify the raw data, the weighted average of each grey band is calculated, as shown in the black curve of Figure 6b. The grey level reaches the maximum at 40 °C, and it has a slight distinction at 39–40 °C. Three experiments were performed to verify the reliability of this measurement method. The percentage increase in the average fluorescence intensity of the fluorescence group relative to the control was calculated as shown in the blue curve in Figure 6b.

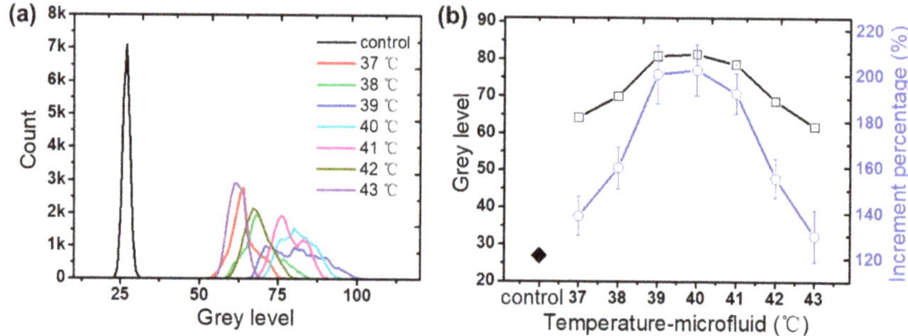

Figure 6. (**a**) Probability distribution of grey scales from 0 to 255. (**b**) Weighted average grey level of each image and increment percentage to control group against the different temperatures of microfluid.

The percent of fluorescence intensity increasing at different temperatures is shown in Table 3. These results show that this measurement method has a good ability to distinguish temperature variation. In previous studies, temperature sensors were placed outside the microfluidic area similar to our method, such as, next to the microfluidic device [39], together with the heater [40], in the reference position [41]. However, the issue in these cases is the precise inside temperature cannot be guaranteed. Therefore, the compensation relationship between the microfluids and the substrate is investigated in this work. In other research, one approach to reducing temperature errors is modifying the control method [42]. Unfortunately, the complex circuit systems and electrode fabricated increase uncertainty. The nanophotonic sensor [43] embedded in a microfluidic chip reaches high precision but is customized. As a more reliable and low-cost sensor, industrial-grade temperature sensors used in our model are capable of providing industrial-grade precisions suitable for various devices to reuse.

Table 3. The percent of fluorescence intensity increased.

Set Temperature (°C)	Fluorescence Intensity Increased (%)
37	138.69 ± 8.51
38	159.82 ± 9.09
38	200.48 ± 12.84
40	202.46 ± 11.11
41	192.22 ± 8.80
42	155.43 ± 8.51
43	130.10 ± 11.40

5. Conclusions

Here in this work, a heat transfer model was presented to investigate the relationship between the microfluids and the glass substrate of a typical microfluidic device. With an intelligent structure design and liquid metal, the millimeter-scale industrial temperature sensor could be utilized for temperature sensing of micro-scale fluids. The method overcomes the limitations of temperature sensing for microfluids. The dynamic linear range of measured temperature is demonstrated from 30 °C to 100 °C, and the uncertainty error is below 0.5 °C. Further, temperature-sensitive nucleic acid amplification experiments have been conducted to clarify the temperature resolution of this method. Therefore, it can be surmised that this method shows high potential for micro–macro interface sensing and is helpful beyond microfluidic applications.

Supplementary Materials: The following supporting information can be downloaded at: https://www.mdpi.com/article/10.3390/mi13050792/s1, Table S1: Numbers of the domain and boundary elements in different element sizes. Table S2: Parameter settings in the fluid-structure interacted simulation model. Figure S1: Discrete grid for the simulation model. Figure S2: Grid independence verification.

Author Contributions: Conceptualization, S.L.; experimental investigation, J.M.; experimental assistant C.Y., C.W. and H.L.; writing—original preparation, numerical simulation, J.M., J.L. and S.L.; writing—review and editing, J.L. and S.D.; supervision, S.L. and S.D.; funding acquisition, S.L. and J.L. All authors have read and agreed to the published version of the manuscript.

Funding: This research was funded by the National Natural Science Foundation of China, grant number 51728502; Fund for Distinguish Young Scholars in Tianjin 2018 3rd Round, grant number 180191; Hebei Science and Technology Foundation, grant number 19271707D; Hebei Natural Science Foundation, grant number E2020202101, E2022202127 and F2021202001; Department of Human Resources and Social Security of Hebei Province, grant number C20200314 and C20210337; and Jiangsu Key Laboratory of Advanced Manufacturing Equipment and Technology, grant number FMZ202016.

Data Availability Statement: The data that support the findings of this study are available from the corresponding author, upon reasonable request.

Conflicts of Interest: The authors declare no conflict of interest.

References

1. Massing, J.; Kähler, C.J.; Cierpka, C. A volumetric temperature and velocity measurement technique for microfluidics based on luminescence lifetime imaging. *Exp. Fluids* **2018**, *59*, 163. [CrossRef]
2. Dinh, T.; Phan, H.P.; Kashaninejad, N.; Nguyen, T.K.; Dao, D.V.; Nguyen, N.T. An on-chip SiC MEMS device with integrated heating, sensing, and microfluidic cooling systems. *Adv. Mater. Interfaces* **2018**, *5*, 1800764. [CrossRef]
3. Jeroish, Z.E.; Bhuvaneshwari, K.S.; Samsuri, F.; Narayanamurthy, V. Microheater: Material, design, fabrication, temperature control, and applications-a role in COVID-19. *Biomed. Microdevices* **2021**, *24*, 3. [CrossRef]
4. Geitenbeek, R.G.; Vollenbroek, J.C.; Weijgertze, H.M.; Tregouet, C.B.; Nieuwelink, A.-E.; Kennedy, C.L.; Weckhuysen, B.M.; Lohse, D.; Van Blaaderen, A.; Van Den Berg, A. Luminescence thermometry for in situ temperature measurements in microfluidic devices. *Lab Chip* **2019**, *19*, 1236–1246. [CrossRef]
5. Abduljabar, A.A.; Clark, N.; Lees, J.; Porch, A. Dual mode microwave microfluidic sensor for temperature variant liquid characterization. *IEEE Trans. Microw. Theory Tech.* **2017**, *65*, 2572–2582. [CrossRef]
6. Wirdatmadja, S.A.; Moltchanov, D.; Balasubramaniam, S.; Koucheryavy, Y. Microfluidic system protocols for integrated on-chip communications and cooling. *IEEE Access* **2017**, *5*, 2417–2429. [CrossRef]
7. Bürkle, F.; Czarske, J.; Büttner, L. Simultaneous velocity profile and temperature profile measurements in microfluidics. *Flow Meas. Instrum.* **2022**, *83*, 102106. [CrossRef]
8. Wang, J.-J.; Wang, T.; Wu, C.-G.; Luo, W.-B.; Shuai, Y.; Zhang, W.-L. Highly precise Ti/Pt/Cr/Au thin-film temperature sensor embedded in a microfluidic device. *Rare Met.* **2019**, *40*, 195–201. [CrossRef]
9. Choi, J.; Bandodkar, A.J.; Reeder, J.T.; Ray, T.R.; Turnquist, A.; Kim, S.B.; Nyberg, N.; Hourlier-Fargette, A.; Model, J.B.; Aranyosi, A.J. Soft, skin-integrated multifunctional microfluidic systems for accurate colorimetric analysis of sweat biomarkers and temperature. *ACS Sens.* **2019**, *4*, 379–388. [CrossRef]
10. Lee, D.; Fang, C.; Ravan, A.S.; Fuller, G.G.; Shen, A.Q. Temperature controlled tensiometry using droplet microfluidics. *Lab Chip* **2017**, *17*, 717–726. [CrossRef]
11. Xu, A.; Li, P.; Khan, F. Microfluidic Device Control System Based on Segmented Temperature Sensor. *Mob. Inf. Syst.* **2021**, *2021*, 9930649. [CrossRef]
12. Deng, Y.; Jiang, Y.; Liu, J. Low-melting-point liquid metal convective heat transfer: A review. *Appl. Therm. Eng.* **2021**, *193*, 117021. [CrossRef]
13. Papadopoulos, V.E.; Kefala, I.N.; Kaprou, G.D.; Tserepi, A.; Kokkoris, G. Modeling heat losses in microfluidic devices: The case of static chamber devices for DNA amplification. *Int. J. Heat Mass Transf.* **2022**, *184*, 122011. [CrossRef]
14. Hollstein, K.; Entholzner, D.; Zhu, G.; Weide-Zaage, K.; Benstetter, G. Developing a micro-thermography system for thermal characterization of LED packages. *Microelectron. Eng.* **2021**, *254*, 111694. [CrossRef]
15. Iwata, N.; Miyazaki, Y.; Yasuda, S.; Ogawa, H. Thermal performance and flexibility evaluation of metallic micro oscillating heat pipe for thermal strap. *Appl. Therm. Eng.* **2021**, *197*, 117342. [CrossRef]
16. Zhao, X.; Gao, W.; Yin, J.; Fan, W.; Wang, Z.; Hu, K.; Mai, Y.; Luan, A.; Xu, B.; Jin, Q. A high-precision thermometry microfluidic chip for real-time monitoring of the physiological process of live tumour cells. *Talanta* **2021**, *226*, 122101. [CrossRef] [PubMed]
17. Fujisawa, H.; Ryu, M.; Lundgaard, S.; Linklater, D.P.; Ivanova, E.P.; Nishijima, Y.; Juodkazis, S.; Morikawa, J. Direct Measurement of Temperature Diffusivity of Nanocellulose-Doped Biodegradable Composite Films. *Micromachines* **2020**, *11*, 738. [CrossRef]

18. Tian, B.; Liu, Z.; Wang, C.; Liu, Y.; Zhang, Z.; Lin, Q.; Jiang, Z. Flexible four-point conjugate thin film thermocouples with high reliability and sensitivity. *Rev. Sci. Instrum.* **2020**, *91*, 045004. [CrossRef]
19. Li, T.; Shi, T.; Tang, Z.; Liao, G.; Duan, J.; Han, J.; He, Z. Real-time tool wear monitoring using thin-film thermocouple. *J. Mater. Processing Technol.* **2021**, *288*, 116901. [CrossRef]
20. Feng, S.; Yan, Y.; Li, H.; He, Z.; Zhang, L. Temperature uniformity enhancement and flow characteristics of embedded gradient distribution micro pin fin arrays using dielectric coolant for direct intra-chip cooling. *Int. J. Heat Mass Transf.* **2020**, *156*, 119675. [CrossRef]
21. Li, C.; Sun, J.; Wang, Q.; Zhang, W.; Gu, N. Wireless thermometry for real-time temperature recording on thousand-cell level. *IEEE Trans. Biomed. Eng.* **2018**, *66*, 23–29. [CrossRef] [PubMed]
22. Gao, Y.; Bando, Y. Carbon nanothermometer containing gallium. *Nature* **2002**, *415*, 599. [CrossRef] [PubMed]
23. Kim, N.-I.; Chang, Y.-L.; Chen, J.; Barbee, T.; Wang, W.; Kim, J.-Y.; Kwon, M.-K.; Shervin, S.; Moradnia, M.; Pouladi, S. Piezoelectric pressure sensor based on flexible gallium nitride thin film for harsh-environment and high-temperature applications. *Sens. Actuators A Phys.* **2020**, *305*, 111940. [CrossRef]
24. Brites, C.D.S.; Lima, P.P.; Silva, N.J.O.; Millán, A.; Amaral, V.S.; Palacio, F.; Carlos, L.D. Thermometry at the nanoscale. *Nanoscale* **2012**, *4*, 4799–4829. [CrossRef]
25. Figueroa, B.; Hu, R.; Rayner, S.G.; Zheng, Y.; Fu, D. Real-time microscale temperature imaging by stimulated raman scattering. *J. Phys. Chem. Lett.* **2020**, *11*, 7083–7089. [CrossRef]
26. Jaque, D.; Vetrone, F. Luminescence nanothermometry. *Nanoscale* **2012**, *4*, 4301–4326. [CrossRef]
27. Uchida, K.; Sun, W.; Yamazaki, J.; Tominaga, M. Role of thermo-sensitive transient receptor potential channels in brown adipose tissue. *Biol. Pharm. Bull.* **2018**, *41*, 1135–1144. [CrossRef]
28. Yang, J.-M.; Yang, H.; Lin, L. Quantum dot nano thermometers reveal heterogeneous local thermogenesis in living cells. *ACS Nano* **2011**, *5*, 5067–5071. [CrossRef]
29. Okabe, K.; Inada, N.; Gota, C.; Harada, Y.; Funatsu, T.; Uchiyama, S. Intracellular temperature mapping with a fluorescent polymeric thermometer and fluorescence lifetime imaging microscopy. *Nat. Commun.* **2012**, *3*, 705. [CrossRef]
30. Mäki, A.-J.; Verho, J.; Kreutzer, J.; Ryynänen, T.; Rajan, D.; Pekkanen-Mattila, M.; Ahola, A.; Hyttinen, J.; Aalto-Setälä, K.; Lekkala, J.; et al. A Portable Microscale Cell Culture System with Indirect Temperature Control. *SLAS Technol.* **2018**, *23*, 566–579. [CrossRef]
31. Cantoni, F.; Werr, G.; Barbe, L.; Porras, A.M.; Tenje, M. A microfluidic chip carrier including temperature control and perfusion system for long-term cell imaging. *HardwareX* **2021**, *10*, e00245. [CrossRef]
32. Tözüm, M.S.; Alay Aksoy, S.; Alkan, C. Development of reversibly color changing textile materials by applying some thermochromic microcapsules containing different color developers. *J. Text. Inst.* **2021**, 1–10. [CrossRef]
33. Fu, J.; Gao, J.; Qin, P.; Li, D.; Yu, D.; Sun, P.; He, Z.; Deng, Z.; Liu, J. Liquid metal hydraulics paradigm: Transmission medium and actuation of bimodal signals. *Sci. China Technol. Sci.* **2021**, *65*, 77–86. [CrossRef]
34. Kim, P.; Kwon, K.W.; Min, C.P.; Lee, S.H.; Suh, K.Y. Soft Lithography for Microfluidics: A Review. *BioChip J.* **2008**, *2*, 1–11.
35. Qin, D.; Xia, Y.; Whitesides, G.M. Soft lithography for micro- and nanoscale patterning. *Nat. Protoc.* **2010**, *5*, 491–502. [CrossRef] [PubMed]
36. Wang, H.; Shi, L.; Zhou, T.; Xu, C.; Deng, Y. A novel passive micromixer with modified asymmetric lateral wall structures. *Asia-Pac. J. Chem. Eng.* **2018**, *13*, e2202. [CrossRef]
37. Corrêa Ribeiro, C.A.; Ferreira, J.R.; Lima e Silva, S.M.M. Thermal influence analysis of coatings and contact resistance in turning cutting tool using COMSOL. *Int. J. Adv. Manuf. Technol.* **2021**, *118*, 275–289. [CrossRef]
38. Wood, A.; Tupholme, G.; Bhatti, M.; Heggs, P. Steady-state heat transfer through extended plane surfaces. *Int. Commun. Heat Mass Transf.* **1995**, *22*, 99–109. [CrossRef]
39. Abeille, F.; Mittler, F.; Obeid, P.; Huet, M.; Kermarrec, F.; Dolega, M.E.; Navarro, F.; Pouteau, P.; Icard, B.; Gidrol, X.; et al. Continuous microcarrier-based cell culture in a benchtop microfluidic bioreactor. *Lab Chip* **2014**, *14*, 3510–3518. [CrossRef]
40. Habibey, R.; Golabchi, A.; Latifi, S.; Difato, F.; Blau, A. A microchannel device tailored to laser axotomy and long-term microelectrode array electrophysiology of functional regeneration. *Lab Chip* **2015**, *15*, 4578–4590. [CrossRef]
41. Lin, J.-L.; Wu, M.-H.; Kuo, C.-Y.; Lee, K.-D.; Shen, Y.-L. Application of indium tin oxide (ITO)-based microheater chip with uniform thermal distribution for perfusion cell culture outside a cell incubator. *Biomed. Microdevices* **2010**, *12*, 389–398. [CrossRef] [PubMed]
42. Maki, A.-J.; Ryynanen, T.; Verho, J.; Kreutzer, J.; Lekkala, J.; Kallio, P.J. Indirect Temperature Measurement and Control Method for Cell Culture Devices. *IEEE Trans. Autom. Sci. Eng.* **2018**, *15*, 420–429. [CrossRef]
43. Habibi, M.; Bagheri, P.; Ghazyani, N.; Zare-Behtash, H.; Heydari, E. 3D printed optofluidic biosensor: NaYF4: Yb3+, Er3+ upconversion nano-emitters for temperature sensing. *Sens. Actuators A Phys.* **2021**, *326*, 112734. [CrossRef]

Article

Thermo-Hydraulic Performance of Pin-Fins in Wavy and Straight Configurations

Mohamad Ziad Saghir [1,*] and **Mohammad Mansur Rahman** [2]

1 Department of Mechanical and Industrial Engineering, Toronto Metropolitan University, Toronto, ON M5B2K3, Canada
2 Department of Mathematics, College of Science, Sultan Qaboos University, Al-Khod PC 123, Muscat, Oman; mansur@squ.edu.om
* Correspondence: zsaghir@ryerson.ca

Abstract: Pin-fins configurations have been investigated recently for different engineering applications and, in particular, for a cooling turbine. In the present study, we investigated the performance of three different pin-fins configurations: pin-fins forming a wavy mini-channel, pin-fins forming a straight mini-channel, and a mini-channel without pin-fins considering water as the working fluid. The full Navier–Stokes equations and the energy equation are solved numerically using the finite element technique. Different flow rates are studied, represented by the Reynolds number in the laminar flow regime. The thermo-hydraulic performance of the three configurations is determined by examining the Nusselt number, the pressure drop, and the performance evaluation criterion. Results revealed that pin-fins forming a wavy mini-channel exhibited the highest Nusselt number, the lowest pressure drop, and the highest performance evaluation criterion. This finding is valid for any Reynolds number under investigation.

Keywords: pin-fins; wavy pin-fins channel; performance criterion; pressure drop; friction factor

Citation: Saghir, M.Z.; Rahman, M.M. Thermo-Hydraulic Performance of Pin-Fins in Wavy and Straight Configurations. *Micromachines* **2022**, *13*, 954. https://doi.org/10.3390/mi13060954

Academic Editors: Ruijin Wang and Junfeng Zhang

Received: 14 May 2022
Accepted: 14 June 2022
Published: 16 June 2022

Publisher's Note: MDPI stays neutral with regard to jurisdictional claims in published maps and institutional affiliations.

1. Introduction

Heat extraction is a desirable topic in engineering. Nowadays, with climate change and the introduction of electrical systems such as vehicles, amongst other applications, researchers focus on extracting and storing heat. Storing heat or energy to be used later is becoming an essential practice in engineering but is in its infancy. More sophisticated means to store a large amount of heat are under investigation. This combines engineers' and architects' efforts in designing a unique system capable of absorbing heat and storing it for later use. In technology, dealing with cooling small area surfaces [1] is of great interest among battery developers in using different types of vehicles. Forced flow is one of these applications through channels with varying widths, up to mini-channels of a few millimeters in width. Straight and wavy tracks are amongst the ones that were investigated recently [2]. However, the users are faced with a pressure drop, thus the need for a higher pumping mechanism. Since we are moving to a smaller scale component, this is becoming an issue. On the other hand, pin-fins have been proposed to extract heat from a surface, and this technology is not new. Forced convection in pin-fins was investigated; the uniqueness of this approach is the low-pressure drop in the system. However, to the authors' knowledge, no studies have been written on replacing channel walls with pin-fin walls or even in a wavy form. Thus, in the present article, an attempt was made to investigate the effectiveness numerically in studying forced convection in wavy wall mini-channels fabricated of pin-fins and compare the performance against straight pin-fin wall mini-channels.

Xu et al. [3] conducted experimental measurements of heat enhancement and pressure drop on a plate with different pin-fin shapes. The shapes used in the experiment were types of round shapes, quadrangle shapes, and streamlined shapes. All forms had an identical nominal diameter. The flow was turbulent, with Reynolds numbers varying from 10,000 to 60,000. Results revealed that, as for heat enhancement, the quadrangle shape performed better than the other two sets of forms. If the pressure varied between the inlet and outlet, the friction factor varied, and thus the streamlined shapes were the best due to the lowest pressure drop. Ye et al. [4] proposed a new test section composed of pin-fins and cutback surfaces. The aim was to investigate the best cooling performance at different flow rates. The pin-fin had a cross-section of round, elliptical, and teardrop shapes having constant heights. The pin-fins were located at various positions, not forming a channel wall. They found that the addition of pin-fins reduces the film coverage in the expanded cutback at a low flow rate. The fin-pins did not show heat enhancement as expected for their particular design. Pan et al. [5] investigated the effectiveness of using pin-fin wall channels compared with solid wall mini-channels. They confirmed that pin-fin walls provide the best heat enhancement. Wang et al. [6] investigated the heat transfer characteristics of a single pin-fins mini-channel. Deng et al. [7] studied the micro pin-fins located on the hot plate. Results revealed that micro pins have a significant heat boiling enhancement.

Niranjan et al. [8] investigated the forced convection numerically on a heated plate with uniformly distributed micro pin-fins; the pin-fins form a channel. They noticed that the presence of pin-fins enhanced heat transfer by 10%. Different fin shapes were investigated by Hirasawa et al. [9], as well as Matsumoto et al. [10] and by Casano et al. [11]. Later, Ahmadian-Elmi et al. [12] continued investigating the effectiveness of pin-fins in heat extraction. Alnaimat et al. [13] used pin-fins inside a channel.

Similarly, Saghir, and Rahman [14] conducted numerical modelling of block pins inside a channel. The channel side was large enough. The numerical results concluded that heat enhancement is evident when the track contains micro-pins. However, because micro-pins act as fins, a non-uniform temperature distribution is detected with a surge in heat extraction at each pin base. Mafeed et al. [15] investigated the importance of fin-pin height in heat extraction. Different pin-fin shapes were studied, and an optimum height was detected for some pin-fins.

Further work was performed by Boyalakuntla et al. [16] but at the larger pin-fin size. Qu and Siu-Ho [17,18] conducted an experimental measurement of the pressure drop, friction coefficient, and Nusselt number for an array of micro pin-fins. They proposed different correlations and discussed other correlations found in the literature. The finding provided an insight into heat enhancement and the dependence of the Nusselt number on the Reynolds number. Recent work by Iasiello et al. [19] and Mauro et al. [20] focused on heat enhancement in the presence of porous media. Olabi et al. [21] focused on using nanofluid in heat extraction.

According to the current review, few studies have investigated the high importance of pin-fins, and no researchers look into the significance of pin-fin distributions. In the present study, we combined two crucial effects missing from the literature review. These are mainly pin-fins forming wavy channels and height variation effects. Thus, the uniqueness of the present work. This novel concept will be investigated compared with traditional, uniformly distributed pin-fin cases and cases with no pin-fins.

2. Problem Description

In the present study, an attempt was made to investigate two different pin-fins mini-channel models. A previous study by the authors [1] demonstrated that wavy channels provide a marginal heat enhancement compared with straight channels. Here we investigated two types of pin-fin mini-channels: a pin-fin forming straight mini-channel and a wavy mini-channel. For both configurations, the spacing between the pins-fins is identical, and the number of rows of pin-fins is also similar. What makes the model unique is the variable pin-fin height. Here, we investigated four pin-fin sizes. Figure 1 shows the two aluminum-fabricated configurations under investigation. The base square surface of the pin-fins is 2 mm on each side, and the pin-fin heights are 2 mm, 4 mm, 6 mm, and 8 mm, respectively. The distance between two pin-fins in the x-direction is 8.05 mm, and 7.67 mm in the y-direction. The base plate of the test section has a square shape of 37.5 mm on each side and a thickness of 3.7 mm in the z-direction. The test section has a square shape of 37.5 mm and a height of 8.7 mm. The reason for such a thick base plate is to act as a step obstacle to deflect the flow toward the pin-fins. We compared these two configurations' performance with the case of no pin-fins to determine whether having pin-fins improves the heat enhancement.

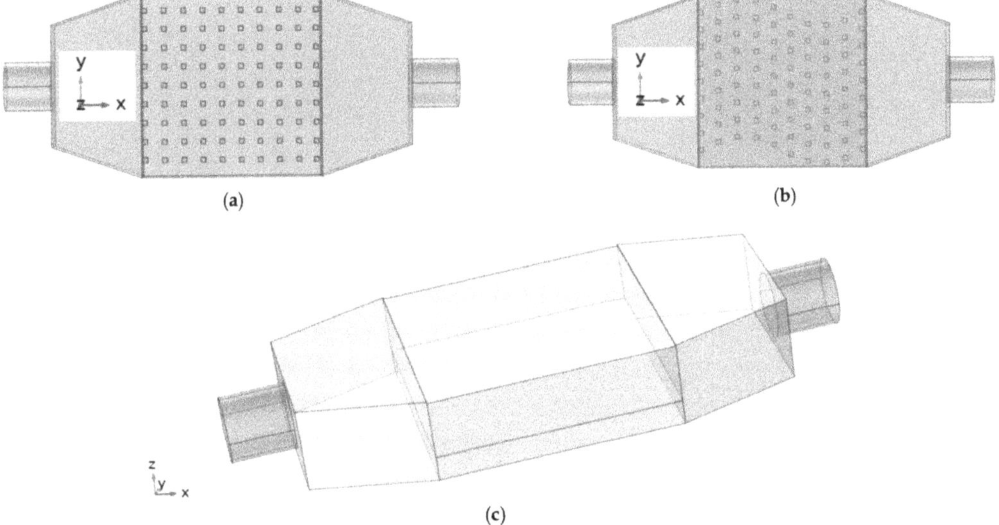

Figure 1. Three different configurations ((**a**) Straight pin-fins configuration, (**b**) Wavy pin-fins configuration, (**c**) No pin-fins configuration).

2.1. Differential Equations and Boundary Conditions

Solving this problem requires three sets of differential equations. The working fluid under investigation is laminar, incompressible, and steady. Since the flow is due to the forced convection, one can neglect the gravity vector from the full Navier–Stokes equations in three dimensions. The energy equation is solved for heat transfer, in which the convective terms couple the heat transfer and the fluid flow. Third, the conduction heat transfer equation is required for the solid boundary of the setup.

The three sets of equations (Navier–Stokes, energy, and heat conduction) were solved numerically using the finite element technique with the aid of the COMSOL software, Newton, USA [17]. The differential equations are omitted for the sake of duplication but are explained in detail in reference [14]. The equations were rendered non-dimensional using the following non-dimensional parameters shown in Equation (1).

$$X = \frac{x}{D}, \; Y = \frac{y}{D}, \; Z = \frac{z}{D}, \; U = \frac{u}{u_{in}}, \; V = \frac{v}{u_{in}}, \; W = \frac{w}{u_{in}}, \; P = \frac{pD}{\mu u_{in}}, \; \theta = \frac{(T - T_{in})k_w}{q'' D} \tag{1}$$

In the non-dimensional parameters shown in Equation (1), D represents the characteristic length equal to 18.97 mm. It is the hydraulic diameter at the larger face of the mixing chamber facing the testing section. Since the study is conducted for different flow rates represented by the Reynolds number, the inlet velocity is u_{in}. Here, k_W is the conductivity, μ is the viscosity and ρ is the density of the fluid. As shown in Figure 2, the red arrows indicate the applied heat flux q'' in watts per m^2. In our analysis, we set the applied heat flux as 50,000 W/m^2. The boundary conditions are set as follows:

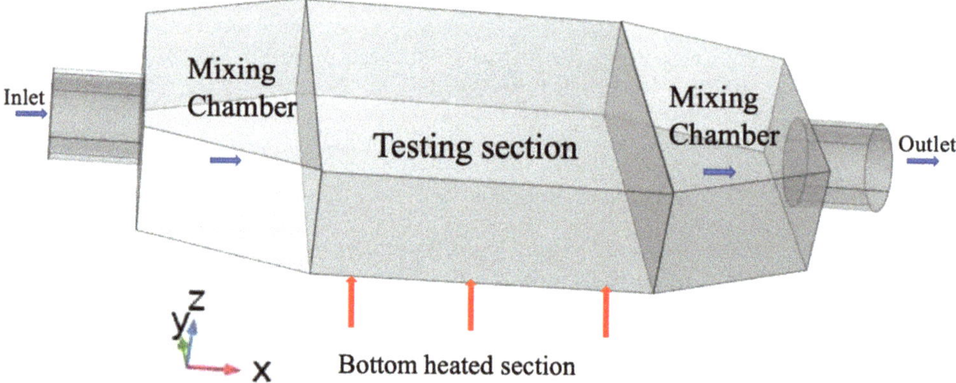

Figure 2. Boundary conditions.

At the inlet: $T = T_{in}$ and $u = u_{in}$, in the non-dimensional form, it becomes $\theta = 0$ and $U = 1$. At the outlet: the boundary is free. At the heated bottom section, the heat flux is $q = q''$, in non-dimensional form; it equals 1. All external surface boundaries are assumed to be insulated for no heat losses. The solid component of the model including the pins-fins are fabricated of Aluminum. Plant and Saghir [2] conducted an experiment and demonstrated minimal heat losses. However, the heater performance with time shows some weakness in delivering the proper heat flux. From time to time, heaters are replaced for accuracy. As shown in Figure 2, the flow moves in the x-direction, and the pin's height varies in the z-direction. The two parameters which control the thermo-hydraulic effect are the Reynolds number defined as $Re = \frac{\rho u_{in} D}{\mu}$ and the Prandtl number defined as $Pr = \frac{c_p \mu}{k_w}$.

In the present study, we investigated the role of different model parameters on the flow and thermal fields. We calculated the temperature distribution at the pin's base where a fluid circulates from the inlet to the outlet. Then, we calculated the Nusselt number at the exact locations where the temperature is measured. The pressure drop is calculated between the inlet and the outlet, along with the friction coefficient for different configurations. Finally, the influence of the Nusselt number and the pressure drop are combined by defining the performance evaluation criterion (PEC). By definition, Nusselt number, $Nu = \frac{hD}{k_w}$ where the heat convection coefficient, $h = \frac{q''}{(T - T_{in})}$. In dimensionless form, it becomes $Nu = \frac{1}{\theta}$. The friction factor in the non-dimensional form is defined as $f = 0.2529 \times \frac{\Delta P}{Re}$, where P is the pressure and Re is the Reynolds number. Therefore, the performance evaluation criterion in non-dimensional form is defined as $PEC = \frac{Nu_{average}}{f^{1/3}}$. These parameters can guide the reader on heat removal effectiveness for various configurations at different conditions. The readers are invited to read reference [14] for the detailed equations used in our model.

2.2. Mesh Sensitivity and Convergence Criteria

A mesh analysis was conducted to ensure the mesh used provides accurate results. Table 1 presents different mesh levels used and the corresponding Nusselt number. As can be seen, using of a normal mesh consisting over 700,000 elements provides accurate results.

Table 1. Mesh sensitivity analysis.

Mesh Name	Number of Elements	Average Nusselt Number
Normal	700,494	7.157
Coarse	599,637	7.153
Coarser	301,134	7.106

In using COMSOL software, we used the default solver segregated method. The reader can refer to reference [22] for more details. In a nutshell, the convergence criteria were set as follows: at every iteration, the average relative error for U, V, W, P, and θ were computed using the following relation:

$$R_c = \frac{1}{n \cdot m} \sum_{i=1}^{i=m} \sum_{j=1}^{j=n} \left| \frac{\left(F_{i,j}^{s+1} - F_{i,j}^{s} \right)}{F_{i,j}^{s+1}} \right| < 1.0 \times 10^{-6} \tag{2}$$

where F represents one of the unknowns, viz. the velocities, pressure, and temperature, s is the iteration number, and (i, j) represents the coordinates on the grid.

3. Results and Discussion

Investigating the effectiveness of heat removal in the presence of a wavy and a straight mini-channel composed of pin-fins is the objective of this current study. Different parameters relating to the flow and thermal fields are examined.

3.1. Thermo-Hydraulic Performance of Different Pin Channels Configuration

The heat enhancement and pressure drop is investigated for various Reynolds number ranges and pin-fin heights. Figure 3 presents the temperature variation near the base of the pin-fins for different configurations and Reynolds numbers. It is observed that regardless of the design under investigation, the temperature profiles are similar, having different intensities. A maximum temperature exists near the end of the flow path for all cases in this figure.

Obviously, as the Reynolds number increases, the heat extraction increases accordingly, resulting in a temperature drop. It is also evident that the development of the flow and thermal boundary layers leads to obstruction in heat removal near the end of the flow path. Interestingly, the temperature variation at the base of the pin-fins is wavy. In particular, it is lower at the pin-fins' bottom. The reason is that pin-fins absorb more heat than the base plate, resulting in waviness. One way to observe this finding is to investigate the average Nusselt number for all cases, as depicted in Figure 4.

By examining Figure 4, and since the average Nusselt number is known as the inverse of the temperature, the wavy pin-fins configuration confirms the previous finding. It is proven here that the pin-fins forming wavy mini-channels exhibit the highest Nusselt number and outperform the case with no pin-fins. The variation in the pin-fins' average Nusselt number with Reynolds number for all heights is not identical. By comparing the two configurations of pin-fins arrangement, the average Nusselt number for fin-pins forming wavy mini-channels case is higher.

As mentioned earlier, the advantage of using pin-fins in mini-channels results in lower pressure drop. The average Nusselt number measuring the performance evaluation criterion for the two pin-fins configurations is depicted in Figure 5. As expected, pin-fins forming wavy mini-channels confirm the earlier conclusion, providing better heat enhancement without additional pumping power. The larger pin-fin sizes demonstrate a more vital performance evaluation criterion for both pin-fins configurations.

Finally, Figure 6 presents the hydrodynamic effect by displaying the flow behavior in different planes when the pin-fin height is 6 mm and the Reynolds number equal to 250. In Figure 6, column A is for straight pin configuration and column B for Wavy pin configuration. Comparing Figure 6a,d it is evident that the flow circulation in the latter one is more effective in removing heat from the bottom of the plate. Flow circulation is more pronounced compared with Figure 6a. What triggered this flow is shown in Figure 6b,e. A step obstacle at the bottom of the mixing chamber forces the flow to jump and penetrate the pin-fins, creating a mixture that remains in the laminar regime. We again re-examined Figure 6a,d by displaying the flow patterns in Figure 6c,f, respectively. Indeed, a more complex flow circulation is evident in Figure 6f; thus, the reason for the better performance evaluation criterion. In Figure 6c, the flow moves in mini-channels; whereas in the wavy shape Figure 6f, it circulates non-uniformly, resulting in better heat extraction. The flow circulates between the pin-fins and the top of the pin-fins, as the pin-fins' height is smaller than the cavity depth. Thus, one can see a counter flow emerging from the bottom of the hot plate to the top of the test section. This helps increase the mixing and heat extraction.

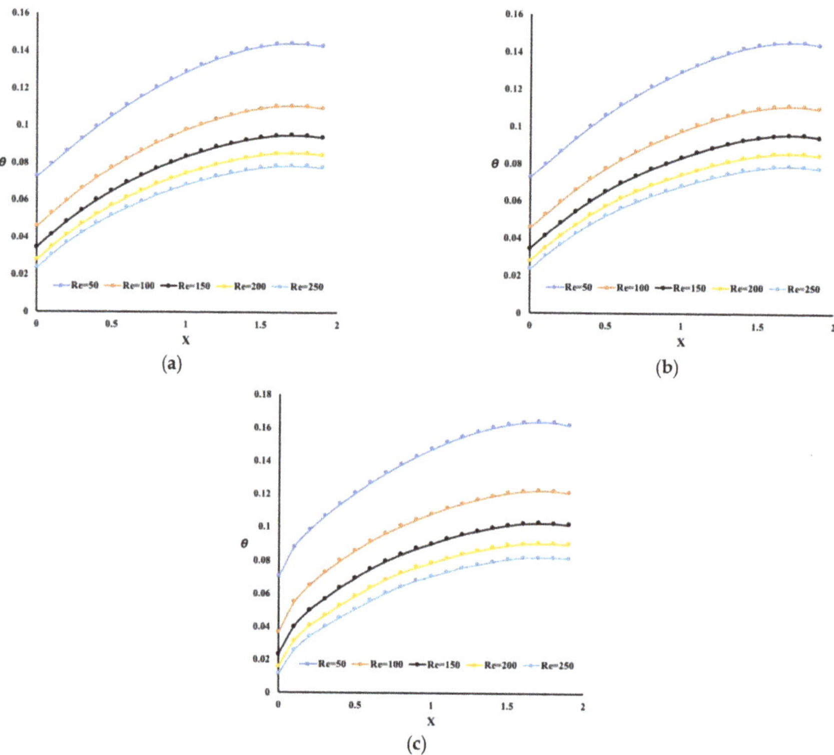

Figure 3. Temperature variation for different configurations (6 mm pin-fins height case, (**a**) Straight pin-fins configuration, (**b**) Wavy pin-fins configuration, (**c**) No pin-fins configuration).

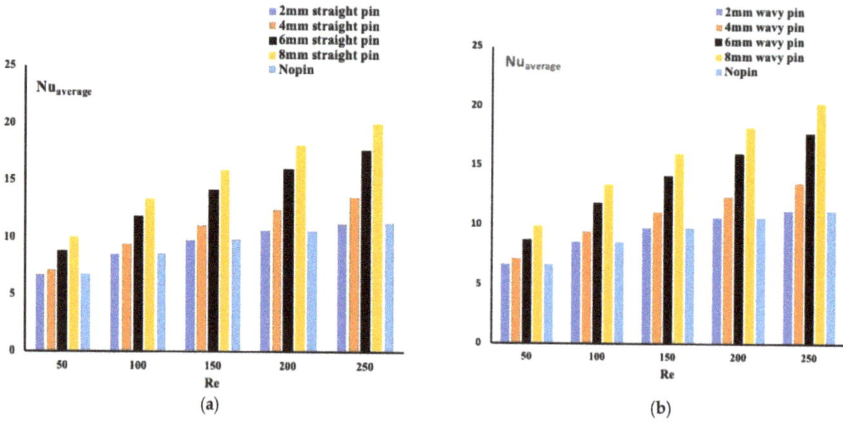

Figure 4. Average Nusselt number for different configuration ((**a**) Straight pin-fins configuration, (**b**) Wavy pin-fins configuration).

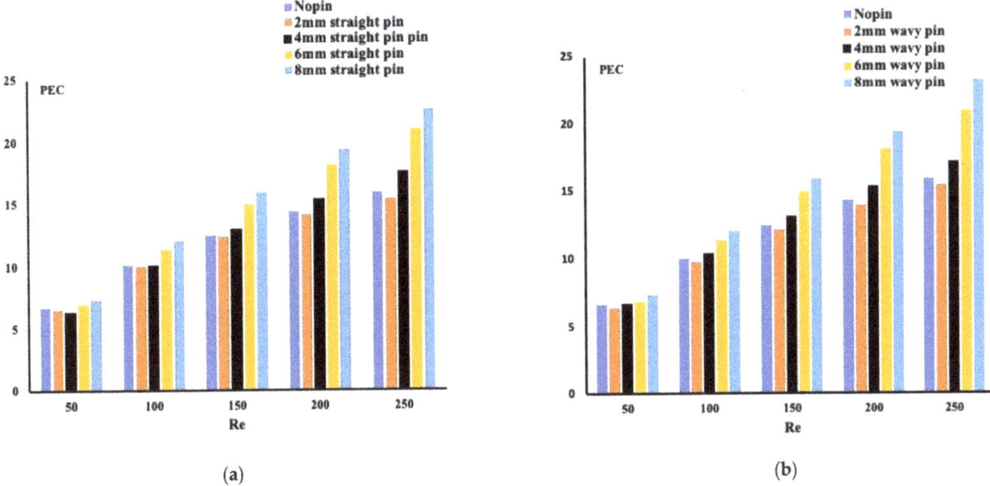

(a)

(b)

Figure 5. Performance evaluation criterion for different configurations ((**a**) Straight pin-fins configuration, (**b**) Wavy pin-fins configuration).

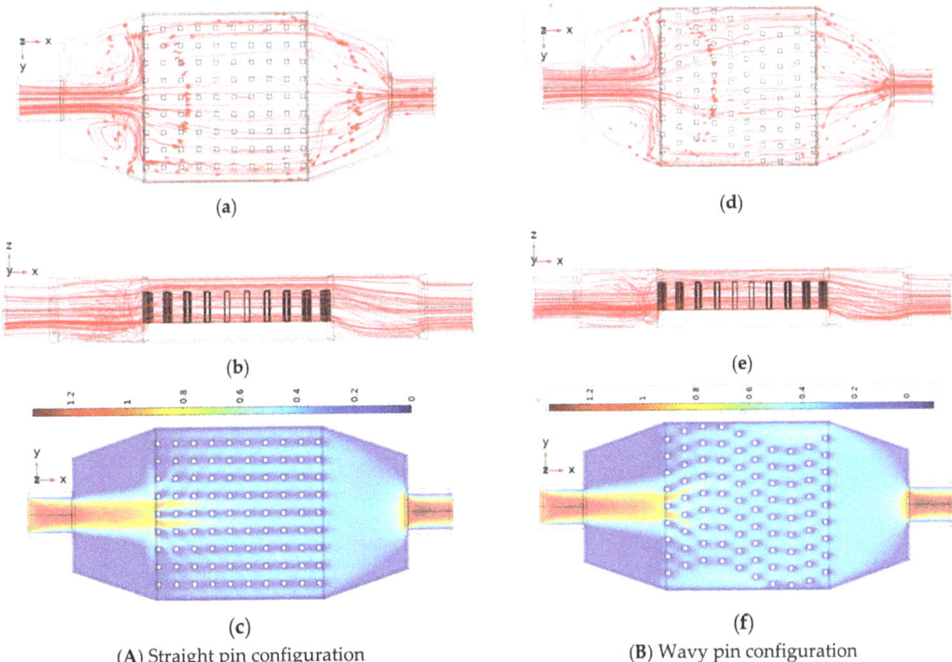

(a)

(d)

(b)

(e)

(c)

(f)

(**A**) Straight pin configuration

(**B**) Wavy pin configuration

Figure 6. Flow distributions for the pin configurations. (Re = 250, pin-fin height 6 mm, (**a**) Streamline in (xy) plane, (**b**) Streamline in (xz) plane, (**c**) Velocity in (xy) plane, (**d**) Streamline in (xy) plane, (**e**) Streamline in (xz) plane, (**f**) velocity in (xy) plane).

3.2. Heat Removal for All Configurations

It is interesting to calculate the amount of heat removed from the system for all configurations. The heat removal is known to be $Q = \dot{m}c_p(T_{out} - T_{in})$, where c_p is the specific heat, T_{in} and T_{out} are the temperatures at the flow inlet and flow outlet, respectively.

The $\dot{m}$ is the mass flow rate which is the product of the flow rate and the density of the water. In the non-dimensional form, the formulation becomes $Re.Pr.\theta_{out}$. The inlet temperature is related to the definition of the non-dimensional temperature presented in Equation (1). Figure 7 illustrates the three configurations' heat removed for different pin-fins' heights. The first observation, which applies to all the cases, is that the absence of pin-fins is the worst case for heat removal. This observation is followed by the fact that pin-fins forming a wavy mini-channel has better heat removal than the straight channel. This confirms all previous findings by showing the temperature, the Nusselt number, or the PEC. For all cases, the non-linear behavior of the heat removal profile suggests that at the Reynolds number of 150, less heat is removed from the system. However, what is evident is that the best heat removal mechanism is when the pin-fins' height is 4 mm. That means flow going through the pin-fins and above the pin-fins plays a significant role in heat extraction.

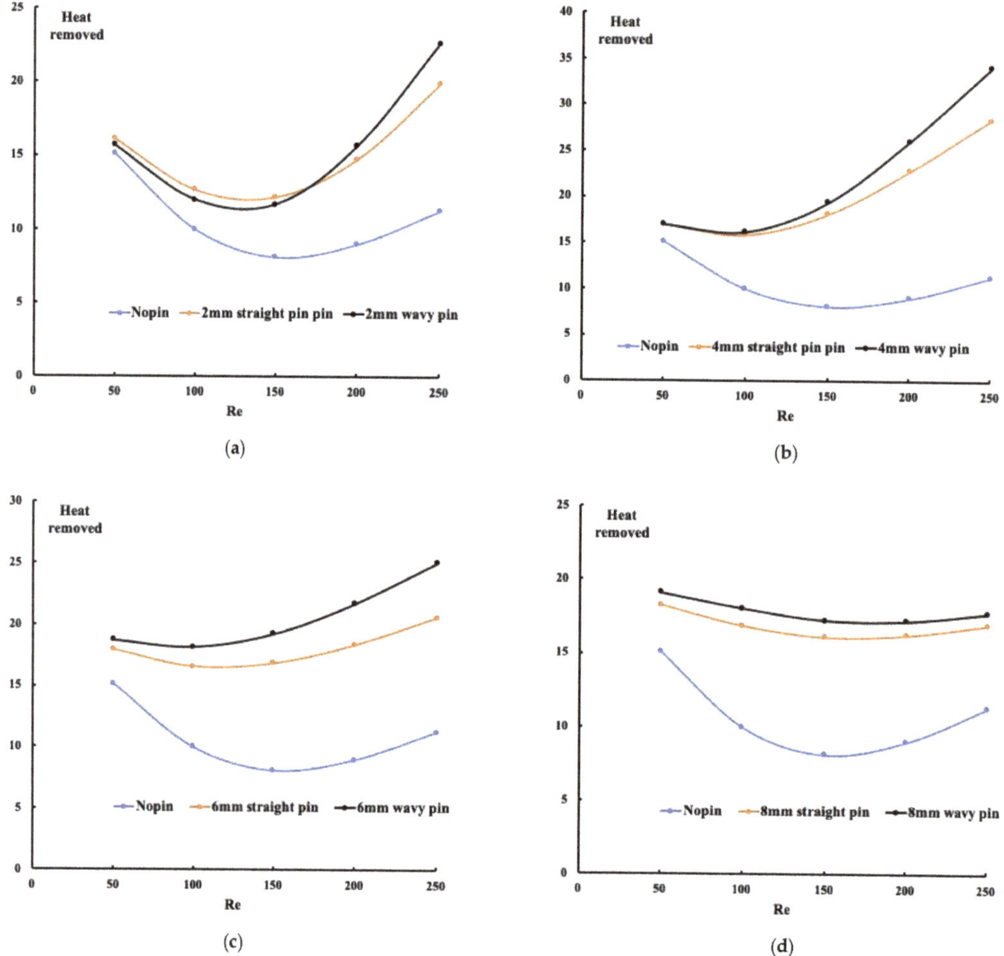

Figure 7. Heat removed for all configurations ((**a**) 2 mm pin height, (**b**) 4 mm pin height, (**c**) 6 mm pin height, (**d**) 8 mm pin height).

3.3. Friction Coefficient and Nusselt Number Correlations

It is worth discussing the relation between the friction factor, the Nusselt number, and the Reynolds number. Qu and Siu-Ho experimented with an array of pin-fins staggered along with the flow and proposed a relation between the friction factor and the Reynolds number. Their experimental setup and the pin-fins distributions are not similar to our proposed model. Figure 8 summarizes our numerical findings of the variation in the friction coefficient with the Reynolds number.

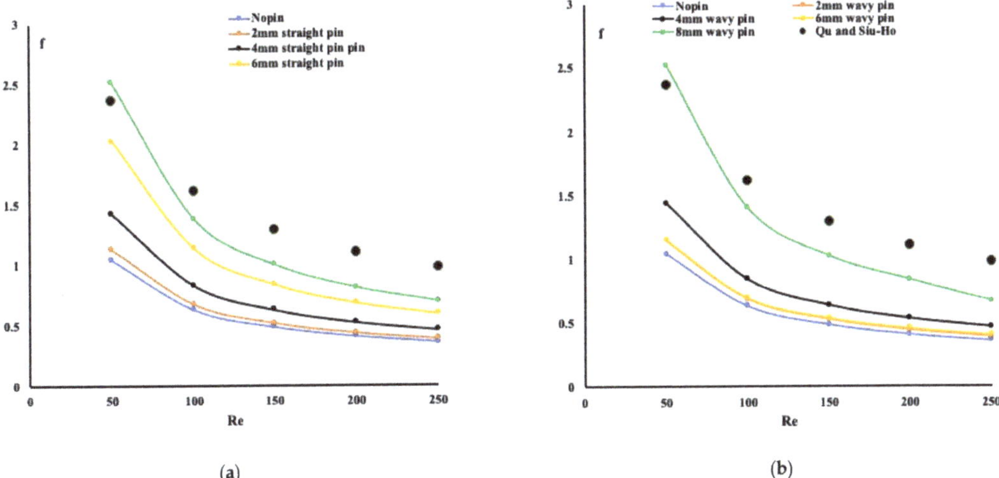

Figure 8. Variation in friction coefficient with Reynolds number ((a) Straight pin-fins mini-channels, (b) Wavy pin-fins mini-channels).

In addition, in the same plot, the experimental findings of Qu and Siu-Ho [17,18] are plotted. As one can observe, the trend for the friction factor is identical, and the friction factor values are also within the obtained values. This indicates that the numerical results obtained are reasonable and correct. As shown in Figure 8, an increased Reynolds number reduces the friction coefficient accordingly. For the case of pin-fins forming a wavy channel, as indicated earlier, greater friction is expected than the pin-fins creating straight channels. A fitting curve for Figure 8 (not including Qu and Siu-Ho curves) is summarized by two analytical relations shown in Equations (3) and (4) as a function of Reynolds number and the pin-fins' height. The friction factor trend variation with the Reynolds number is well-presented in these two equations having a 97% accuracy.

For straight mini-channels:

$$f = 59.355 \text{Re}^{-0.733}\left(\frac{H}{D}\right)^{0.488} \tag{3}$$

For wavy mini-channels:

$$f = 38.49 \text{Re}^{-0.707}\left(\frac{H}{D}\right)^{0.353} \tag{4}$$

Furthermore, a correlation between the average Nusselt number as a function of the Reynolds number and the pin-fins' height was determined for the two cases of pin-fins distribution in the mini-channels. Equations (5) and (6) display the obtained relations for different Reynolds numbers and heights. It is essential to indicate that these relations are valid for a range of Reynolds numbers from 0 to 250 and for a range of pin-fins size from 2 mm to 8 mm with an accuracy of 97%.

For straight mini-channels:

$$Nu_{average} = 2.848 Re^{0.398} \left(\frac{H}{D} \right)^{0.358} \tag{5}$$

For wavy mini-channels:

$$Nu_{average} = 2.82 Re^{0.402} \left(\frac{H}{D} \right)^{0.359} \tag{6}$$

It is evident from these two relations that the pin-fins forming wavy channels exhibit a higher Nusselt number.

4. Conclusions

In the present study, we investigated the thermo-hydraulic performance of a set of pin-fins forming wavy and straight mini-channels. For this purpose, the Navier–Stokes equations and the energy equations were solved numerically using the finite element technique. Different variables were examined in this study, mainly the variation in Reynolds number and the height of the pin-fins. Thus, two flows were involved in the test section, primarily the first flow circulating between the pin-fins and the other one spreading above the pin-fins. This setup's thermal and hydraulic performance was compared with the case of pin-fins forming straight mini-channels and having no pin-fins. Results revealed;

1. The pin-fins forming wavy mini-channels exhibit more significant heat enhancement and less pressure drop;
2. The performance evaluation criterion is found to be higher for the wavy mini-channel configuration compared with the straight pin-fins configuration;
3. The flow above the pin-fins is found to reduce the pressure drop.

Author Contributions: Conceptualization, M.Z.S. and M.M.R.; methodology, M.Z.S.; software, M.Z.S. and M.M.R.; validation, M.Z.S. and M.M.R.; formal analysis, M.Z.S. and M.M.R.; investigation, M.Z.S. and M.M.R.; resources, M.Z.S.; data curation, M.Z.S. and M.M.R.; writing—original draft preparation, M.Z.S. and M.M.R. All authors have read and agreed to the published version of the manuscript.

Funding: The funding of this research is from the National Science and Engineering Council Canada (NSERC) and the Sultan Qaboos University for funding through the research grant IG/SCI/MATH/20/03.

Institutional Review Board Statement: Not applicable.

Informed Consent Statement: Not applicable.

Data Availability Statement: Not applicable.

Acknowledgments: We would like to thank the anonymous referees for their valuable comments for the further improvement of the paper. M.Z.S. is grateful to National Science and Engineering Council Canada (NSERC), M.M.R. is grateful to Sultan Qaboos University for funding through the research grant IG/SCI/MATH/20/03.

Conflicts of Interest: The authors declare no conflict of interest.

References

1. Elsafy, K.M.; Saghir, M.Z. Forced convection in wavy microchannels porous media using TiO_2 and Al_2O_3-Cu nanoparticles in water base fluids: Numerical results. *Micromachines* **2021**, *12*, 654. [CrossRef] [PubMed]
2. Plant, R.D.; Saghir, M.Z. Numerical and experimental investigation of high concentration aqueous alumina nanofluids in a two and three channel heat exchanger. *Int. J. Thermofluids* **2021**, *9*, 100055. [CrossRef]
3. Xu, J.; Zhang, K.; Duan, J.; Lei, J.; Wu, J. Systematic comparison on convective heat transfer characteristics of several pin fins for turbine cooling. *Crystals* **2021**, *11*, 977. [CrossRef]
4. Ye, L.; Liu, C.L.; Zhang, F.; Li, J.C.; Zhou, T.L. Effect of the multiple rows of pin-fins on the cooling performance of cutback trailing-edge. *Int. J. Heat Mass Transf.* **2021**, *170*, 120992. [CrossRef]
5. Pan, Y.; Zhao, R.; Nian, Y.; Cheng, W. Study on the flow and heat transfer characteristics of pin-fin manifold microchannel heat sink. *Int. J. Heat Mass Transf.* **2022**, *183*, 122052. [CrossRef]

6. Wang, Y.; Nayebzadeh, A.; Yu, X.; Shin, J.H.; Peles, Y. Local heat transfer in a microchannel with a pin fin—experimental issues and methods to mitigate. *Int. J. Heat Mass Transf.* **2017**, *106*, 1191–1204. [CrossRef]

7. Deng, D.; Wan, W.; Qin, Y.; Zhang, J.; Chu, X. Flow boiling enhancement of structured microchannels with micro pin fins. *Int. J. Heat Mass Transf.* **2017**, *105*, 338–349. [CrossRef]

8. Niranjan, R.S.; Singh, O.; Ramkumar, J. Numerical study on thermal analysis of square micro pin fins under forced convection. *Heat Mass Transf.* **2022**, *58*, 263–281. [CrossRef]

9. Hirasawa, S.; Fujiwara, A.; Kawanami, T.; Shirai, K. Forced convection heat transfer coefficient and pressure drop of diamond-shaped fin-array. *J. Electron. Cool. Therm. Control* **2014**, *4*, 78. [CrossRef]

10. Matsumoto, N.; Tomimura, T.; Koito, Y. Heat transfer characteristics of square micro pin fins under natural convection. *J. Electron. Cool. Therm. Control* **2014**, *4*, 59. [CrossRef]

11. Casano, G.; Collins, M.W.; Piva, S. Experimental investigation of the fluid dynamics of a finned heat sink under operating conditions. *J. Electron. Cool. Therm. Control* **2014**, *4*, 86. [CrossRef]

12. Ahmadian-Elmi, M.; Mashayekhi, A.; Nourazar, S.S.; Vafai, K. A comprehensive study on parametric optimization of the pin-fin heat sink to improve its thermal and hydraulic characteristics. *Int. J. Heat Mass Transf.* **2021**, *180*, 121797. [CrossRef]

13. Alnaimat, F.; Al Hamad, I.M.; Mathew, B. Heat transfer intensification in MEMS two-fluid parallel flow heat exchangers by embedding pin fins in microchannels. *Int. J.* **2021**, *9*, 100048. [CrossRef]

14. Saghir, M.Z.; Rahman, M.M. Forced convection in multichannel configuration in the presence of blocks insert along the flow path. *Int. J. Thermofluids* **2021**, *10*, 100089. [CrossRef]

15. Mafeed, M.P.; Salman, A.M.; Prabin, C.; Ramis, M.K.; Baig, A.; Khan, S.A. Optimum length for pin fins used in electronic cooling. In *Applied Mechanics and Materials*; Trans Tech Publications Ltd.: Stafa-Zurich, Switzerland, 2012; Volume 110, pp. 1667–1673.

16. Boyalakuntla, D.S.; Murthy, J.Y.; Amon, C.H. Computation of natural convection in channels with pin fins. *IEEE Trans. Compon. Packag. Technol.* **2004**, *27*, 138–146. [CrossRef]

17. Qu, W.; Siu-Ho, A. Liquid single-phase flow in an array of micro-pin-fins-Part II: Pressure drop characteristics. *J. Heat Transf.* **2008**, *130*, 124501. [CrossRef]

18. Qu, W.; Siu-Ho, A. Liquid single-phase flow in an array of micro-pin-fins-Part I: Heat transfer characteristics. *J. Heat Transf.* **2008**, *130*, 122402. [CrossRef]

19. Iasiello, M.; Bianco, N.; Chiu, W.K.S.; Naso, V. The effects of variable porosity and cell size on the thermal performance of functionally-graded foams. *Int. J. Therm. Sci.* **2021**, *160*, 10669. [CrossRef]

20. Mauro, G.M.; Iasiello, M.; Bianco, N.; Chiu, W.K.S.; Naso, V. Mono-and Multi-Objective CFD Optimization of Graded Foam-Filled Channels. *Materials* **2022**, *15*, 968. [CrossRef] [PubMed]

21. Olabi, A.G.; Wilberforce, T.; Sayed, E.T.; Elsaid, K.; Rahman, S.A.; Abdelkareem, M.A. Geometrical effect coupled with nanofluid on heat transfer enhancement in heat exchangers. *Int. J.* **2021**, *10*, 100072. [CrossRef]

22. *COMSOL User Manual*, Version 5.6; Newton, MA, USA, 2021.

MDPI

St. Alban-Anlage 66

4052 Basel

Switzerland

Tel. +41 61 683 77 34

Fax +41 61 302 89 18

www.mdpi.com

Micromachines Editorial Office

E-mail: micromachines@mdpi.com

www.mdpi.com/journal/micromachines

9 783036 549682